Heterocycles - Synthesis and Biological Activities

Edited by B. P. Nandeshwarappa and Sadashiv S. O.

Published in London, United Kingdom

IntechOpen

Supporting open minds since 2005

Heterocycles - Synthesis and Biological Activities
http://dx.doi.org/10.5772/intechopen.78709
Edited by B. P. Nandeshwarappa and Sadashiv S. O.

Contributors
Sundaram Singh, Savita Kumari, Dinesh Kumar Jangid, Surbhi Dhadda, Rajesh Y B R D, Avula Srinivas, Malladi Sunitha, Dattatraya Navnath Pansare, Rohini Shelke, Rajendra Sukhadeorao Dongre, Roshan Nasare, Mohammad Idrees, Satish Kola, N V Anil Kumar, Hillemane Venkatachalam, Daniel Plano, Sandra Ramos-Inza, Carlos Aydillo, Carmen Sanmartín, Rohit Singh, Lyudmila Kayukova, Kaldybay Praliyev, Sridhar Goud, Pardeep Kumar, Rose Dawn Bharath, Belakatte Parameshwarappa Nandeshwarappa

First published in London, United Kingdom, 2020 by IntechOpen
IntechOpen is the global imprint of INTECHOPEN LIMITED, registered in England and Wales, registration number: 11086078, 7th floor, 10 Lower Thames Street, London, EC3R 6AF, United Kingdom
Printed in Croatia

British Library Cataloguing-in-Publication Data
A catalogue record for this book is available from the British Library

Additional hard and PDF copies can be obtained from orders@intechopen.com

Heterocycles - Synthesis and Biological Activities
Edited by B. P. Nandeshwarappa and Sadashiv S. O.
p. cm.
Print ISBN 978-1-83969-003-7
Online ISBN 978-1-83880-624-8
eBook (PDF) ISBN 978-1-83880-625-5

Meet the editors

Dr. B.P. Nandeshwarappa holds an M.Sc. and Ph.D. from Kuvempu University, Karnataka, India. He was a Postdoctoral fellow at the Institute of Fine Chemicals, East China University of Science and Technology, Shanghai. His research has been largely in organic chemistry. He has published many scientific papers in reputed journals. He has presented his research papers at national and international conferences, seminars, and workshops. He has edited five books, authored seven specialized books, and five book chapters. He has guided fifty-one dissertation/project theses for postgraduates and is currently guiding four Ph.D. scholars. He has fourteen years of teaching experience and nearly three years of industry experience. He is currently working as an Associate Professor at the Department of Studies in Chemistry, Davangere University, Davangere.

Dr. Sadashiv S. O. is presently an Assistant Professor in the Department of Food Technology, Davangere University, Davanagere, Karnataka. He completed his M.Phil and Ph. D. in microbiology from Karnataka University, Dharwad, Karnataka, India. He has published various research articles in national and international peer-reviewed journals and written six book chapters, two books/monographs, and edited four books. He has secured the Young Scientists Award twice, one Gold Medal award for the research contributions in the field of Veterinary Microbiology from various academic organizations, and also received an Appreciation Award from the Zoological Society of India (ZSI). He has participated and presented research papers in many conferences/seminars/workshops/training sessions at national and international levels. His main areas of research are veterinary microbiology, dairy microbiology, environmental microbiology, and food technology.

Contents

Preface

Heterocyclic systems are important building blocks for new materials possessing interesting electronic, mechanical, or biological properties. For more than a century, many heterocycles have constituted the largest areas of research in organic chemistry. Heterocycles play an important role in biochemical processes because the side groups of the most typical and essential constituents of living cells, DNA and RNA, are based on aromatic heterocycles. The presence of heteroatoms results in significant changes in the cyclic molecular structure due to the availability of unshared pairs of electrons and the difference in electronegativity between heteroatoms and carbon.

Many synthetic methods have been developed for the preparation of heterocycles. The current book project covers the basics and advanced utility of heterocycles in biochemistry, industrial chemistry, environmental sciences, pharmacy, microbiology, biotechnology, and related disciplines. Also, it will fulfill the requirement for new approaches in experimental skills and solving problems obtained during functional group transformation. This book provides a broad range of solutions to the problems of graded difficulties from organic reactions, rearrangements, and reagents along with their isolation and synthetic applications. The main goal of this book is to build on the foundation of synthesis and characterization. Special emphasis has been placed on the synthesis and mechanism of the reactions of different classes of heterocycles. Furthermore, it will bridge the gap between teaching, research, and practice. Finally, this book project provides a concise and comprehensive overview of the methods and biological importance of heterocyclic compounds synthesis.

B.P. Nandeshwarappa
Davangere University,
India

Sadashiv S.O.
Garden City University,
India

Section 1

Synthetic Section

Chapter 1

Introductory Chapter: Synthesis and Antimicrobial Activities of Dihydroazeto[2′,3′:4,5]seleno[2,3-*b*]quinolines

B.P. Nandeshwarappa, G.K. Prakash and S.O. Sadashiv

1. Introduction

Literature survey reveals that sulfur- and selenium-containing molecules have attracted great importance in synthetic organic chemistry; particularly, aromatic five- and six-membered heterocycles fused or bridged to quinoline ring in linear fashion are found in many natural products due to their great pharmacological importance [1–7].

Substituted 2-azetidinone is an important class of compound for its importance in β-lactam antibiotic synthesis [8–10]. β-Lactam drugs in heterocycles are still the most widely prescribed antibiotics used in medicine [11]. The discovery of penicillin 2-azetidinone-based heterocycles have been one of the main classes of drugs with wide therapeutic activities, viz. anticonvulsant [12], anti-inflammatory [13], antibacterial [14], herbicidal [15], and also functioning as enzyme inhibitor [16] and are effective on central nervous system.

The conversion of aryl quinolines into dihydroazeto[2′,3′:4,5]seleno[2,3-*b*]quinolines is of vital importance in synthetic organic chemistry. Seleno[2,3-*b*]quinolines are prepared via their corresponding halogenated derivatives. In the current approach, we have planned to synthesize title compounds by efficient methods for the synthesis of seleno[2,3-*b*]quinolines, starting from easily accessible dichloro substituents by a reaction with sodium hydrogen selenide in water media with quantitative yield under remarkably soft conditions. Also, we came to know that good results were achieved using sodium hydrogen selenide (NaHSe). Here, we wish to examine the feasibility and efficiency of an approach to synthesis of some new seleno[2,3-*b*]quinolines.

In continuation of our research program directed toward the studies on Sulfur Chemistry [17–30] and synthesis of new potentially bioactive molecules, we were in need of a medicinal, bioorganic, industrial, cost-effective and commercial method for the synthesis of quinoline-based sulfur and selenium compounds. Also, the extensive biological properties and pharmaceutical applications have attracted interests in development of such sulfur and selenium-containing analogs.

2. Results and discussions

In this contribution, we focused our attention on the fast and efficient synthesis of dihydroazeto[2′,3′:4,5]seleno[2,3-*b*]quinolines. At first, the key intermediate 3-formyl-2-chloroquinoline [31] and azitidones [32] have been prepared from available reported methods.

In the current investigation very interesting result was observed in the reaction of **1a–e** on subjected to ring cyclisation with sodium hydrogen selenide in water offered seleno quinolines **2a–e**. As expected, the ^{1}H NMR spectrum exhibited two peaks at δ 5.71 ppm and δ 4.91 ppm of two protons present in azetidinone ring, i.e., -N-CH-C- and -Se-CH- of newly formed thieno ring. The aromatic protons resonate as multiplets at δ 7.26–8.50 ppm. The structure was further confirmed by recording its mass spectra. It gave the molecular ion peak at m/z 351(M^+) which corresponds to molecular formula $C_{18}H_{12}N_2OSe$. Also, halogen test was used to confirm the absence of chlorine.

3. Experimental

IR spectra were taken on a Perkin Elmer 157 Infrared spectrophotometer. ^{1}H NMR spectra (300 MHz) were recorded on a Bruker supercon FT-NMR instrument using TMS as internal standard and mass spectra on a Jeol JMS-D 300 mass spectrometer operating at 70 eV. Melting points were determined in open capillary and are uncorrected. Purity of the compounds was checked by TLC on silica gel and the compounds were purified by column chromatography.

3.1 Preparation of sodium hydrogen selenide

A mixture of 1 g of selenium powder and 25 ml of water was taken in a 500 ml beaker. The heat obtained was controlled by keeping this mixture at ice cold condition. A calculated amount of sodium borohydride of 0.026 moles was added in stepwise with constant stirring. During this, immediate liberation of foaming takes place because of the formation of hydrogen gas. Once the addition of sodium borohydride is over, approximately 25 ml of water was added along the side of the beaker and stirring was continued over 15 min. During this, a colorless, deep, reddish NaHSe formed and thus the obtained result was used without further any purification.

3.2 Preparation of dihydroazeto[2′,3′:4,5]seleno[2,3-*b*]quinolines (2a: E)

About 0.01 mole of azitidone **1a**, and 0.01 mole sodium hydrogen selenide, and 50 ml of water were taken in a 500-ml round-bottom flask. The contents of the flask were refluxed over 10–15 min on a water bath. The crystalline solid **2a** was precipitated in the flask. The contents of the flask was poured into a beaker containing 500 mL ice cold water, stirred, filtered, and finally, washed with ethanol. The compound obtained was dried, and recrystallized from ethyl acetate. In the same way the compounds, **2b–e** were prepared (**Figure 1**).

2a. 1-Phenyl-2a, 7b-dihydroazeto[2′,3′:4,5]seleno[2,3-*b*]quinoline

Solid, mp. 280°C; ^{1}H NMR (300 MHz, DMSO-*d6*) *δ* (ppm): 4.91 (1H, s, -Se-CH-), 5.71 (1H, s, -N-CH-C), 7.26–8.50 (10H, m, Ar-H); IR (KBr) *ν* (cm^{-1}): 1735.81 (C=O azetidinone), 1653. [M+], 351. Calcd. (%) for $C_{18}H_{12}N_2OSe$: C; 61.55, H; 3.44, N; 7.98, Se; 22.48, Found: C; 61.08, H; 3.44, N; 7.95, Se; 22.43.

2b. 1-(4-Methylphenyl)-2a,7b-dihydroazeto[2′,3′:4,5]seleno[2,3-*b*]quinoline

Solid, mp. 272°C; ^{1}H NMR (300 MHz, DMSO-*d6*) *δ* (ppm): 2.60 (s, -CH_3), 4.87 (1H, s, -Se-CH-), 5.89 (1H, s, -N-CH-C), 7.20–8.49 (9H, m, Ar-H); IR (KBr) *ν* (cm^{-1}): 1735.61 (C=O azetidinone), [M+], 365. Calcd. (%) for $C_{19}H_{14}N_2OSe$: C; 62.47, H; 3.86, N; 7.67, Se; 21.62, Found: C; 62.45, H; 3.83, N; 7.64, Se; 21.65.

2c. 1-(4-Methoxyphenyl)-2a,7b-dihydroazeto[2′,3′:4,5]seleno[2,3-*b*]quinoline

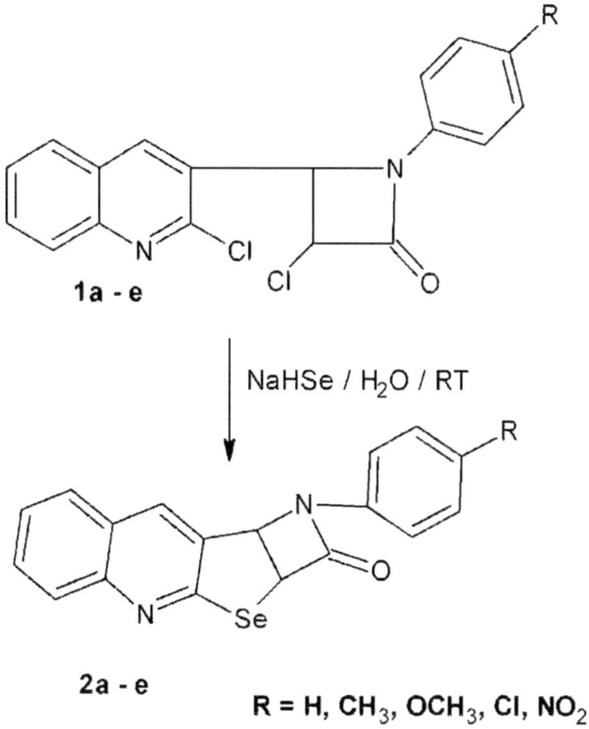

Figure 1.
*General synthetic procedure for dihydroazeto[2′,3′:4,5]seleno[2,3-*b*]quinolines.*

Solid, mp. 287°C; ^{1}H NMR (300 MHz, DMSO-*d6*) δ (ppm): 3.90 (s, -OCH_3), 4.92 (1H, s, -Se-CH-), 5.79 (1H, s, -N-CH-C), 7.23–8.42 (9H, m, Ar-H); [M+], 381. Calcd. (%) for $C_{19}H_{14}N_2O_2Se$: C; 59.85, H; 3.70, N; 7.35, Se; 20.71. Found: C; 59.87, H; 3.73, N; 7.39, Se; 20.73.

2d. 1-(4-Chlorophenyl)-2a,7b-dihydroazeto[2′,3′:4,5]seleno[2,3-*b*]quinoline

Solid, mp. 292°C; ^{1}H NMR (300 MHz, DMSO-*d6*) δ (ppm): 4.79 (1H, s, -Se-CH-), 5.70 (1H, s, -N-CH-C), 7.29–8.55 (9H, m, Ar-H); [M+], 385. Calcd. (%) for $C_{18}H_{11}ClN_2OSe$: C; 56.05, H; 2.87, N; 7.26, Se; 20.47, Found: C; 56.09, H; 2.89, N; 7.23, Se; 20.43.

2e. 1-(4-Nitrophenyl)-2a,7b-dihydroazeto[2′,3′:4,5]seleno[2,3-*b*]quinoline

Solid, mp. 288°C; ^{1}H NMR (300 MHz, DMSO-*d6*) δ (ppm): 4.70 (1H, s, -Se-CH-), 5.68 (1H, s, -N-CH-C), 7.34–8.60 (9H, m, Ar-H). [M+], 396. Calcd. (%) for $C_{18}H_{11}N_3O_3Se$: C; 54.56, H; 2.80, N; 10.60, Se; 19.93, Found: C; 54.52, H; 2.83, N; 10.63, Se; 19.95.

4. Antimicrobial activity

The in vitro antimicrobial activity was carried out against 24 h old cultures of three bacteria by disk diffusion method [33] using ampicillin as the reference. Compounds **2a–e** were tested against Gram-positive bacteria (*Staphylococcus aureus, Micrococcus roseus*) and Gram-negative bacteria (*Escherichia coli*). The compounds were tested at a concentration of 0.001 mol/ml in DMF against all organisms. The zone of inhibition was compared with the standard drug after 24 h of incubation at 25°C and measured in mm. Results are reported in **Table 1**, and it was found that compounds **2d** and **2e** were highly active against *S. aureus* and *M. roseus*

Compound number	Microorganisms		
	S. aureus	***M. roseus***	***E. coli***
Ampicillin	20	22	22
2a	5	7	5
2b	4	4	4
2c	4	5	4
2d	13	16	10
2e	14	15	9

Zone of inhibition was expressed in mm.
Highly active +++ (inhibition zone >12 mm); moderately active ++ (inhibition zone 9–12 mm); slightly active + (inhibition zone 6–9 mm); and inactive (inhibition zone <6 mm).

Table 1.
*Antimicrobial activity tests of dihydroazeto[2′,3′:4,5]seleno[2,3-*b*]quinolines (**2a–e**).*

(Gram-positive) and moderately active against *E. coli* (Gram-negative), and compound **2c** was slightly active against *M. roseus* and *E. coli*. Compound **2e** was slightly active against *S. aureus* and *M. roseus*, and compounds **2a** and **2b** were slightly active against *M. roseus.* Other compounds were all inactive against these three pathogenic microorganisms. Hence, further studies in these compounds are planned to obtain clinically useful agents.

Author details

B.P. Nandeshwarappa[1]*, G.K. Prakash[2] and S.O. Sadashiv[3]

1 Department of PG Studies and Research in Chemistry, Shivagangothri, Davangere University, Davangere, Karnataka, India

2 Department of Chemistry, Bapuji Institute of Engineering and Technology, Davangere, India

3 Department of Food Science and Technology, Shivagangothri, Davangere University, Davangere, Karnataka, India

*Address all correspondence to: belakatte@gmail.com

References

[1] Thomys H. Chemical constituents of the rutaceae. Berichte der Deutschen Pharmazeutischen Gesellschaft. 1923;**33**:68

[2] Nandeshwarappa BP, Arun Kumar DB, Kumaraswamy MN, Ravikumar YS, Bhojya Naik HS, Mahadevan KM. Microwave assisted synthesis of some novel thiopyrano[2,3-b]quinolines as a new class of antimicrobial agent. Phosphorous, Sulfur, and Silicon and the Related Elements. 2006;**181**:1545

[3] Bhojya Naik HS, Ramesha MS, Shwetha BV, Roopa TR. A facile synthesis of novel 9-methyl[1, 2, 3] selenadiazoles[4, 5-b]quinoline and 9-methyl[1, 2, 3]thiadiazole[4, 5-b] quinoline as a new class of antimicrobial agents. Phosphorous, Sulfur, and Silicon and the Related Elements. 2006;**81**:533

[4] Nandeshwarappa BP, Arun Kumar DB, Bhojya Naik HS, Mahadevan KM. An efficient microwave-assisted synthesis of thieno[2,3-b]quinolines under solvent-free conditions. Journal of Sulfur Chemistry. 2005;**26**(4-5):373

[5] Raghavendra M, Bhojya Naik HS, Sherigara BS. Microwave induced synthesis of thieno[2,3-b]quinoline-2-carboxylic acids and alkyl esters and their antibacterial activity. Journal of Sulfur Chemistry. 2006;**27**(4):1

[6] Raghavendra M, Bhojya Naik HS, Sherigara BS. One pot synthesis of some new 2-hydrazino-[1,3,4]thiadiazepino [7,6-b]quinolines under microwave irradiation conditions. Arkivoc. 2006;**15**:153

[7] Michael JP, Natural Products Reports. 1997;**14**:605. b) Balasubramanian M, Keay JG. Comprehensive Heterocyclic Chemistry, II, Vol. 5. In: Katrizky AR, Rees CW, Scriven EFV, editors. Oxford: Pergamon Press, Vol. 06, Chapter 5; 1996. p. 245

[8] Mukerjee AK, Singh AK. β-Lactams: retrospect and prospect. Tetrahedron. 1978;**34**:1731

[9] Kametani T. Synthesis of carbapenem antibiotics. Heterocycles. 1982;**17**:463

[10] Nagahara T, Kametani T. Enantioselective syntheses of carbapenem antibiotics. Heterocycles. 1987;**25**:729

[11] Laila MG, Fatema DE. Mansoura Journal of Pharmaceutical Sciences/ Chemical Abstrcts. 1996;**125**:114552

[12] Koniya T. LRES Lab, Suiisawa Pharma Co. Ltd, Osaka (Japan) Kugaku No Ryoiki Zokon. Chemical Abstracts. 1946/1977;**29/86**:112/1656a

[13] Maffii G. Lepetit S P A, Milan Italy, Farmaco (Pavia)/Chemical Abstracts. 1954/1959;**14/53**:176/20553B

[14] Shah SK, Peter BL. European Patent Application, Ep 360/Chemical Abstracts. 1986/1989;**110**:173037a

[15] Cooper and Robin David Gray. European Patent Application, Ep 252/Chemical Abstracts. 1988/1989;**744/10**:4934

[16] Firestone RA, Barker PL, Pisano JM, Ashe BM, Dahlgren ME. Monocyclic β-lactam inhibitors of human leukocyte elastase. Tetrahedron. 1990;**46**:2255

[17] Nandeshwarappa BP, Aruna Kumar DB, Bhojya Naik HS, Mahadevan KM. An efficient microwave-assisted synthesis of thieno [2, 3-b] quinolines under solvent-free conditions. Journal of Sulfur Chemistry. 2005;**26**(4-5):373

[18] Nandeshwarappa BP, Aruna Kumar DB, Kumaraswamy MN, Ravi Kumar YS, Naik HSB, Mahadevan KM. Microwave assisted synthesis of some novel thiopyrano[2,3-b]quinolines

as a new class of antimicrobial agent. Phosphorus, Sulfur, and Silicon and the Related Elements. 2006;**81**:1545

[19] Nandeshwarappa BP, Aruna Kumar DB, Bhojya Naik HS, Mahadevan KM. A fast and large-scale synthesis of 3-formyl-2-mercaptoquinolines. Phosphorus, Sulfur, and Silicon and the Related Elements. 2006;**181**:1997

[20] Kiran BM, Nandeshwarappa BP, Vaidya VP, Mahadevan KM. Chemistry of substituted quinolines: thieno [2, 3-b] and thiopyrano [2, 3-b] quinolines. Phosphorus, Sulfur, and Silicon and the Related Elements. 2007;**182**:969

[21] Kiran BM, Nandeshwarappa BP, Prakash GK, Vaidya VP, Mahadevan K. Synthesis of new seleno-substituted quinolines. Phosphorus, Sulfur, and Silicon and the Related Elements. 2007;**182**:993

[22] Nandeshwarappa BP. Solvent free, inorganic solid supported synthesis of furoquinolines under microwave irradiation: A potent antimicrobial agent. Journal of Chemistry and Chemical Sciences. 2017;**7**(1):15

[23] Nandeshwarappa BP, Manjunatha Swamy HM. New and efficient synthesis of [(3-formylquinolin-2-yl)thio] acetic acids. Journal of Chemistry and Chemical Sciences. 2017;**7**(2):131

[24] Nandeshwarappa BP. Novel synthesis of new thieno [2,3-b] quinoline-2-carboxylates. Journal of Chemistry and Chemical Sciences. 2017;**7**(3):230

[25] Nandeshwarappa BP. A facile synthesis of sulfur containing condensed quinolines. Journal of Chemistry and Chemical Sciences. 2017;**7**(3):222

[26] Nandeshwarappa BP. Novel approach towards synthesis of ethyl [(3-formylquinolin-2-yl)thio]acetate. Journal of Chemistry and Chemical Sciences. 2017;**7**(3):237

[27] Nandeshwarappa BP. An efficient synthesis of ethyl {[3-(hydroxymethyl) quinolin-2-yl]thio}acetate. Journal of Chemistry and Chemical Sciences. 2017;**7**(3):242

[28] Nandeshwarappa BP. Synthesis of {[3-(hydroxymethyl)quinolin-2-yl] thio} acetic acid. Journal of Chemistry and Chemical Sciences. 2017;**7**(3):247

[29] Nandeshwarappa BP. Synthesis of ethyl {[3-(chloromethyl)quinolin-2-yl] sulfanyl}acetate. Journal of Chemistry and Chemical Sciences. 2017;**7**(3):256

[30] Nandeshwarappa BP. Rapid synthesis of 5H-[1,4]oxathiepino[5,6-b] quinolin-3(2H)-ones. Journal of Chemistry and Chemical Sciences. 2017;**7**(3):251

[31] Meth-Kohn O, Narin B, Tarnowski B, Hayes R, Keyzad A, Rhousati S, et al. A versatile new synthesis of quinolines and related fused pyridines. Part 9. Synthetic application of the 2-chloroquinoline-3-carbaldehydes. Journal of the Chemical Society, Perkin Transactions. 1981;**1**:2509

[32] Nandeshwarappa BP, Manjappa S, Kishore B. A novel approach toward the synthesis of azetidinones derivatives. Journal of Sulfur Chemistry. 2011;**32**(5):475

[33] Finegold SM, Martin WJ. Diagnostic Microbiology. 6th ed. London: Mosby; 1982. p. 450

Chapter 2

Phenacyl Bromide: An Organic Intermediate for Synthesis of Five- and Six-Membered Bioactive Heterocycles

Dinesh Kumar Jangid and Surbhi Dhadda

Abstract

An environmentally friendly, economic synthetic protocol was advanced for synthesis of biologically and pharmacologically vital five- and six-membered heterocycles containing nitrogen, sulphur and oxygen as heteroatom. A series of thiazole derivatives was prepared by the reaction of substituted phenacyl halides and phenyl thiourea in the presence of TiO_2 nanoparticles (NPs) as nanocatalyst in DCM. Similarly, another series of six-membered heterocyclic compounds were synthesized by the reaction of phenacyl halides with phenylenediamine, 2-amino-phenol, 2-aminobenzenethiol to produce corresponding products (1,4-quinoxaline, benzoxazine, benzothiazine) under catalytic effect of TiO_2nanocatalyst. Analytical and spectral (FTIR, ^{1}H and ^{13}C NMR and SEM) techniques were employed for the structural elucidation of the synthesized compounds.

Keywords: environmentally friendly, thiazole derivatives, nanocatalyst, 1,4-quinoxaline, benzoxazine, benzothiazine

This chapter is divided into two sections:

1. Synthesis of five-membered heterocylces from phenacyl halides

2. Synthesis of six-membered heterocylces from phenacyl halides

1. Synthesis of five-membered heterocylces from phenacyl halides

1.1 Introduction

Cyclic compounds which contain one or more hetero atoms besides carbon are called heterocyclic compounds. Most commonly nitrogen, sulphur and oxygen are present as hetero atoms. Phosphorous, tin, boron, silicon, etc. are other less common hetero atoms. Numerous heterocyclic compounds have three to six atoms in the ring, but only those compounds which have five- or six-membered ring are by far most significant. Heterocyclic compounds are broadly circulated in nature and are predominantly important because of the extensive variety of physiological activities related with this course of substances. Several of the important compounds contain heterocyclic rings, e.g. most of the members of alkaloids, vitamin B complex, chlorophyll, antibiotics, other plants pigments, dyes, amino acids, enzymes, the genetic material, DNA, drugs, etc. These biologically active molecules always drawn the attention of chemist over the years specifically because of their biological significance.

One striking structural article characteristic to heterocycles, which continue to be exploited to great benefit by the drug industry, lies in their capability to manifest substituents around a core scaffolds in sharp three-dimensional representations [1]. In early studies of chemistry, nitrogen and sulphur containing heterocyclic compounds contained predominantly and they were thoroughly associated with the enlargement of organic chemistry which was concerned with the study of materials separated from living sources and are widely used as structural motif in drug discovery [2].

Heterocycles form by far the leading classical splits of organic chemistry and are of enormous prominence in the biological and industrial field. One of the major causes for the extensive use of heterocyclic compounds is their structures that can be precisely manipulated to attain the required alteration in function. Another important feature embraced by heterocycles is the possibility of incorporating functional groups either as substituents or as the part of ring system itself. They are also the integral part of the wide range of drugs, most of the vitamins, biomolecules, many natural products, and biologically active compounds, including antifungal, antitumor, antimicrobial [3], antibiotic, anti-inflammatory, antidepressant, antimalarial [4] antibacterial, antiviral, herbicidal, anti-HIV, antidiabetic, insecticidal and fungicidal agents. Further, most of the heterocycles possess vital applications in materials science such as dyestuff, fluorescent sensor, plastics, information storage, brightening agents, and analytical reagents. In addition, they have applications in polymer and supramolecular chemistry, especially in conjugated polymers. Moreover, they act as organic light-emitting diodes (OLEDs), organic conductors, light harvesting systems, photovoltaic cells, optical data carriers [5], chemically controllable switches, semiconductors, molecular wires, and liquid crystalline compounds. Thus consideration has been given to advanced effective new methods to synthesize heterocycles.

Now, nanotechnologies are broadly considered to have the potential to bring assistances in area as diverse as water contamination, drug development, information and communication technologies and the production of lighter and strong materials. Nanotechnologies include the conception and manipulation of materials at the nanometre scale, either by refining or reducing bulk materials or by scaling up from single groups of atoms. Nanoparticles (1–100 nm size) have a distinctive place in nanoscience and nanotechnology, not only because of their specific properties subsequent from their reduced dimensions, but also because they are auspicious building blocks for more complex nanostructures. Nanoparticles with the diameter of less than 10 nm have created extreme curiosity over the past decade due to their developed potential application in area such as nanoscale electronics, sensors, optics and catalysis. Due to this importance of nanoparticles so many efforts have been devoted to the synthesis of nanoparticles from last few years.

Furthermore, the α-halogenation of ketones is an important conversion in synthetic organic chemistry [6]. Due to high reactivity of α-bromoketones, they react with a large number of nucleophiles which provide a range of biologically active compounds [7]. α-bromoacetophenone derivatives have been examined for their active contribution in the inhibition of protein tyrosine phosphatase such as PTP1B and SHP-1 [8]. Bromination of 1,3-keto compounds at the reactive position increases bioactivity, mainly cytotoxicity against breast cancer 1A9 cells, with respect to the unsubstituted compound [9].

For several conversions employed in organic and pharmaceutical synthesis, especially, α-bromo carbonyl compounds have become a significant structural motif for the development of numerous biologically active compounds for instancethiazolidin-4-one, quinoxalines [10], cyclohexanone derivatives, thiophene, pyrazolo[1,5-α][1,3,5]triazine, imidazo[1,2-a][1,3,5]triazin, pyrazolines, imidazo[2,1-b] benzothiazoles [11], thiadiazine and triazolo[3,4-b][1,3,4]thiadiazine. Additionally, they are adaptable building blocks for the retro-synthesis of natural products. α-bromoalkanones were prepared by direct method from α-bromination of carbonyl compounds, which has fascinated significant consideration in the synthetic organic chemistry [12]. The brominated products are important intermediates for the synthesis of various useful molecules such as pharmaceuticals, surfactants, pesticides and biologically active heterocyclic compounds [13]. α-Bromination is also a crucial step for introducing a functional group into a molecule for further conversions. α-Halogenated carbonyl compounds are broadly used in organic synthesis as appreciated reaction intermediates and they show versatile uses in organic conversions.

Thiazole ring containing heterocyclic systems are a significant structural entity for several bioactive molecules [14]. Thiazole has been used in the preparation of imperative drugs essential for antibacterial treatment, inflammation and possesses immunosuppressant activity [15]. It also possesses inhibitor's activity against antiallergies, enzyme cyclo-independent kinase, antitumor and schizophrenia [16]. Some of the thiazole derivatives prepared as fungicide as well as preventing in vivo growth of *Xanthomonas* and anti-arthritis. Amino-thiazoles act as an oestrogen receptor and as a potent class of adenosine receptor antagonists. Development of heterocyclic chemistry is still in advance phase where lot of scope is accessible for researcher. In continuation of this research, various publications on development of biologically active heterocycles containing nitrogen and sulfur as heteroatom in recent years have come into light.

Previously many synthetic methods have been used to synthesize α-halo carbonyl compounds and various reagents have been applied for halogenation of active α-hydrogen of carbonyl compounds such as bromine has been previously used as a elementary brominating reagent for the α-bromination of carbonyl compounds but bromine is very harmful chemical to use. To overcome this limitation, several different reagents such as copper(II) bromide [17], tribromoacetophenone [18], 1,4-dioxane bromooxonium bromide [19], pyridium and tetrabutylammonium tribromide have also been employed as substitutions to bromine. The most commonly used reagents for a-bromination of ketones include molecular bromine [20], N-bromosuccinimide (NBS) [21]. Recently, various methods have been reported using NBS-NH_4OAc [22], NBS-photochemical [23], NBS-PTSA [24], NBS-silica supported sodium hydrogen sulphate [25], NBS-Amberlyst-15 [26], NBS-Lewis acids [27], NBS-ionic liquids [28], $MgBr_2$-(hydroxy(tosyloxy)iodo)benzene-MW [29], N-methylpyrrolidin-2-one hydrotribromide (MPHT) [30], $(CH_3)_3SiBr$-KNO_3 [31], BDMS, NaBr [32].

Here we are reporting a new efficient synthetic procedure for the synthesis of α-halo acetophenoes and thiazole derivatives using heterogeneous catalyst TiO_2 NPs. Presented route is more advanced, eco-friendly and more efficient.

1.2 Experimental details

All the required chemicals were purchased from Sigma Aldrich, Alpha Aesar and used without further purification. The melting points were checked in open capillary tubes in melting point apparatus and are uncorrected. The completion of the reaction was checked on TLC plates coated with silica gel-G in the n-hexane-EtOAc (v/v = 7:3) and visualised by exposure in UV chamber. The IR spectra were recorded on Shimadzu IR-435 spectrophotometer (ν_{max} in cm^{-1}). ^{1}H NMR, and ^{13}C NMR spectra were recorded using a JEOL RESONANCE Spectrometer at 400.0 and 100.0 MHz respectively (δ in ppm) using TMS (d = 0.0) as an internal standard for ^{1}H NMR, and $CDCl_3$ was used as internal standard (d = 77.0) for ^{13}C NMR. Chemical shifts are reported in parts per million (ppm) as follows: chemical shift, multiplicity (s = singlet, d = doublet, t = triplet, q = quartet, m = multiplet, br = broad). The elemental analysis (C, H and N) were performed using vario-III analyser. The nanoparticles are characterized by FTIR and SEM.

1.2.1 General procedure for the synthesis of TiO_2 NPs

TiO_2 NPs were prepared by sol-gel method [33], using titanium(IV) isopropoxide. For the synthesis of TiO_2 NPs, $[Ti\{OPr^i\}_4]$ (1.75 g) was taken in round bottom flask with dry isopropanol (~35 ml). 2–3 drops of water-isopropanol mixture (1,1) was added to the above mentioned clear solution and magnetically stirred for 2 h then sol formation occurred immediately. To ensure complete hydrolysis, excess of water ~10 ml {stoichiometric amount (0.22 g)}, in small lots with continuous stirring for ~4 h was added. The mixture was again stirred for 1 h, till a gel is formed. The synthesized gel was dried in an oven (100°C) and then washed properly with acetone then an off-white powder was obtained. This powder was sintered at 600°C for 4 h to yield a white powder, which was characterized by FTIR and SEM as pure TiO_2.

1.2.2 Characterization of TiO_2 NPs

The TiO_2 nanocatalyst was prepared using sol-gel method and characterized by various techniques using FT-IR and Scanning Electron Microscopy (SEM). The FT-IR spectrum of TiO_2 NPs is given in **Figure 1**. The absorbance bands at around 3235–3550 cm^{-1} were proved to the adsorbed water and hydroxyl group in nano sized TiO_2 (**Figure 1**). The band observed at 720 cm^{-1} is due to Ti-O-Ti while absorbance bands at 460 cm^{-1} show stretching vibration due to Ti-O, which is customary with the reported IR spectra for nano TiO_2 [33]. The SEM images of this oxide are revealed in **Figure 2**. The scales that are shown in **Figure 2** are of 500 nm come into sight to specify formation of agglomerates granular morphology, constituted by nano-sized crystallites.

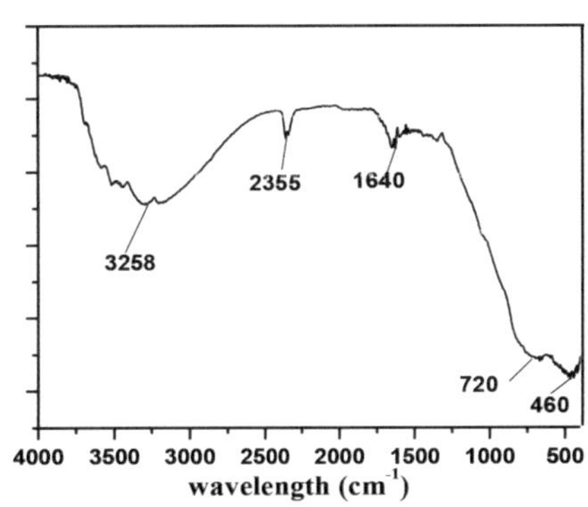

Figure 1.
The FT-IR spectra of TiO_2 NPs.

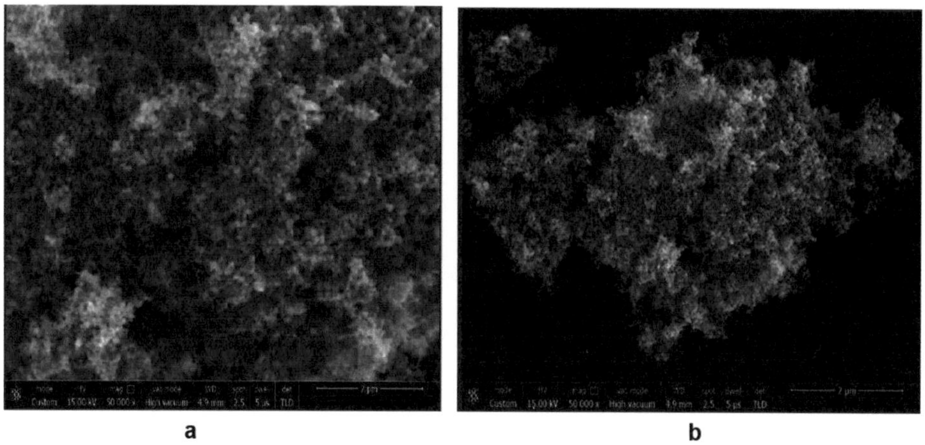

Figure 2.
(a and b). The SEM image of the TiO_2 NPs.

*1.2.3 General procedure for the synthesis of substituted thiazoles (**3a–e**)*

In a 20 ml round bottom flask phenacyl bromide (**1**) (0.5 mmol) and substituted phenyl thioureas (**2**) (0.5 mmol) were added in 5 mL DCM. A catalytic amount of TiO_2 NPs (5 mol%) is added to reaction mixture. Thereafter, the reaction mixture was allowed to stir at magnetic stirrer at 50°C for 20–30 min. The progress of reaction was monitored by TLC, the solid separated was filtered, washed with Hypo solution and recrystallized with ethanol (**Figure 3**) The detailed mechanism of the synthesis is shown in **Figure 4** (**Table 1**).

1.3 Spectral data of substituted thiazole (3a–e)

*1.3.1 [4-(4-Bromo-phenyl)-thiazole-2-yl]-(4-chloro-phenyl)-amine (**3a**)*

IR (cm^{-1}, KBr): 3315, 1611, 1462, 1370, 762, 644; ^1H NMR ($CDCl_3$, 400 MHz): δ 4.03 (s, NH), 7.17 (d, 2H, Ar–H), 7.34 (d, 2H, Ar–H), 7.42 (d, 2H, Ar–H), 7.55 (d, 2H, Ar–H); ^{13}C NMR ($CDCl_3$, 100.4 MHz): δ 106.2, 118.5, 119.5, 128.4, 130.5, 131.0, 131.3, 138.6, 149.4, 175.3; HRMS; *m/z* 365.94 (M^+); $C_{15}H_{10}BrClN_2S$: calcd. C, 49.27; H, 2.76; N, 7.66; found C, 49.25; H, 2.75; N, 7.69.

*1.3.2 [4-(4-Chloro-phenyl)-thiazole-2-yl]-phenyl-amine (**3b**)*

IR (cm^{-1}, KBr): 3349, 1646, 1434, 1389, 779, 667; ^1H NMR ($CDCl_3$, 400 MHz): δ 4.12 (s, NH), 7.18 (d, 2H, Ar–H), 7.42 (d, 2H, Ar–H), 7.45 (d, 2H, Ar–H), 7.57 (d, 2H, Ar–H); ^{13}C NMR ($CDCl_3$, 100.4 MHz): δ 107.3, 119.3, 120.7, 130.3, 131.5, 131.9, 132.3, 140.6, 150.3, 176.9; HRMS; *m/z* 286.03 (M^+); $C_{15}H_{11}ClN_2S$: calcd. C, 62.82; H, 3.87; N, 9.77; found C, 62.83; H, 3.89; N, 9.78.

*1.3.3 [4-(4-Methoxy-phenyl)-thiazole-2-yl]-phenyl-amine (**3c**)*

IR (cm^{-1}, KBr): 3317, 1633, 1467, 1379, 767, 648; ^1H NMR ($CDCl_3$, 400 MHz): δ 4.07 (s, NH), 7.18 (d, 2H, Ar–H), 7.39 (d, 2H, Ar–H), 7.48 (d, 2H, Ar–H), 7.56(d, 2H, Ar–H); ^{13}C NMR ($CDCl_3$, 100.4 MHz): δ 109.5, 112.5, 120.6, 130.7, 131.8, 131.9, 132.4, 140.6, 150.7, 177.5; HRMS; *m/z* 282.08 (M^+); $C_{16}H_{14}N_2OS$: calcd. C, 68.06; H, 5.00; N, 9.92; found C, 68.08; H, 5.02; N, 9.93.

*1.3.4 [4-(4-Fluoro-phenyl)-thiazole-2-yl]-phenyl-amine (**3d**)*

IR (cm^{-1}, KBr): 3340, 1623, 1470, 1389, 766, 665; ^1H NMR ($CDCl_3$, 400 MHz): δ 4.09 (s, NH), 7.21 (d, 2H, Ar–H), 7.37 (d, 2H, Ar–H), 7.45 (d, 2H, Ar–H), 7.58(d, 2H, Ar–H); ^{13}C NMR ($CDCl_3$, 100.4 MHz): δ 109.8, 121.7, 123.8, 130.6, 131.6, 131.9, 132.1, 139.7, 151.5, 179.8; HRMS; *m/z* 270.06 (M^+); $C_{15}H_{11}FN_2S$: calcd. C, 66.65; H, 4.10; N, 10.36; found C, 66.67; H, 4.11; N, 10.37.

*1.3.5 [4-(2-Chloro-phenyl)-thiazole-2-yl]-phenyl-amine (**3e**)*

IR (cm^{-1}, KBr): 3325, 1632, 1472, 1379, 768, 647; ^1H NMR ($CDCl_3$, 400 MHz): δ 4.06 (s, NH), 7.21 (d, 2H, Ar–H), 7.39 (d, 2H, Ar–H), 7.46 (d, 2H, Ar–H), 7.59(d, 2H, Ar–H); ^{13}C NMR ($CDCl_3$, 100.4 MHz): δ 110.5, 119.5, 121.6, 131.6, 131.9, 132.2, 132.5, 141.9, 152.7, 179.3; HRMS; *m/z* 286.03 (M^+); $C_{15}H_{11}ClN_2S$: calcd. C, 62.82; H, 3.87; N, 9.77; found C, 62.83; H, 3.89; N, 9.78.

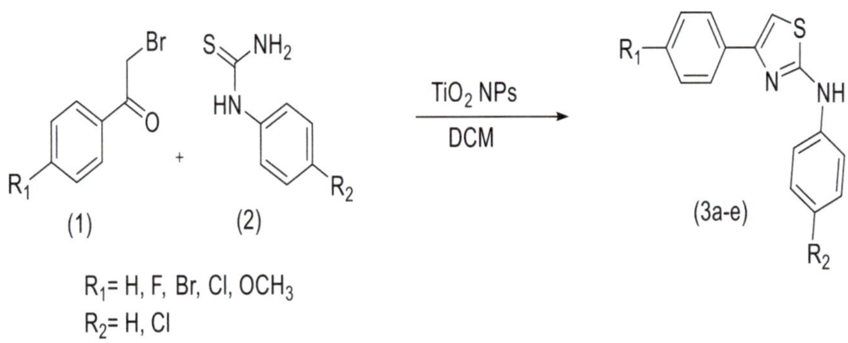

Figure 3.
Synthesis of various substituted thiazole (3a-e).

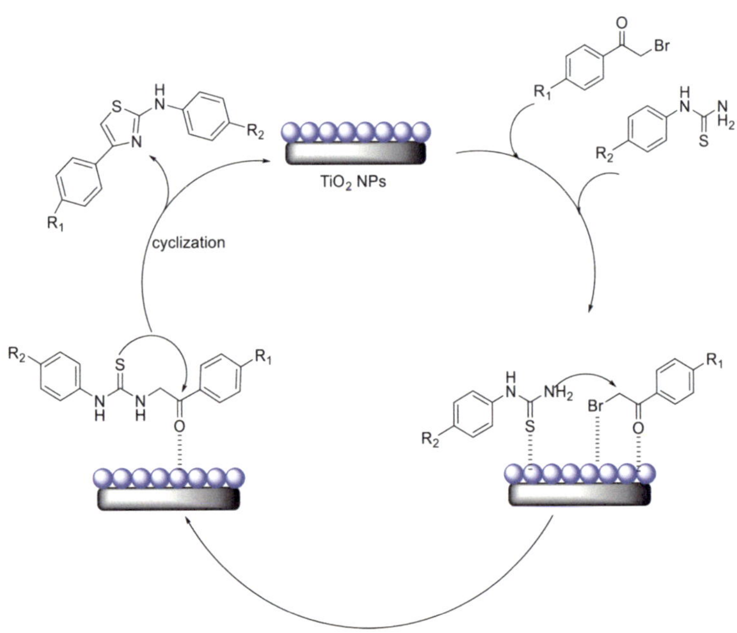

Figure 4.
Proposed mechanism for TiO_2 NPs catalysed thiazole synthesis.

DOI: http://dx.doi.org/10.5772/intechopen.88243

Entry	Phenacyl bromide (1)	Phenyl thiourea (2)	Product (3a–e)	Time (min)
1	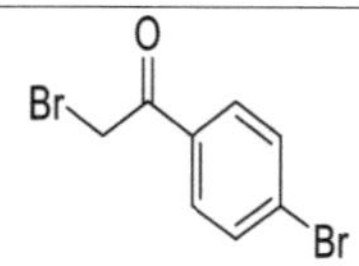	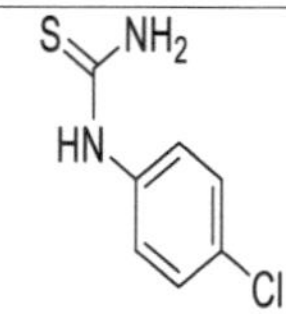	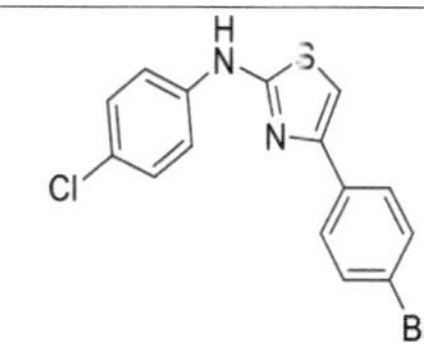 [4-(4-Bromo-phenyl)-thiazole-2-yl]-phenyl-amine	5
2	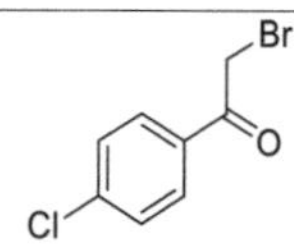	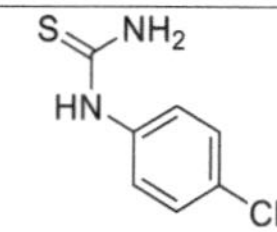	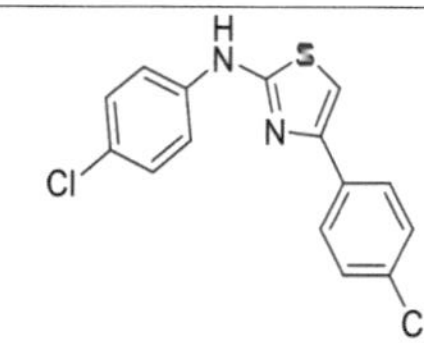 [4-(4-Chloro-phenyl)-thiazole-2-yl]-phenyl-amine	6
3	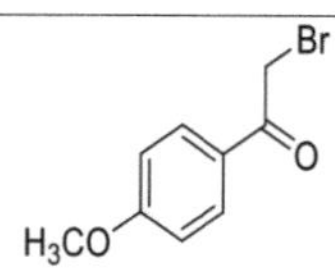	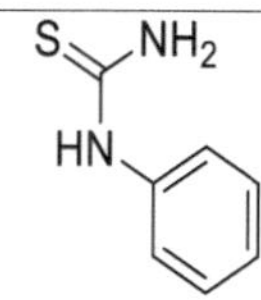	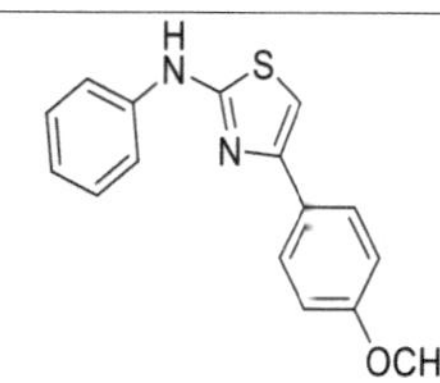 [4-(4-Methoxy-phenyl)-thiazole-2-yl]-phenyl-amine	10

Entry	Phenacyl bromide (1)	Phenyl thiourea (2)	Product (3a–e)	Time (min)
4	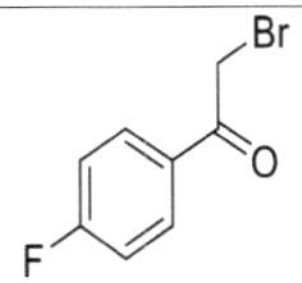	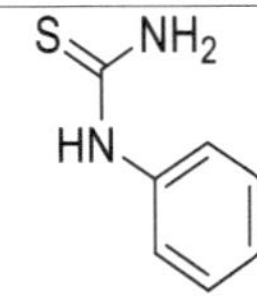	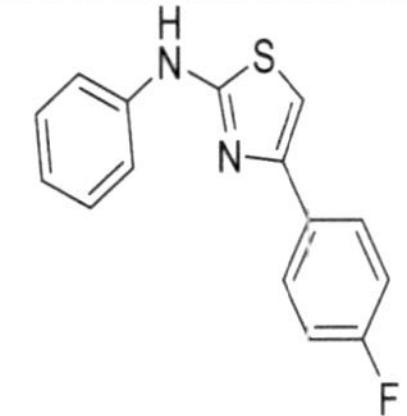 [4-(4-Floro-phenyl)-thiazole-2-yl]-phenyl-amine	9
5	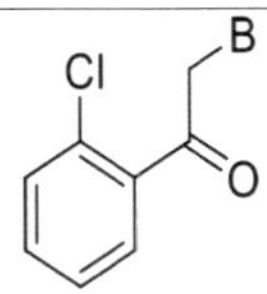	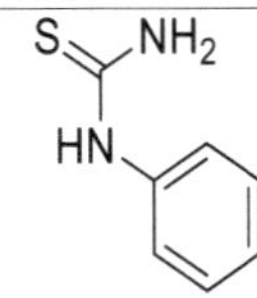	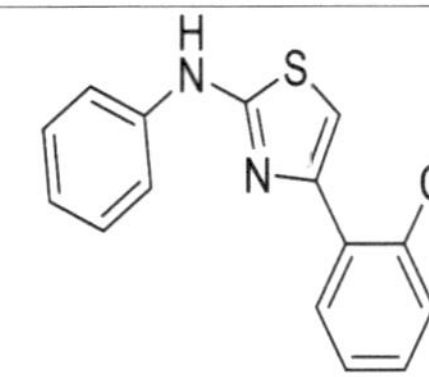 [4-(2-Chloro-phenyl)-thiazole-2-yl]-phenyl-amine	12

Table 1.
Synthesis of various substituted thiazole (3a–e).

2. Synthesis of six-membered heterocylces from phenacyl halides: (1,4-quinoxaline, 1,4-benzoxazines, 1,4-benzothiazines)

2.1 Introduction

Quinoxalines are important class of nitrogen containing heterocycles, possessing nitrogen atom at 1,4 position. These heterocycles possess various pharmacological [34–38] and biological properties such as antibiotic (echinomycin, bleomycin), anticancer [39] anti-viral [40], anti-bacterial, antibiotic, and anti-inflammatory. The compounds of quinoxalines used to develop organic semiconductors [41, 42], dehydroannulenes [43], and also used in dyes [44]. Various methods have been reported in literature for the synthesis of quinoxalines, i.e. condensation of 1,2-diketone with phenylene diamine to yield the desired quinoxaline under reflux condition at ambient temperature with various solvents such as benzene, ethanol [45] with use of different catalyst like molecular iodine, copper(II) sulphate, indium(III) chloride, o-iodoxybenzoic acid, ceric ammonium nitrate, silica gel, gallium(III) triflate phosphorus oxychloride, oxidative coupling of epoxides with ene-1,2-diamines [46], 1,4-addition of 1,2-diamines to diazenylbutenes [47], cyclization-oxidation of phenacyl bromides with 1,2-diamines by $HClO_4$-SiO_2 [48] and by using solid phase synthesis [49, 50]. Quinoxaline has also been synthesized by the chemical reaction of phenylene diamine and different substituted phenacyl bromides via solid phase [49, 50], synthesis by using different catalyst like 1,4-diazabicyclo [2,2,2]octane, trimethylsilyl chloride, perchloric acid supported on silica, KF-alumina, β-cyclodextrin.

1,4-Benzoxazines are important moiety of heterocyclic compounds having considerable biological [51], pharmaceutical [52] and wide range of synthetic utilities. Therefore, new methods should be developed for an efficient protocol for their synthesis. In addition of above information these compounds also served as precursors for the synthesis of many medicinally important drugs [53]. The skeleton of these type of structures are synthesized by the direct intramolecular reductive cyclization of appropriate nitroketones or by intramolecular annulation of 2-aminophenoxy ketones [54]. Benzoxazines are also prepared by the condensation reaction of 2-aminophenols with substituted phenacyl halides [55]. Although, these reported procedures are not specific and general because involvement of more than one-steps, requirement of high temperature, give low to moderate yields and use of commercially unavailable starting material. Hence the discovery of new protocols which leads to an efficient synthetic procedure for synthesis of 1,4-benzoxazine and their derivatives.

There are many reported methods in literature for the efficient synthesis for multicomponent reactions (**MCRs**) for the natural products, these methods are of great advantage at atom and step economy level, and these are also environmental friendly. 4*H*-1,4-benzothiazines (having nitrogen and sulphur heteroatom at 1,4 position) are a family of heterocycles possessing number of important biological and pharmacological properties [56]. A compound containing thiazine ring namely 2-benzoyl-7-chloro-3-methyl-5-trifluoromethyl-4*H*-1,4-benzothiazine [57] possessing numerous biological activities like antiemetics, neuroleptics, antihistaminics, antipsychotics, antibacterial, tranquilizers, sedatives, and anticarcinogen. This type of heterocycles has attracted considerable interest of researchers, which leads to the development of synthetic strategies. Therefore, the development and use of new MCRs have been an interesting topic in the areas of various branches of chemistry like synthetic organic, medicinal and pharmachemistry.

Although, these reported methods suffered from various limitations such as toxic nature of reagents, excess loading of catalyst, need of high temperature, expensive reagents and complicated work-up to complete the reaction. In present era development of green and sustainable protocols attract the attention of scientists because the use of these above reagents causes many allergic diseases. In this connection of research, various researchers have considerable attention on use of non-hazardous reagents like nanoparticles as heterogeneous catalysts for organic transformations. So, keeping in view these facts of green technology we have tried our effort to develop, a new synthetic strategy for the synthesis of 1,4-quinoxaline, 1,4-benzoxazines, 1,4-benzothiazines catalysed by **TiO_2** nanoparticles (**NPs**).

This method is considered to be environment friendly because of use of solid heterogeneous catalyst that provides many advantages such as, ease of handling, non-corrosiveness, high yield, low cost and reusability of the used nanocatalyst.

2.2 General procedure for the synthesis of six-membered heterocycles

*2.2.1 General procedure for the synthesis of 1,4-quinoxaline (**3a** and **b**)*

In a 50 ml round bottom flask we took 1,2-phenylenediamine (**1**) (1 mmol) and substituted phenacyl bromide (**2**) (1 mmol) and dissolved both in dichloromenthane-DCM (5 mL). Now add catalytic amount of TiO_2 nanoparticles and stirring is continuous at 50°C for appropriate time limit. After completion of reaction, the whole content was filtered fir the removal of nanocatalyst and it was well washed with ethyl acetate (10 mL). This obtained filtrate was concentrated and purified by column chromatography by using hexane/ethylacetate (15% ethyl acetate in hexane) as an eluent to yield desired quinoxaline derivatives in appropriate yields. The nanocatalyst can be recovered after thoroughly washing with ethyl acetate, air dried, and activation at 80°C for 3 h and can reused for further cycles (**Figure 5**).

2.2.2 Spectral data of synthesized compounds

*2.2.2.1 2-Phenylquinoxaline (**3a**)*

Dark yellow solid; M.P: 75–78.3°C; ^{1}H NMR (400 MHz, $CDCl_3$, TMS = 0 PPM): δ = 7.50–7.58 (m, 3H, ArH), 7.70–7.82 (m, 2H, ArH), 8.14–8.28 (m, 4H, ArH), 9.43 (s, 1H, C3-H) ppm; ^{13}C NMR (100.4 MHz, $CDCl_3$, TMS = 0 PPM): δ = 127.1, 129.19, 129.26, 129.5, 129.6, 130.5, 130.8, 136.4, 141.9, 142.8, 143.6, 152.8 ppm; LCMS (ESI-MS): m/z calcd. for $C_{14}H_{10}N_2$ (M+): 206.24; found: 207.1 (M + H).

*2.2.2.2 2-(3-bromophenyl)quinoxaline (**3b**)*

Light brown solid; M.P: 132–133.8°C;^{1}H NMR (400 MHz, $CDCl_3$, TMS = 0 PPM): δ = 7.44–7.48 (t, 1H, ArH), 7.66–7.68 (dd, 1H, ArH), 7.77–7.85 (m, 2H, ArH), 8.14–8.19 (m, 3H, ArH), 8.32–8.40 (t, 1H, ArH), 9.50 (s, 1H, C3-H) ppm; ^{13}C NMR (100 MHz, $CDCl_3$, TMS = 0 PPM): δ = 124.4, 126.8, 129.8, 130.6, 130.9, 131.5, 131.6,

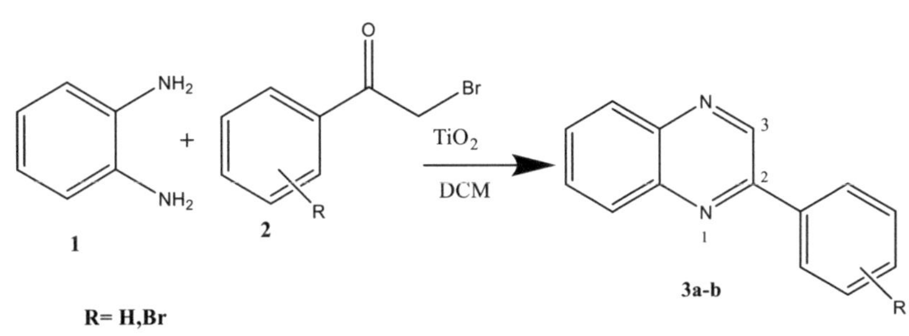

Figure 5.
Synthesis of 1,4-quinoxaline.

138.9, 141.8, 142.8, 143.9, 153.2 ppm; LCMS (ESI-MS): m/z calcd. for $C_{14}H_9BrN_2$ (M+): 285.14; found: 287.0 (M + 2H).

*2.2.3 General procedure for the synthesis of 1,4-benzoxazines (**3c** and **d**)*

A catalytic amount of nanoparticles (TiO_2) added to the stirring mixture of o-aminophenol (**1**) (1 mmol), substituted phenacyl bromide (**2**) (1 mmol) and triethyl amine (Et_3N) (1.1 mmol). After the completion of the reaction the whole content was extracted with Et_2O (5X2, 10 mL). After extraction the organic layer was washed with brine (20 mL), and dried over the layer of anhydrous Na_2SO_4, concentrated and purified by column chromatography by using silica gel and EtOAc/hexane (1:20) (**Figure 6**).

*2.2.3.1 2-Phenyloxazine (**3c**)*

Brown solid; M.P:88–89.5°C; ^{1}H NMR (400 MHz, $CDCl_3$, TMS = 0 PPM): δ = 7.50–7.62 (m, 3H, ArH), 7.76–7.80 (m, 2H, ArH), 8.15–8.29 (m, 4H, ArH), 4.88 (s, 2H, C3-H) ppm; 13C NMR (100.4 MHz, $CDCl_3$, TMS = 77.0 PPM): δ = 127.5, 129.0, 129.2, 129.5, 129.6, 130.1, 130.8, 136.8, 141.5, 142.8, 143.3, 151.8 ppm; LCMS (ESI-MS): m/z calcd for $C_{14}H_{10}N_2$ (M+): 206.24; found: 207.1 (M + H).

*2.2.3.2 2-(3-bromophenyl) oxazine (**3d**)*

Light yellow solid, M.P: 142–143.8°C;^{1}H NMR (400 MHz, $CDCl_3$, TMS = 0 PPM): δ = 7.51–7.58 (t, 1H, ArH), 7.66–7.70 (dd, 1H, ArH), 7.74–7.78 (m, 2H, ArH), 8.16–8.19 (m, 3H, ArH), 8.36–8.40 (t, 1H, ArH), 4.98 (s, 2H, C3-H) ppm; ^{13}C NMR (100 MHz, $CDCl_3$, TMS = 77.0 PPM): δ = 74.5, 126.9, 128.8, 130.8, 131.2, 131.5, 131.9, 139.9, 141.9, 142.4, 143.9, 153.2 ppm; LCMS (ESI-MS): m/z calcd for $C_{14}H_{10}BrN_2$ (M+): 285.14; found: 287.0 (M + 2H).

2.2.4 General procedure for the synthesis of 1,4-benzothiazines (4e and f)

In a round bottom flask the 2-aminobenzenethiol (**1**) (1 mmol), aromatic aldehyde (**2**) (1 mmol), and substituted phenacyl bromide (**3**) (1 mmol), were added in the stirring solution of the DABCO (0.2 mmol) in (Et_3N), and stirring was continued for 6 h in an oil bath at 65°C. After the completion of the reaction the whole reaction mixture was cooled and diluted with DCM (20 mL) and then washed with water. The obtained residue was purified by column chromatography on silica gel (300–400 mesh) with EtOAc and petroleum ether (1:20, v/v) as the eluent to yield the desired product (**Figure 7**).

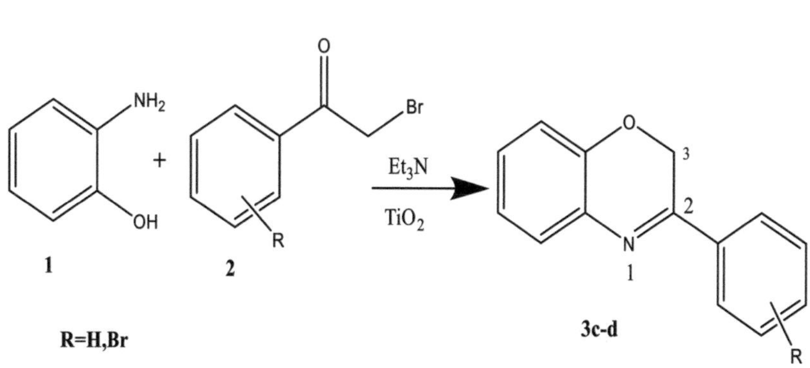

Figure 6.
Synthesis of 1,4-benzoxazines.

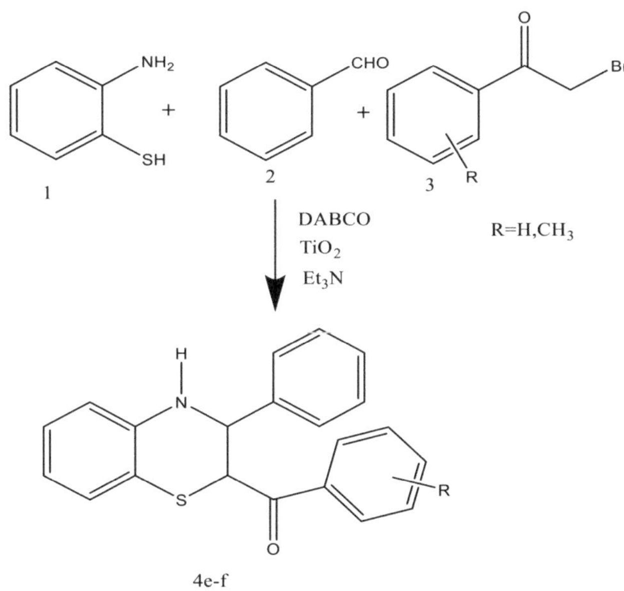

Figure 7.
Synthesis of 1,4-benzothiazines.

*2.2.4.1 Phenyl (3-phenyl-3,4-dihydro-4H-benzo[b][1,4]thiazin-2-yl) methanone (**4e**)*

Light yellow solid; M.P.: 124–126°C; ^{1}H NMR (400 MHz, $CDCl_3$): δ 7.86 (d, J = 7.5 Hz, 2H), 7.52 (t, J = 7.1 Hz, 1H), 7.41–7.26 (m, 7H), 7.08 (q, J = 7.5 Hz, 2H), 6.70 (t, J = 8.9 Hz, 2H), 5.12 (d, J = 4.8 Hz, 1H), 4.72 (d, J = 5.4 Hz, 1H), 4.45 (s, 1H) ppm; ^{13}C NMR (100.4 MHz, $CDCl_3$) δ47.6, 57.9, 113.8, 115.2, 118.4, 127.4, 127.8, 128.2, 128.4, 128.6, 128.6, 128.7, 133.8, 135.4, 142.4, 142.6, 194.4 ppm; HRMS (ESI) calcd for [$C_{21}H_{17}NOS$ + H] + 332.1109, found 332.1104.

*2.2.4.2 (3-Phenyl-3,4-dihydro-2H-benzo[b][1,4]thiazin-2-yl)(p-tolyl) methanone (**4f**)*

Brown solid; M.P.: 118–120°C; ^{1}H NMR (400 MHz, $CDCl_3$): δ 7.80 (d, J = 7.5 Hz, 1H), 7.56 (t, J = 7.8 Hz, 1H), 7.40 (t, J = 7.2 Hz, 2H), 7.26 (t, J = 8.0 Hz, 3H), 7.10–7.04 (m, 4H), 6.68–6.60 (m, 2H), 5.02 (d, J = 5.7, 1H), 4.58 (d, J = 6.3 Hz, 1H), 4.34 (s, 1H), 2.26 (s, 3H) ppm; ^{13}C NMR (75 MHz, $CDCl_3$) δ 47.1, 57.6, 113.6, 115.0, 118.2, 127.4, 127.8, 128.2, 128.4, 128.5, 128.6, 128.7, 133.8, 135.2, 142.2, 142.6, 194.2, 21.5 ppm; HRMS (ESI) calcd. for [$C_{22}H_{19}NOS$ + H] + 346.1266, found 346.1260.

3. Results and discussion

In this chapter we have tried to develop an efficient protocol for the synthesis of five-membered disubstituted derivatives (**thiazole, 3a–e, Figure 3**) by using TiO_2 NPs in moderate to excellent yields from the starting materials phenacyl halides and thioureas. Similarly, six-membered nitrogen containing heterocycles (**quinoxaline derivatives, 3a, b, Figure 5**) from o-phenylenediamines and substituted phenacyl bromides in the presence of TiO_2/DCM at 50°C. In the similar way, six-membered nitrogen and oxygen containing heterocycles (**benzoxzines, 3c** and **d, Figure 6**) was synthesized under the set of conditions Et_3N/TiO_2 at room temperature by annulation of o-aminophenols with substituted phenacyl bromides via one pot process. 1,4-benzothiazines are prepared by the reaction of the benzaldehyde, phenacyl halides and 2-aminothiophenols in the presence of set of conditions DABCO, TiO_2, Et_3N to yield benzothiazine (**4e** and **f, Figure** 7). The TiO_2 NPs was characterised by FTIR and SEM images which confirmed the synthesis of TiO_2 NPs in the nano range.

4. Conclusion

In conclusion, we have developed a green and economic procedure for the synthesis of bioactive five- and six-membered heterocycles. This synthetic methodology allowed us to synthesize products in good to excellent yields, which is irrespective to the functional groups which are present in the starting material. The used protocol is mild and environmental friendly. There are many merits of the used protocol like, low cost of green catalyst, obtaining high yield of products, operational simplicity, and the catalyst can be reused without any significant loss in catalytic property up to four catalytic cycle. These outstanding features of this method make it environmentally friendliness.

Author details

Dinesh Kumar Jangid* and Surbhi Dhadda
Department of Chemistry, University of Rajasthan, Jaipur, Rajasthan, India

*Address all correspondence to: dinu.jangid@gmail.com

References

[1] Gupta RR, Kumar M, Gupta V. Nomenclature of heterocycles. Journal of Heterocyclic Chemistry. 1998;**1**:3-38. DOI: 10.1007/978-3-642-72276-9_2

[2] Joshi RC. Mild and efficient one pot synthesis of imidazolines and benzimidazoles from aldehydes. E-Journal of Chemistry. 2007;**4**:606-610

[3] Fox HH, Bogert MT. A review of biologically active benzothiazole, thiazole and pyridine derivatives. Journal of the American Chemical Society. 1993;**61**:2013-2017

[4] Burger A, Sawhney SN. Antimalarials. III. Benzothiazole amino alcohols. Journal of Medicinal Chemistry. 1968;**11**:270-273

[5] Correndor CC, Huang ZL, Belfield KD, Morales AR, Bondar MV. Photochromic polymer composites for two-photon 3D optical data storage. Chemistry of Materials. 2007;**19**:5165-5173

[6] Erian AM, Sherif SM, Gaber HM. The chemistry of α-haloketones and their utility in heterocyclic synthesis. Molecules. 2003;**8**:793-865

[7] Bolton R. In: Price D, Iddon B, Wakefield BJ, editors. Bromine Compounds: Chemistry and Applications. Amsterdam: Elsevier; 1988. p. 151

[8] Arabaci G, Guo XC, Beebe KD, Coggeshall KM, Pei D. α-haloacetophenone derivatives as photoreversible covalent inhibitors of protein tyrosine phosphatases. Journal of the American Chemical Society. 1999;**121**:5085-5086

[9] Ishida J, Ohtsu H, Tachibana Y, Nakanishi Y, Bastow KF, Nagai M, et al. Antitumor agents. Part 214: Synthesis and evaluation of curcumin analogues as cytotoxic agents. Bioorganic & Medicinal Chemistry. 2002;**10**:3481-3487

[10] Meshram HM, Ramesh P, Kumar GS, Reddy BC. One-pot synthesis of quinoxaline-2-carboxylate derivatives using ionic liquid as reusable reaction media. Tetrahedron Letters. 2010;**51**:4313-4316. DOI: 10.1016/j.tetlet.2010.05.099

[11] Kumbhare RM, Vijay Kumar K, Janaki Ramaiah M, Dadmal T, Pushpavalli SN, Mukhopadhyay D, et al. Synthesis and biological evaluation of novel Mannich bases of 2-arylimidazo [2, 1-b] benzothiazoles as potential anti-cancer agents. European Journal of Medicinal Chemistry. 2011;**46**:4258-4266. DOI: 10.1016/j.ejmech.2011.06.031

[12] Larock RC. Comprehensive Organic Transformations. 3rd ed. New York: VCH; 1999. p. 709

[13] Talegaonkar J, Mukhija S, Boparai KS. Determination of thiosemicarbazones by reaction with ω-bromoacetophenone. Talanta. 1982;**29**:327-328

[14] Lewis JR. Nitrile-containing natural products. Natural Product Reports. 1999;**16**:389-416

[15] Lednicer D, Mitscher LA, Georg GI. Organic Chemistry of Drug Synthesis. New York: Wiley; 1990. pp. 95-97

[16] Ei-Subbagh HI, AI-Obaid AM. 2,4-Disubstituted thiazoles II. A novel class of antitumor agents, synthesis and biological evaluation. European Journal of Medicinal Chemistry. 1996;**31**:1017-1021

[17] King LC, Ostrum GK. Selective bromination with Copper(II) bromide. The Journal of Organic Chemistry. 1964;**29**:3459-3461

[18] Krohnke F, Ellegast K. Selektives Bromieren von Methylketonen und Bromierungen mit Bromketonen. Chemische Berichte. 1953;**86**:1556-1562

[19] Yanovskaya LA, Terentev AP, Belen LI. Chemical Abstracts. 1953;**47**:8032

[20] Boyd RE, Rasmussen CR, Press JB. Regiospecific synthesis of unsymmetrical α-bromoketones. Synthetic Communications. 1995;**25**:1045-1051

[21] Coats SJ, Wasserman HH. The conversion of R-bromo-α-dicarbonyls to vicinal tricarbonyls using dimethyldioxirane and base. Tetrahedron Letters. 1995;**36**:7735-7738

[22] Arbuj SS, Waghmode SB, Ramaswamy AV. Photochemical α-bromination of ketones using *N*-bromosuccinimide: A simple, mild and efficient method. Tetrahedron Letters. 2007;**48**:1411-1415

[23] Pravst I, Zupan M, Stavber S. Halogenation of ketones with *N*-halosuccinimides under solvent-free reaction conditions. Tetrahedron Letters. 2008;**64**:5191-5199

[24] Das B, Venkateswarlu K, Mahender G, Mahender I. A simple and efficient method for α-bromination of carbonyl compounds using *N*-bromosuccinimide in the presence of silica-supported sodium hydrogen sulfate as a heterogeneous catalyst. Tetrahedron Letters. 2005;**46**:3041-3044

[25] Meshram HM, Reddy PN, Sadashiv K, Yadav JS. Amberlyst-15®-promoted efficient 2-halogenation of 1,3-keto-esters and cyclic ketones using N-halosuccinimides. Tetrahedron Letters. 2005;**46**:623-626. DOI: 10.1016/j.tetlet.2004.11.140

[26] Yang D, Yan YL, Lui B. Mild α-halogenation reactions of 1,3-dicarbonyl compounds catalyzed by Lewis acids. The Journal of Organic Chemistry. 2002;**67**:7429-7431

[27] Meshram HM, Reddy PN, Vishnu P, Sadashiv K, Yadav JS. Mild α-halogenation reactions of 1,3-dicarbonyl compounds catalyzed by Lewis acids. The Journal of Organic Chemistry. 2006;**47**:991-995. DOI: 10.1016/j.tetlet.2005.11.141

[28] Lee JC, Park JY, Yoon SY, Bae YH, Lee SJ. Efficient microwave induced direct α-halogenation of carbonyl compounds. Tetrahedron Letters. 2004;**45**:191-193

[29] Bekaert A, Provot O, Rasolojaona O, Alami M, Brion JD. *N*-Methylpyrrolidin-2-one hydrotribromide (MPHT) a mild reagent for selective bromination of carbonyl compounds: Synthesis of substituted 2-bromo-1-naphtols. Tetrahedron Letters. 2005;**46**:4187-4191

[30] Prakash GKS, Ismail R, Garcia J, Panja C, Rasul G, Mathew T, et al. α-Halogenation of carbonyl compounds: halotrimethylsilane-nitrate salt couple as an efficient halogenating reagent system. Tetrahedron Letters. 2011;**52**:1217-1221

[31] Khan AT, Ali MA, Goswami P, Choudhury LH. A mild and regioselective method for α-bromination of β-keto esters and 1,3-diketones using bromodimethylsulfonium bromide (BDMS). The Journal of Organic Chemistry. 2006;**71**:8961-8963

[32] Kim EH, Koo BS, Song CE, Lee K. Halogenation of aromatic methyl ketones using oxone® and sodium halide. Synthetic Communications. 2001;**31**:3627-3632

[33] Venkatachalam N, Palanichamy M, Murugesan V. Sol-gel preparation and characterization of alkaline earth metal doped nano TiO_2: Efficient photocatalytic degradation of 4-chlorophenol. Journal of Molecular Catalysis A: Chemical. 2007;**273**:177-185

[34] Ali MM, Ismail MMF, El-Gabby MSA, Zahran MA, Ammar YA. Synthesis and antimicrobial activities of some novel quinoxalinone derivatives. Molecules. 2000;**5**:864-873

[35] Sakata G, Makino K, Kurasawa Y. Recent progress in the quinoline chemistry. Synthesis and biological activity. Heterocycles. 1998;**27**:2481-2515

[36] Gomtsyan A, Bayburt EK, Schmidt RG, Zheng GZ, Perner RJ, Didomenico S, et al. Novel transient receptor potential vanilloid 1 receptor antagonists for the treatment of pain: Structure-activity relationships for ureas with quinoline, isoquinoline, quinazoline, phthalazine, quinoxaline, and cinnoline moieties. Journal of Medicinal Chemistry. 2005;**48**:744-752

[37] Seitz LE, Suling WJ, Reynolds RC. Synthesis and antimycobacterial activity of pyrazine and quinoxaline derivatives. Journal of Medicinal Chemistry. 2002;**45**:5604-5606

[38] Discovery of a new series of centrally active tricyclic isoxazoles combining serotonin (5-HT) reuptake inhibition with α_2-adrenoceptor blocking activity. Journal of Medicinal Chemistry. 2005;**48**:2019

[39] Lindsley CW, Zhao Z, Leister WH, Robinson RG, Barnell SF, Defeozones D, et al. Allosteric Akt (PKB) inhibitors: Discovery and SAR of isozyme selective inhibitors. Bioorganic & Medicinal Chemistry Letters. 2005;**15**:761-764

[40] Loriga M, Piras S, Sanna P, Paglietti G. 2-[aminobenzoates] and 2-[aminobenzoylglutamate]-quinozalines as classical antifolate agents. Synthesis and evaluation of in vitro anticancer, anti-HIV and antifungal activity. Farmaco. 1997;**52**:157-166

[41] Dailey S, Feast JW, Peace RJ, Sage IC, Till S, Wood EL. Synthesis and device characterization of side-chain polymer electron transport materials for organic semiconductor applications. Journal of Materials Chemistry. 2001;**11**:2238-2243

[42] O'Brien D, Weaver MS, Lidzey DG, DDC B. Use of poly (phenyl quinoxaline) as an electron transport material in polymer light-emitting diodes. Applied Physics Letters. 1996;**69**:881-883

[43] Ott S, Faust R. Quinoxalinodehydroannulenes: A novel class of carbon-rich materials. Synlett. 2004:1509-1512

[44] Brock ED, Lewis DM, Yousaf TI, Harper HH. WO 9951688. USA: The Procter & Gamble Company; 1999

[45] Brown DJ. The Chemistry of Heterocyclic Compounds Quinoxalines: Supplement II. New Jersey: John Wiley & Sons; 2004

[46] Antoniotti S, Dunach E. Direct and catalytic synthesis of quinoxaline derivatives from epoxides and ene-1,2-diamines. Tetrahedron Letters. 2002;**43**:3971-3973. DOI: 10.1016/S0040-4039(02)00715-3

[47] Aparicio D, Attanasi OA, Filippone P, Ignacio R, Lillini S, Mantellini F, et al. Straightforward access to pyrazines, piperazinones, and quinoxalines by reactions of 1,2-diaza-1,3-butadienes with 1,2-diamines under solution, solvent-free, or solid-phase conditions. The Journal of Organic Chemistry. 2006;**71**:5897-5905. DOI: 10.1021/jo060450v

[48] Das B, Venkateswarlu K, Suneel K, Majhi A. An efficient and convenient protocol for the synthesis of quinoxalines and dihydropyrazines via cyclization–oxidation processes using $HClO_4 \cdot SiO_2$ as a heterogeneous recyclable catalyst. Tetrahedron Letters. 2007;**48**:5371-5374. DOI: 10.1016/j.tetlet.2007.06.036

[49] Wu Z, Ede NJ, et al. Solid-phase synthesis of quinoxalines on SynPhase™ Lanterns. Tetrahedron Letters. 2001;**42**:8115-8118. DOI: 10.1016/S0040-4039(01)01733-6

[50] Singh SK, Gupta P, Duggineni S, Kundu B. Solid phase synthesis of quinoxalines. Synlett. 2003;**14**:2147-2150. DOI: 10.1055/s-2003-42065

[51] Yang W, Wang Y, Ma Z, Golla R, Stouch T, Seethala R, et al. Synthesis and structure-activity relationship of 3-arylbenzoxazines as selective estrogen receptor b agonists. Bioorganic & Medicinal Chemistry Letters. 2004;**14**:2327-2330. DOI: 10.1016/j.bmcl.2004.01.099

[52] Das BC, Madhukumar AV, Anguiano J, Mani S. Design, synthesis and biological evaluation of 2*H*-benzo[b][1,4]oxazine derivatives as hypoxia targeted compounds for cancer therapeutics. Bioorganic & Medicinal Chemistry Letters. 2009;**19**:4204-4206. DOI: 10.1016/j.bmcl.2009.05.110

[53] Satoh K, Inenaga M, Kanai K. Synthesis of a key intermediate of levofloxacin via enantioselective hydrogenation catalyzed by iridium(I) complexes. Tetrahedron: Asymmetry. 1998;**9**:2657-2662. DOI: 10.1016/S0957-4166(98)00276-6

[54] Chiccara F, Novellino E. Some new aspects on the chemistry of 1,4-benzoxazines. Journal of Heterocyclic Chemistry. 1985;**22**:1021-1023. DOI: 10. 1002/jhet.5570220418

[55] Sabitha G, Subba Rao AV. Synthesis of 3-arylcoumarins, 2-aroylbenzofurans and 3-aryl-2*H*-1,4-benzoazines under phasetransfer catalysis conditions. Synthetic Communications. 1987;**17**:341-354. DOI: 10.1080/00397918708077315

[56] Hari A, Miller BL. Rapid and efficient synthesis of 2-amino-4H-benzothiazines. Organic Letters. 2000;**2**:3667-3670. DOI: 10.1021/o1006580m

[57] Wu J, Shang Y, Wang C, He X, Yan Z, Hu M, et al. Synthesis of 3,4-dihydro-2H-1,4-benzo[b]thiazine derivatives via DABCO-catalyzed one-pot three-component condensation reactions. RSC Advances. 2013;**3**(14):4643-4651. DOI: 10.1039/C3RA00123G

Chapter 3

Amidoxime Derivatives with Local Anesthetic, Antitubercular, and Antidiabetic Activity

Lyudmila Kayukova, Umirzak Jussipbekov and Kaldybay Praliyev

Abstract

Our task in the field of new derivatives of amidoximes was the elaboration for new medication with increased activity and lower toxicity than medications used in practice. Here are the results of the search for new painkillers and antitubercular and antidiabetic drugs in the class of amidoxime derivatives. Nitrous derivatives of α-chloro-α-isonitrosoacetone, *O*-aroyl-β-aminopropioamidoximes, and 3-[β-(piperidine-1-yl)]ethyl-5-aryl-1,2,4-oxadiazoles were tested for conduction, infiltration, and terminal anesthesia. Among them hit compounds were discovered. The search for new anti-TB drugs is executed in the world. Salts and bases of *O*-aroylation products of β-(thiomorpholin-1-yl) and β-(4-methylpiperazin-1-yl) propioamidoximes during *in vitro* antitubercular screening for DS, DR, and MDR strains of *M. tuberculosis* manifest themselves as highly active competitive compounds. In the series of the derivatives of β-aminopropioamidoximes, a search for new antidiabetic drugs was done. The compounds with pronounced antidiabetic properties were revealed. The obtained data of the most promising samples with a preliminary assessment of their average toxic dose in animals can be used in further *in vivo* testing of infiltration anesthesia conditions, of antidiabetic properties, and at the development of doses and new treatment regimens for TB.

Keywords: nitrous derivatives of α-chloro-α-isonitrosoacetone, bases and salts of *O*-aroyl-β-aminopropioamidoximes, 3-(β-amino)ethyl-5-aryl-1,2,4-oxadiazoles, local anesthetics, *in vitro* antitubercular, antidiabetic screening

1. Introduction

First of all, researchers' interest in amidoximes is due to the possibility of their synthetic modification according to the reaction groups NOH and NH_2. The largest number of derivatives was obtained as a result of acylation reactions at the O-atom of the NOH group and subsequent transformations involving the NH_2 fragment to 1,2,4-oxadiazoles [1]. In most cases, amidoxime derivatives, including heterocyclic radicals, under standard conditions are stable, allowing their structural identification, and withstand storage and biological screening. Arrays of data were obtained on their diverse biological activity: antitubercular, local anesthetic, antidiabetic, antioxidant, etc. [2, 3].

The rational use of drugs is one of the urgent problems of modern medicine. A doctor of any profile most often faces the need to eliminate and prevent pain.

With pain of varying intensities, adequate pain relief reduces the patient's tension and fear, prevents him from forming a negative attitude to medical manipulations, and protects the nervous system of the doctor and patient, providing better medical care. The search for new painkillers with increased activity and lower toxicity than painkillers used in practice is one of the tasks of modern medical chemistry. We developed new β-aminopropioamidoximes and studied their neurotropic properties. Herein we present results from a study of the local anesthetic activities of three chemical groups of new amidoxime derivatives [4].

Tuberculosis (TB) is the leading cause of death and morbidity in more than one third of the world's population. Of the 56.4 million deaths worldwide due to the 10 leading causes in 2016, tuberculosis ranked 10th, from which 1.4 million people died [5].

In May 2014, the World Health Organization (WHO) approved a new global TB control strategy "End TB". This strategy marks a critical shift from tuberculosis control to ending the epidemic by 2035. The "End TB" strategy emphasizes the need for innovation to accelerate progress by optimizing existing ones in the short term and introducing new innovative modes in the long term [6]. In order to reduce the duration of treatment, the rapid development of drug resistance and toxic and side effects of existing anti-TB drugs, and to reduce the cost of extremely expensive treatment of TB (DS, MDR, XDR), the world is searching for new anti-TB drugs. We have synthesized the salts and bases of the *O*-aroylation products of β-(thiomorpholin-1-yl) and β-(4-methylpiperazin-1-yl)propioamidoximes, containing in the β-position pharmacophore fragments of 1-methylpiperazine and thiomorpholine. *In vitro* antitubercular screening of β-aminopropioamidoxime derivatives in the DS, DR, and MDR strains of *M. tuberculosis* revealed highly active competitive compounds which are less toxic than rifampicin and isoniazid with activity significantly exceeding the activity of the reference preparations. It is assumed that these compounds may be the subject of subsequent trials in the development of doses and new treatment regimens for TB [7, 8].

Diabetes is on the rise across the globe. Presently every 7 seconds someone is estimated to die from diabetes or its complications. This is against the background of a global diabetes prevalence of 8.8% of the world population in 2017. The prevalence is expected to further increase to 9.9% by the year 2045. In total numbers, this reflects a population of 424.9 million people with diabetes worldwide in 2017 with an estimate of a 48% increase to 628.6 million people for the year 2045 [9]. Due to the urgency of the problem of diabetes in the world, a search is underway for new antidiabetic drugs. The antidiabetic activity of amidoxime derivatives is known [10, 11]. We conducted *in vitro* testing of derivatives of β-aminopropioamidoximes: bases and pharmacologically acceptable salts of *O*-aroyl-β-(morpholin-1-yl) propioamidoxime and 5-aryl-3-β-(piperidin-1-yl and morpholin-1-yl)ethyl-1,2,4-oxadiazoles with respect to their ability to inhibit the activity of α-amylase and α-glucosidase enzymes. Identified compounds with pronounced antidiabetic properties must be noted; a series of 3,5-disubstituted 1,2,4-oxadiazoles is more active than a series of *O*-aroyl-β-aminopropioamidoximes [12].

The data obtained can be used in further *in vivo* testing of the antidiabetic properties of the most promising samples with a preliminary assessment of their average toxic dose in animals.

2. Local anesthetic activity of new amidoxime derivatives

Herein we present results from a study of the local anesthetic activities of three chemical groups of new amidoxime derivatives (**1–10**) [4].

The first group includes derivatives of α-chloro-α-isonitrosoacetone such as 3-acetyl-5,5-bis(hydroxymethyl)-5,6-dihydro-4*H*-1,2,4-oxadiazine (**1**) and the *anti*-isomer of *N*-(4-methylphenyl)acetylformamidoxime thiosemicarbazone (**2**). The second group includes the hydrochlorides of *O*-aroyl-β-aminopropioamidoximes with piperidine (**3**), morpholine (**4–6**), and benzimidazole (**7** and **8**) in the β-position. The third group consists of 3-[β-(piperidin-1-yl)]ethyl-5-*p*-tolyl-1,2,4-oxadiazole (**9**) and 3-[β-(piperidin-1-yl)]ethyl-5-*m*-chlorophenyl-1,2,4-oxadiazole (**10**).

The local anesthetic activity of **1–10** was studied in three types of anesthesia, i.e., infiltration, conduction, and terminal. The reference drugs were trimecaine, lidocaine, novocaine, and kazcaine [hydrochloride of 1-(2-ethoxyethyl)-4-ethynyl-4-benzoyloxypiperidine] (**Tables 1–3**).

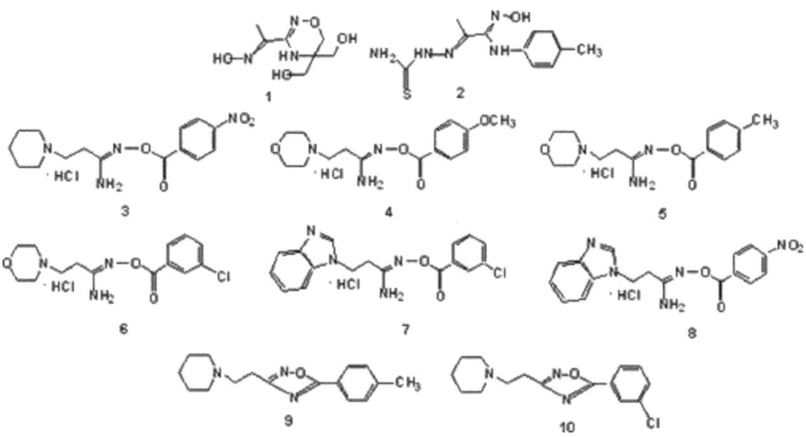

Compound	Anesthesia index (M ± m)	Duration of complete anesthesia (M ± m), min	Total duration of anesthesia (M ± m), min
1	22.6 ± 1.4*	10.0 ± 2.5**	40.0 ± 2.7*
2	28.0 ± 2.2**	14.4 ± 0.4***	29.1 ± 2.2**
3	28.8 ± 3.7****	15.0 ± 0.0*	28.0 ± 4.9***
4	31.4 ± 1.4**	18.3 ± 7.7	55.0 ± 1.8****
5	30.6 ± 1.3*	20.8 ± 2.4****	55.8 ± 2.1**
6	21.0 ± 2.4*	8.3 ± 2.7	34.1 ± 0.8*
7	31.0 ± 1.2****	20.0 ± 2.8*	45.8 ± 2.5****
8	34.0 ± 1.15*	25.0 ± 2.8**	55.8 ± 2.1**
9	34.1 ± 0.7**	25.0 ± 0.3*	58.3 ± 2.7*
10	36.0 ± 0.0*	85.0 ± 0.8****	125.0 ± 1.8**
Trimecaine	34.1 ± 0.5	30.0 ± 1.7	44.1 ± 1.7
Lidocaine	32.3 ± 2.3	25.8 ± 0.8	54.5 ± 2.3
Novocaine	30.0 ± 0.2	10.0 ± 0.0	22.0 ± 0.1
Kazkain	31.1 ± 1.2	25.0 ± 2.5	75.0 ± 0.7

**Compared to trimecaine.*
***Compared with lidocaine.*
****In comparison with novocaine.*
*****Compared to cascaine.*

Table 1.
*Activity and duration of action of compounds **1–10** (0.5% concentration) for infiltration anesthesia.*

Compound	Anesthesia index (M ± m)	Duration of complete anesthesia (M ± m), min	Total duration of anesthesia (M ± m), min
1	329.0 ± 20.0*	10.0 ± 0.0***	64.0 ± 1.5*
2	301.0 ± 5.3**	15.0 ± 0.0*	72.0 ± 4.0*
3	319.7 ± 5.6***	45.0 ± 0.0**	69.3 ± 3.0**
4	427.0 ± 44.0*	48.0 ± 0.0****	80.6 ± 2.0***
5	310.0 ± 43.7*	20.0 ± 0.0*	65.0 ± 3.1*
6	242.9 ± 4.7**	10.0 ± 0.8*	61.2 ± 1.2*
7	425.7 ± 15.6*	84.0 ± 2.6**	144.0 ± 3.5****
8	534.0 ± 12.0****	88.0 ± 0.0***	118.0 ± 3.1*
9	591.0 ± 34.0*	90.0 ± 2.4*	105.0 ± 5.9*
10	600.0 ± 0.0*	70.4 ± 1.1***	90.0 ± 3.1**
Trimecaine	324.0 ± 14.0	20.0 ± 0.0	63.0 ± 1.3
Lidocaine	366.8 ± 94.8	10.0 ± 0.0	68.0 ± 2.8
Novocaine	310.0 ± 43.7	10.0 ± 0.0	60.0 ± 0.0
Kazkain	600.0 ± 0.0	208.9 ± 7.3	280.0 ± 0.0

*Compared to trimecaine.
**Compared with lidocaine.
***In comparison with novocaine.
****Compared to cascaine.

Table 2.
Activity and duration of action of compounds 1–10 (1% concentration) for conduction anesthesia.

Compound	Regnier index (M ± m)	Duration of complete anesthesia (M ± m), min	Total duration of anesthesia (M ± m), min
1	85.6 ± 5.0	0.0	14.4 ± 1.5
2	242.5 ± 16.4	0.0	33.1 ± 1.6
3	186.3 ± 9.7	0.0	28.0 ± 2.4
4	150.6 ± 16.2	0.0	27.5 ± 1.9
5	281.6 ± 18.5	0.0	38.0 ± 1.9
6	13.0 ± 0.0	0.0	0.0
7	103.0 ± 10.5	0.0	22.0 ± 0.9
8	373.4 ± 37.3	0.0	43.1 ± 4.0
9	601.5 ± 32.7	0.0	62.0 ± 2.3
10	430.0 ± 14.4	0.0	48.75 ± 2.1
Dikain	1300.0 ± 0.0	65.0 ± 0.0	120.0 ± 0.0

Table 3.
Activity and duration of action of compounds 1–10 (1% concentration) for terminal anesthesia.

The experimental results indicated that all compounds **1–10** were effective to different degrees in infiltration anesthesia (**Table 1**). The most active compound was **10**, which induced the maximum deep anesthesia (anesthesia index 36.0) and exceeded statistically that of the reference drugs with the exception of lidocaine.

This compound also turned out to be more active than the other tested compounds. The anesthesia indices of **8** and **9** were almost the same as that for trimecaine and were slightly greater than those for lidocaine, novocaine, and kazcaine. The strength of the anesthesia induced by **4** and 7 was greater than that of novocaine, equal to that of kazcaine, and less than that of trimecaine and lidocaine. The anesthesia indices of **1–3** and **4** were less than those of the reference drugs. Compound **10** had a longer duration of conduction anesthesia than the other tested compounds (including the reference drugs).

Compounds 5 and 7–**9** had longer durations of action than novocaine, shorter than trimecaine, and essentially the same as lidocaine and kazcaine. The duration of total anesthesia of **4** was longer than that of novocaine and slightly shorter than that of the other reference drugs. The durations of total anesthesia for **1** and **4** (10.0 and 8.3 min) were comparable with that of novocaine.

Table 2 presents results from a study of conduction anesthesia by **1–10**.

Like in the preceding series of tests, **10** had the highest activity. Its anesthesia index was greater than those of trimecaine, lidocaine, and novocaine and equal to that of kazcaine. Compounds **1**–**5** and 7–**9** were rather active. Their anesthesia indices were greater than those of trimecaine, novocaine, and lidocaine. However, they were less than that of kazcaine. Compound **6** was less active than the reference drugs. Like in the preceding series of tests, the durations of total anesthesia of the studied compounds were compared. **Table 2** shows that all compounds **1–10** had total anesthesia duration indices that were shorter than that of kazcaine although **3**, **4**, and 7–**10** had durations of action longer than those of novocaine, lidocaine, and trimecaine.

Compounds **1–10** in terminal anesthesia were weaker and shorter acting than dicaine (**Table 3**). However, not one of these compounds exhibited an irritating effect.

Thus, it was shown that amidoxime derivatives **1–10** exhibited anesthetic effects that were greater than those of the reference drugs in conduction and infiltration anesthesia. The 1,2,4-oxadiazoles **9** and **10** and to a lesser extent *O*-aroylaminopropioamidoximes with a β-benzimidazole substituent 7 and **8** had longer durations of action than the reference drugs.

3. Search for new antitubercular drugs among the salts and bases of *O*-aroylation products of β-(thiomorfolin-1-yl)- and β-(4-methylpiperazin-1-yl)propioamidoximes

A search for qualitatively new antitubercular drugs with the requirements of reducing the duration of treatment, eliminating of the rapid drug resistance development and toxic side effects of the existing antitubercular drugs, and reducing the cost of extremely expensive treatment of TB (DS, MDR, XDR) is being conducted in the world.

1,5-Diphenylpyrroles have been identified as a class of compounds with high *in vitro* antitubercular activity. Replacing of the methylpiperazine substituent for thiomorpholine and replacing the chlorine atom in position 4 of the N-phenyl moiety with the fluorine atom, as well as varying the aromatic substituents at the C-2 atom of the pyrrole ring during the transition from *p*-CH_3 (BM221) to *p*-CH_3O (BM233) and to *p*-CH_3S (BM579) in 1,5-(4-chlorophenyl)-2-methyl-3-(4-methylpiperazin-1-yl)methyl-1H-pyrrole (BM212), leads to an increase in *in vitro* antitubercular activity on *M. tuberculosis* H37Rv strains [13, 14].

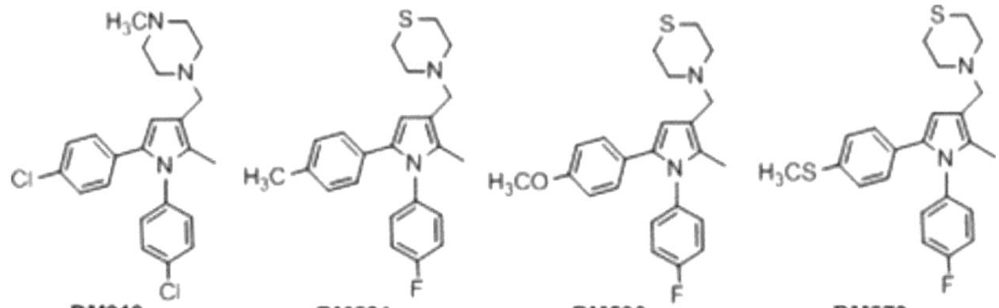

BM212 MIC: 5 μM | BM521 MIC: 0,16 μM (0,4 μg/mL) | BM533 MIC: 0,20 μM | BM579 MIC: 0,16 μM

Taking into account the above examples, we synthesized compounds of the β-aminopropioamidoxime series containing in the β-position fragments of 1-methylpiperazine and thiomorpholine (**11–21**).

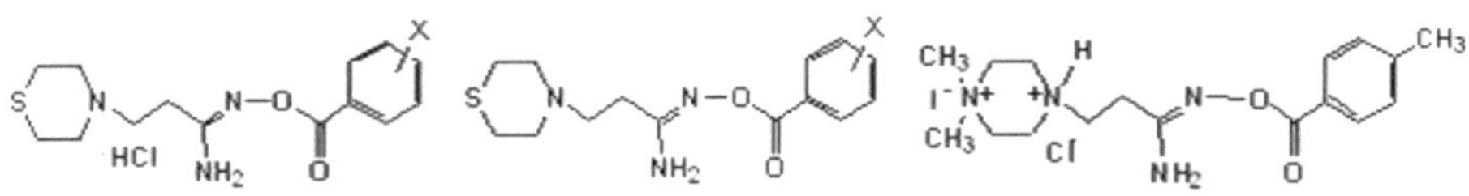

11–15 | 16–20 | 21

X = p-CH_3O (11, 16), p-CH_3 (12, 17), H (13, 18), p-Br (14, 19), m-Cl (15, 20)

In vitro antitubercular screening of a series of *O*-aroyl-β-aminopropioamidoximes (**11–21**) on DS museum H37Rv and wild* I MTB strains and two wild DR and MDR strains of MTB II and III on Shkolnikova liquid medium found that compounds **11–21** in varying degrees have antitubercular activity from >100 to 0.01 μg/ml (**Table 4**).

Thus, on the DS strains of MTB *O*-benzoyl-β-(thiomorpholin-1-yl) propioamidoxime (**18**) and hydrochloride, iodomethylate of *O*-*p*-toluoyl-β-(1-methylpiperazin-1-yl)propioamidoxime (**21**) showed the highest activity at 0.01 μg/ml; compound **19** had an average antitubercular activity with MBC >20 μg/ml; the remaining compounds **11–17** and **20** had MBC from 100 to >100 μg/ml.

The highest activity in 0.1 μg/ml on DR and MDR strains of MTB II and III was shown by hydrochloride, iodomethylate of *O*-*p*-toluoyl-β-(1-methylpiperazin-1-yl) propioamidoxime (**21**) (**Table 4**).

The acute toxic effect of rifampicin, isoniazid, and compounds **18** and **21** (LD_{50}) was determined on white mice of both sexes weighing 17–23 g when administered subcutaneously. The toxicity of rifampicin SV is 267.6 ± 7.2 mg/kg; of isoniazid 62.5 ± 12.8 mg/kg; and of compounds **18** and **21**, respectively, 325.0 ± 17.8 and 1750.0 ± 35.6 mg/kg.

Thus, hydrochloride, iodomethylate of *O*-*p*-toluoyl-β-(4-methylpiperazin-1-yl) propioamidoxime, is by 100 times more active against DS strains than rifampicin SV and by 10 times more active than isoniazid; it is by 20 times more active against DR strains than rifampicin SV and by 10 times more active than isoniazid. Hydrochloride, iodomethyl *O*-*p*-toluoyl-β-(4-methylpiperazin-1-yl)propioamidoxime, is less toxic than rifampicin SV by 6.5 times and by 28 times less toxic than isoniazid.

O-Benzoyl-β-(thiomorpholin-1-yl)propioamidoxime is by 100 times more active against DS strains than rifampicin SV and by 10 times more than isoniazid; it is less toxic than rifampicin SV by 1.2 times and by 5.2 times less toxic than isoniazid. These data are protected by the patents of the Republic of Kazakhstan [7, 8].

№ comp.	MBC on the *M. tuberculosis* strains, μg/ml				
	H37Rv	I	II	III	LD_{50}, mg/kg
11	>100	>100	100	100	—
12	100	100	100	100	—
13	100	100	100	100	—
14	>100	>100	100	100	—
15	100	100	100	100	—
16	100	100	100	100	—
17	>100	>100	>100	>100	—
18	0.01	0.01	100	100	325.0 ± 17.8
19	>20	>20	100	100	—
20	100	100	100	100	—
21	0.01	0.01	0.1	0.1	1750.0 ± 35.6
Rifampicin	1	1	2	2	267.6 ± 7.2
Isoniazid	0.1	0.1	1	1	62.5 ± 12.8

Wild strains of M. tuberculosis I, II, and III were isolated from the patients and typed in the RSE "National Scientific Center for Phthisiopulmonology of the Republic of Kazakhstan" of the Ministry of Health of the Republic of Kazakhstan: I, DS (drug-sensitive) to anti-TB drugs; II, DR (drug-resistant) to rifampicin; III, MDR (multidrug-resistant) to rifampicin, isoniazid, and ethambutol.

Table 4.
*Bactericidal activity and average subcutaneous toxicity of O-aroyl-β-(thiomorpholin-1-yl)propioamidoximes (**11–20**) and double salt of O-p-toluoyl-(4-methylpiperazin-1-yl)propioamidoxime (**21**) on DS and DR strains of M. tuberculosis.*

Based on the high priority requirements of increasing the effectiveness and safety of treatment in the development of new antitubercular drugs, it can be argued that *O*-benzoyl-β-(thiomorpholin-1-yl)propioamidoxime and hydrochloride, iodomethylate of *O*-*p*-toluoyl-β-(4-methylpiperazine-1-yl)propioamidoxime, are competitive because they are less toxic and more active than the basic tuberculostatics used in practice: isoniazid and rifampicin.

4. Inhibition of α-amylase and α-glucosidase by new β-aminopropioamidoxime derivatives

The urgency of discovering effective medicines to treat diabetes and information about the antidiabetic activity of amidoxime derivatives [9–11] prompted us to test β-aminopropionamidoxime bases and pharmacologically acceptable salts of *O*-aroyl-β-(morpholin-1-yl)propionamidoximes and 5-aryl-3-β-(piperidin-1-yl- and morpholin-1-yl)ethyl-1,2,4-oxadiazoles for *in vitro* inhibitory activity against the enzymes α-amylase and α-glucosidase, which determine the supply level of glucose from the gastrointestinal tract into the blood pool.

Herein, results from *in vitro* screening of new β-aminopropionamidoximes (**22–26**) for antidiabetic activity are now reported. The series of β-aminopropionamidoximes included bases and pharmacologically acceptable salts (hydrochloride, acetate, oxalate, citrate, and methyl iodide) of *O*-aroyl-β-(morpholin-1-yl) propionamidoximes **22–24** and 5-(*p*-, *m*-substituted phenyl)-3-(β-piperidin-1-yl- and morpholin-1-yl)-1,2,4-oxadiazoles **25** and **26**.

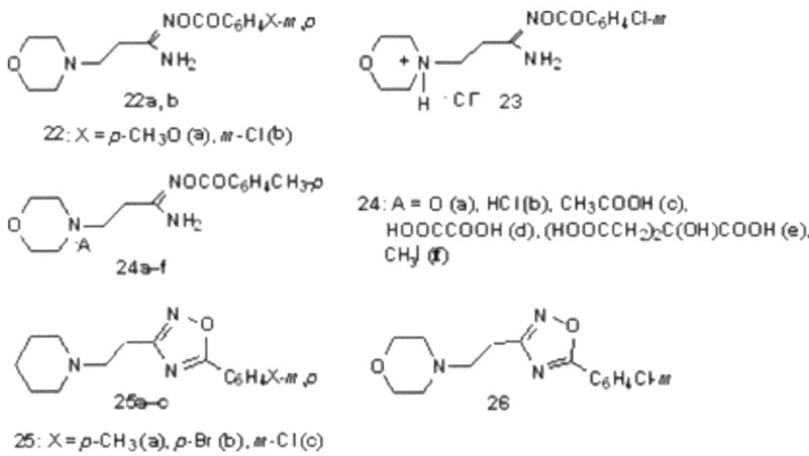

Compounds **22–24a** and **b**, **25**, and **26** were described [12, 15, 16].

Compounds **24b–f** were derived from the base of *O-p*-toluoyl-β-(morpholin-1-yl)propionamidoxime **24a** and were prepared in one step by adding of equivalent amounts of organic acids (acetic, oxalic, citric) and methyl iodide in various solvents. Acetate **24c** was prepared by reacting **24a** with a twofold excess of glacial AcOH in refluxing in EtOH.

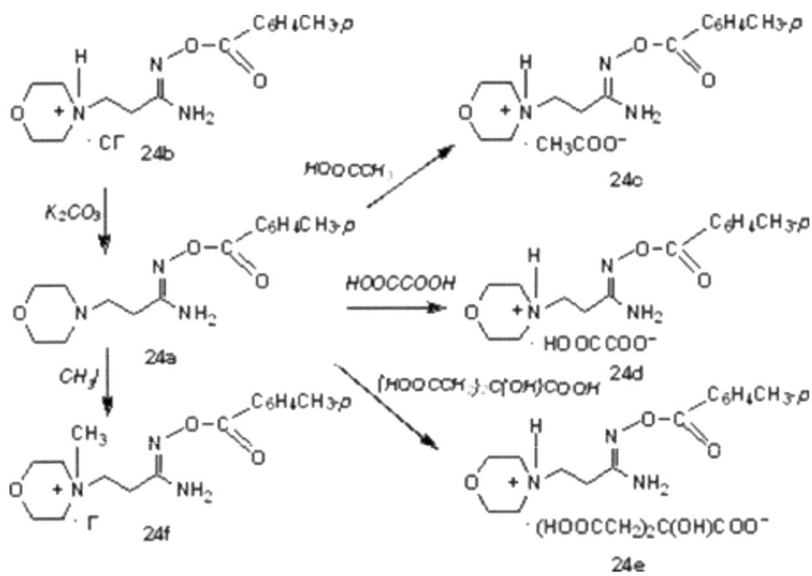

The *in vitro* activity of **22–26** for inhibition of α-amylase and α-glucosidase was tested using two series of experiments. **Table 5** presents the screening results using acarbose as the standard in both instances.

The greatest inhibitory activities (~50%) for α-amylase were found for O-*m*-chlorobenzoyl-β-(morpholin-1-yl)propionamidoxime (**22b**, 48%); 5-(*p*-bromophenyl)-3-[(β-piperidin-1-yl)ethyl]-1,2,4-oxadiazole (**25b**, 51%); and 5-(*m*-chlorophenyl)-3-[(β-morpholin-1-yl)ethyl]-1,2,4-oxadiazole (**26**, 48%). Moderate activity for α-amylase (from 27 to 43%) was found for O-*m*-chlorobenzoyl-β-(morpholin-1-yl)propionamidoxime hydrochloride (**23**, 35%); base *O-p*-toluoyl-β-(morpholin-1-yl)propionamidoxime (**24a**, 32.5%); citrate of *O-p*-toluoyl-β-

Compound		22a	22b	23	24a	24b	24c	24d	Acarbose
Inhibition, %	α-Amylase	—	48.0 ± 5.8	35.0 ± 0.6	32.5 ± 0.22	27.0 ± 5.5	25.6 ± 0.26	—	71.0 ± 2.7
	α-Glucosidase	78.7 ± 0.91**	23.0 ± 0.84	45.1 ± 1.99	22.8 ± 0.09	34.7 ± 1.36	27.4 ± 0.15	—	75.0 ± 1.32

Compound		24e	24f	25a	25b	25c	26	Acarbose
Inhibition, %	α-Amylase	37.0 ± 3.4	—	43.0 ± 3.0	51.0 ± 9.1	—	48.0 ± 5.9	71.0 ± 2.7
	α-Glucosidase	—	78.1 ± 4.41**	67.2 ± 0.82	68.7 ± 1.81	67.2 ± 1.79	61.7 ± 2.26	75.0 ± 1.32

*Activity absent (−).
**$p > 0.05$ vs. acarbose.

Table 5.
*Inhibitory activity of **22–26** for α-amylase and α-glucosidase, %*.*

(morpholin-1-yl)propionamidoxime (**24e**, 37%); and 5-(*p*-toluoyl)-3-[(β-piperidin-1-yl)ethyl]-1,2,4-oxadiazole (**25a**, 43%).

The highest inhibitory activities against α-glucosidase were exhibited by O-*p*-anisoyl-β-(morpholin-1-yl)propionamidoxime (**22a**, 78.7%); iodine methylate of O-*p*-toluoyl-β-(morpholin-1-yl)propionamidoxime (**24f**, 78.1%); and 5-(*m*-chlorophenyl)-3-[(β-morpholin-1-yl)ethyl]-1,2,4-oxadiazole (**26**, 61.7%).

Moderate inhibitory activity for α-glucosidase was manifested by O-*m*-chlorobenzoyl-β-(morpholin-1-yl)propionamidoxime (**22b**, 23%) and its hydrochloride (**23**, 45.1%).

The reference compound acarbose exhibited the standard inhibitory activity against α-amylase and α-glucosidase of 71.0 and 75.5%, respectively.

In conclusion, it is noteworthy that bases and pharmacologically acceptable salts of *O*-aroyl-β-aminopropionamidoximes and 5-substituted phenyl-3-β-(piperidin-1-yl and morpholin-1-yl)ethyl-1,2,4-oxadiazoles (**22–26**) showed more pronounced inhibitory activity for α-glucosidase than for α-amylase. Both **22a** and **24f** had α-glucosidase activity comparable with that of the standard acarbose.

A structure–activity relationship for two series of screening experiments found that, as a rule, 3,5-disubstituted 1,2,4-oxadiazoles exhibited greater inhibition of α-amylase and α-glucosidase than their chemical precursors, i.e., bases and pharmacologically acceptable salts of *O*-aroyl-β-aminopropionamidoximes.

Acknowledgements

Sincere gratitude is expressed to organic chemists and to the biological activity testers who provided support and understanding in resolving the practical issues addressed in this work: I.S. Zhumadildaeva, A.L. Ahelova, M.O. Orazbaeva, G.I. Gapparova, G.P. Baitursynova, A.B. Uzakova, G.T. Dyusembaeva, G.M. Pichkhadze, D.M. Kadyrova, G.S. Mukhamedzhanova, R.A. Agzamova, V.L. Bismilda, L.T. Chingisova, B.T. Toksanbaeva, A.E. Gulyaev, Z.T. Schulgau, and Sh. D. Sergazu.

The publication of this chapter was made possible thanks to the financial support for the basic research program on the topic "Physico-chemical fundamentals of creating inorganic, organic, polymer compounds, systems and materials with desired properties" and the award from the Ministry of Education and Science of the Republic of Kazakhstan No. 207 of 03/19/2018.

Author details

Lyudmila Kayukova*, Umirzak Jussipbekov and Kaldybay Praliyev
JSC "A. B. Bekturov Institute of Chemical Sciences", Almaty, Kazakhstan

*Address all correspondence to: lkayukova@mail.ru

References

[1] Kayukova LA. Mint: Synthesis of 1,2,4-oxadiazoles (a review). Pharmaceutical Chemistry Journal. 2005;**39**(10):539-547. DOI: 10.1007/s11094-006-0017-7

[2] Fylaktakidou KC, Hajipavlou-Litina DJ, Litinas KE, Varella EA, Nicolaides DN. Mint: Recent developments in the chemistry and in the biological applications of amidoximes. Current Pharmaceutical Design. 2008;**14**(10):1001-1047. DOI: 10.2174/138161208784139675

[3] Khalilova SF, Poplavskaya IA, Blonskaya LI, Blagodarnyi YA. Mint: N-arylacetylformamidoxime thiosemicarbazones and their tuberculostatic properties. Pharmaceutical Chemistry Journal. 1986;**20**(12):846-848. DOI: 10.1007/BF00763717

[4] Kayukova LA, Praliev KD, Akhelova AL, Kemel'bekov US, Pichkhadze GM, Mukhamedzhanova GS, et al. Mint: Local anesthetic activity of new amidoxime derivatives. Pharmaceutical Chemistry Journal. 2011;**45**(8):468-471. DOI: 10.1007/s11094-011-0657-0

[5] Fact Sheets WHO. The top 10 causes of death, 24 May 2018. [Internet]. 2018. Available from: https://www.who.int/news-room/fact-sheets/detail/the-top-10-causes-of-death [Accessed: 4 September 2019]

[6] WHO. Implementing the end TB strategy: The essentials. 2015/2016. 130p. ISBN: 978924150993; WHO/HTM/TB/2015.31. http://www.who.int/tb/publications/2015/The_Essentials_to_End_TB/en/ [Accessed: 4 September 2019]

[7] Kayukova LA, Orazbaeva MA, Bismilda VL. Highly active on sensitive strains of *M. tuberculosis* derivative of β-(thiomorpholin-1-yl)propioamidoxime. Pat. № 21544 RK. Bull. No. 12. (In Russian). 2011. Available from: https://www.kazpatent.kz/ru; https://gosreestr.kazpatent.kz/ [Accessed: 5 September 2019]

[8] Kayukova LA, Bismilda VL. Derivative of β-(piperazin-1-yl) propioamidoxime as a bactericidal agent against sensitive, resistant and multidrug-resistant strains of *M. tuberculosis*. Pat. № 21543 RK. Bull. No. 12. (In Russian). 2011. Available from: https://www.kazpatent.kz/ru; https://gosreestr.kazpatent.kz/ [Accessed: 5 September 2019]

[9] International Diabetes Federation. IDF Diabetes Atlas. 8th ed. Brussels, Belgium: International Diabetes Federation; 2017. Available from: https://diabetesatlas.org/resources/2017-atlas.html [Accessed: 4 September 2019]

[10] Peter LN, Zoltan S, Kalman T, Laszlo T, Kalman T, Jozsef M, et al. Reducing overweight or obese. Pat. № 2443417 Russia. Patentee: N-GIN Laboratories Research, Inc. (US). 2012. Available from: http://www.freepatent.ru/images/patents/15/2443417/patent-2443417.pdf [Accessed: 4 September 2019]

[11] Zbigniew M, Katarzhina R, Kludkevich D, Daniel S, Krzysztof K, Katarzyna M, et al. The novel derivatives of 3-phenylpropionic acid. Pat. № 2369602 Russia. Patentee: Adamed JV. Z OO. 2009. Available from: http://webcache.googleusercontent.com/search?q=cache:http: //bd.patent.su/2369000-2369999/pat/servl/servlet6a95.html [Accessed: 4 September 2019]

[12] Kayukova LA, Uzakova AB, Baitursynova GP, Dyusembaeva GT, Shul'gau ZT, Gulyaev AE, et al. Mint: Inhibition of α-amylase and

α-glucosidase by new β-aminopropionamidoxime derivatives. Pharmaceutical Chemistry Journal. 2019;**53**(2):129-133. DOI: 10.1007/s11094-019-01966-5

[13] Poce G, Bates RH, Alfonso S, Cocozza M, Porretta GC, Ballell L, et al. Mint: Improved BM212 MmpL3 inhibitor analogue shows efficacy in acute murine model of tuberculosis infection. PLoS ONE. 2013;**8**(2):e56980. DOI: 10.1371/journal.pone.0056980

[14] Biava M, Porretta GC, Poce G, Battilocchio C, Alfonso S, De Logu A, et al. Mint: Identification of a novel pyrrole derivative endowed with antimycobacterial activity and protection index comparable to that of the current antitubercular drugs streptomycin and rifampin. Bioorganic & Medicinal Chemistry. 2010;**18**(22): 8076-8084. DOI: 10.1016/j.bmc.2010.09.006

[15] Kayukova LA, Praliev KD, Zhumadil'daeva IS. Mint: Cyclization of *O*-benzoyl-β-piperidinopropionamidoximes to form 5-phenyl-3-(β-piperidino)ethyl-1,2,4-oxadiazoles. Russian Chemical Bulletin. 2002;**51**(11):2100-2105. DOI: 10.1023/A:1021628430346

[16] Kayukova LA. Mint: Conditions for the heterocyclization of *O*-aroyl-β-morpholinopropioamidoximes to 5-aryl-3-(β-morpholino)ethyl-1,2,4-oxadiazoles. Chemistry of Heterocyclic Compounds. 2003;**39**(2):223-227. DOI: 10.1023/A:1023724626003

Chapter 4

Recent Developments of Target-Based Benzimidazole Derivatives as Potential Anticancer Agents

Nerella Sridhar Goud, Pardeep Kumar and Rose Dawn Bharath

Abstract

Cancer is one of the major life burdens and around 18.1 million new cancer cases and 9.6 million deaths have been estimated in 2018 globally. Recent reports of the World Health Organization (WHO) stated that about one in six death cases globally is mainly due to cancer. Hence, the development of efficacious drugs with novel mechanisms is necessary for various cancer types. The chemotherapy drug resistance and non-selectivity toward targets have turned the current cancer research on to the highly emerging selective targets for the development of potential anticancer agents. Benzimidazole is regarded as an essential pharmacophore of the cancer research because of wide anticancer potentials with versatile mechanisms to inhibit the tumor progression and also facile synthetic strategies for an easy synthesis of various benzimidazole derivatives. The selective anticancer potentials also depend on the substitution of the benzimidazole nucleus. Therefore, it would lead to providing a path for the development of novel target-specific and highly effective benzimidazole-based anticancer agents.

Keywords: benzimidazole, cancer, specific targets, synthetic strategies

1. Introduction to cancer

Cancer is one of the dreadful diseases in the world and mainly characterized by uncontrolled cell proliferation. Worldwide, one in six women and one in five men develop cancer during their lifetime, and one in eleven women and one in eight men die from the disease. Global data clearly show that nearly half of the new cases and more than half of the cancer deaths worldwide in 2018 are estimated to occur in Asian countries because the region has nearly 60% of the global population and it is estimated to have a rise of over 21.4 million new cases per year, with 13.2 million cancer deaths, by 2030. The top three cancer types *viz.* breast, lung, and colorectal are responsible for one-third of the cancer incidence and mortality burden worldwide [1, 2]. Behavioral risk factors such as tobacco usage and smoking; physical risk factors such as exposure to ionizing radiations and asbestos; and genetic predominant factors are the main contributors to cancer. Even though utmost care has been taken, the disease still causes the death of millions of people globally [3].

Although scientific advances have focused on knowing the exact pathophysiology of the disease and tremendous efforts have been made on early diagnosis of cancer, the overall mortality rate has not subsided. Moreover, the cancer survival rate tends to be extremely low in some developing countries. This is due to the combination of both late-stage detection and limited access to time and qualitative treatment [4].

Radiotherapy, surgery, and chemotherapy are the usual cancer treatment strategies [5]. Among these, chemotherapy is considered as one of the efficient and first-line strategies in suppressing tumor prognosis and eradication. Most of the chemotherapeutic drugs target the key cellular mechanisms and inhibit the cell division and thereby prevent cancer cell multiplication. Current clinical anticancer drugs usually act on metabolically effective or fast replicating cells and show drawbacks such as poor selectivity between cancer cells and healthy cells [6]. Cancer cells generally disturb the cell signaling pathways and tissue morphogenesis for the neoplastic propagation of tumors. Therefore, targeting these cell pathways by cytotoxic agents has been a proven therapeutic approach to subside tumor growth and disease progression. Unfortunately, most of the cytotoxic drugs cause side effects due to the poor selectivity and specificity toward cancer cells. However, the higher toxic profiles and poor tolerance of the present chemotherapeutic drugs are major obstacles to the effective treatment of cancer [7, 8]. Therefore, it is highly pertinent to design and synthesize new anticancer agents with improved efficiency and reduced side effects to complement the present

DNA interacting agents	
Alkylating agents	**Alkylation of DNA bases** **Procarbazine, dacarbazine, and temozolomide**
DNA cleaving agents	Cause strand scission at the binding site-Bleomycin
Cross-linking agents	Binding to DNA results in intra- and inter-strand cross-linking Platinum complexes-carboplatin, cisplatin, oxaliplatin Nitrogen mustards-cyclophosphamide, ifosfamide
Intercalating agents	Stacking between DNA base pairs Anthracyclines-doxorubicin, epirubicin Mitoxantrone and actinomycin-D
Topoisomerase inhibitors	Topoisomerase I-camptothecins Topoisomerase II-Anthracyclines, etoposide
Anti-metabolites	
Purine analogues	Mercaptopurine
Pyrimidine analogues	5-Fluorouracil
DHFR inhibitors	Methotrexate
Antitubulin agents	
Taxol	Paclitaxel, Docetaxel
Vinca alkaloids	Vincristine, Vinblastine, Vinorelbine
Tyrosine kinase inhibitors	
Small molecule	Imatinib (Gleevec): inhibits ABL, c-Kit kinase, PDGFR Gefitinib (Iressa): inhibits EGFR
Monoclonal antibody	Trastuzumab: inhibits EGFR2, HER2
Angiogenesis/ Metastasis inhibitors	
Monoclonal antibody	Bevacizumab (Avastin): targets VEGF

Table 1.
Common anticancer drugs along with their mechanisms of action.

chemotherapeutic approaches. Identifying new drugs and drug combinations for cancer treatment is essential to combat this lethal disease. Hence, further research that emphasizes mainly on the development of efficient chemotherapeutic agents is an emerging area of research in the field of medicinal chemistry. The list of various available chemotherapeutic agents has been shown in **Table 1** [9].

2. Introduction to Benzimidazole

Benzimidazole heterocyclic nucleus can be termed as "Master Key" due to its overwhelming biological profile and synthetic applications in medicinal chemistry. It is among the top five most common five-membered aromatic nitrogen heterocycles in U.S. FDA-approved pharmaceutical drugs [10]. Benzimidazoles are structural isosteres of nucleobases due to the fused nitrogen nuclei and they readily interact with biomolecular targets and elicit many biological activities such as anticancer [11], anti-inflammatory [12], antiulcer [13], anti-hypertensive [14], and anthelmintic [15]. Akhtar et al. in his recent review described the therapeutic evolution of benzimidazole scaffolds during the last quinquennial period [16]. This nitrogen-containing heterocycle was present in a number of well-established clinical drugs with diverse therapeutic activities. For instance, drugs like rabeprazole (**1**) and omeprazole (**2**) are benzimidazole-containing drugs, act as proton pump inhibitors, and are, therefore, used in the treatment of stomach ulcers [17]. Albendazole (**3**) and thiabendazole (**4**) are anthelmintic drugs that act by the inhibition of tubulin polymerization and impair the uptake of glucose, eventually leading to the death of the parasites [18]. Nocodazole (**5**) is a well-recognized antineoplastic agent that mainly acts by tubulin polymerization inhibition. Candesartan (**6**) is a benzimidazole-based orally active potent angiotensin II receptor antagonist that is used for the treatment of hypertension [19]. Bendamustine (7) is nitrogen mustard which belongs to alkylating agents, a class of chemotherapeutic agent and used in the treatment of chronic lymphomas [20]. Dovotininb (**8**) is the orally active benzimidazole quinolinone compound with potential antineoplastic activity (**Figure 1**). It strongly binds to the fibroblast growth receptor 3 (FGFR3) and inhibits its phosphorylation and induces tumor cell death [21].

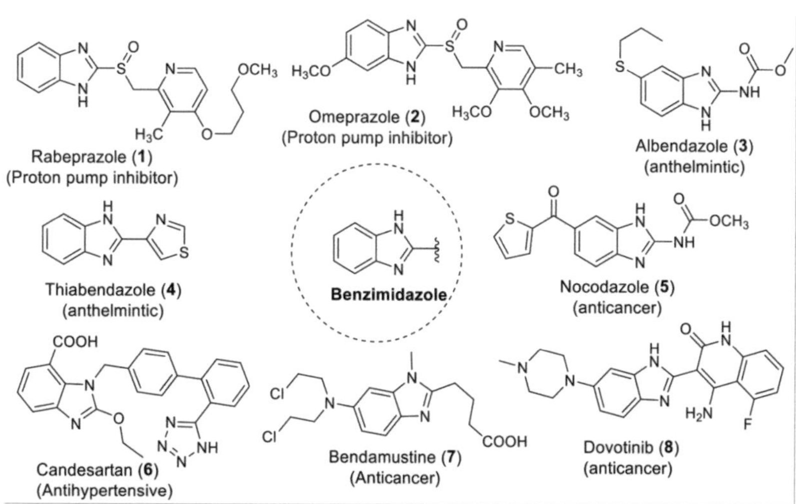

Figure 1.
Examples of drugs and other bioactive molecules containing benzimidazole motif.

In 1954, Tamm, Folkers, and co-workers first reported the synthesis and antiviral activities of halogenated benzimidazole nucleosides [22]. They found that 5,6-dichloro-1-β-D-ribofuranosyl benzimidazole (DRB) has multiple biological activities including activity against RNA and DNA viruses. DRB inhibits cellular RNA polymerase II, thus affecting the multiple cellular processes so that it is more cytotoxic than antiviral. Slayden et al. found that albendazole (**3**) and thiabendazole (**4**) known tubulin inhibitors interfered and delayed the *Mtb* cell division processes [23]. Later, Kumar et al. proposed that the benzimidazole core would be a novel FtsZ inhibitor, which will have activity against both drug-sensitive and drug-resistant *Mtb* [24]. This molecular framework displays numerous biological properties and is usually present in various drug compositions. Benzimidazoles tethered to various bioactive pharmacophores have also displayed potent antitumor activities.

Benzimidazoles have revolutionized the drug discovery process by their diverse range of biological activities, which make this scaffold an indispensable anchor for the innovation of novel therapeutic agents. Thus, the therapeutic potential of the benzimidazole and related drugs has attracted researchers to design and synthesize more potent derivatives with a wide range of pharmacological activities. Owing to the immense synthetic value and extended bioactivities exhibited by benzimidazoles and their derivatives, efforts have been made from time to time to create libraries of these compounds.

3. Target-based benzimidazole derivatives

3.1 Galectin-1 inhibitors

Galectin-1 (Gal-1) is expressed in various normal and pathological conditions and has multiple functions with a wide range of biological activity. Gal-1, a human homodimeric lectin protein of 14KDa, is implicated in many signaling pathways, immune responses associated with cancer progression, neurological conditions, and immune disorders [25]. Gal-1 has a carbohydrate recognition domain (CRD), which is selective toward β-galactosides in the body. Inhibition of human Gal-1 has been regarded as one of the potential therapeutic approaches for the treatment of cancer, as it plays a major role in tumor development and metastasis by modulating various biological functions viz. angiogenesis, apoptosis, migration, and cell immune escape [26]. The overexpression of Gal-1 has been reported in many cancer types like the brain, breast, osteosarcoma, lung, prostate, melanoma, etc. [27]. Gal-1 can mediate neoplastic transformation by interacting with oncogenes, such as H-Ras and promote Ras-mediated signal transduction involving RAF1 and extracellular signal-regulated kinase (ERK). Gal-1 multivalently mediates tumor cell-ECM adhesion at the primary site by cross-linking cell surface glycoproteins, such as integrins, and glycosylated proteins in the ECM, such as laminin and fibronectin [28]. Hence, Gal-1 is regarded as a promising molecular target for the development of new therapeutic drugs for cancer.

Recently, a new series of 1-benzyl-1H-benzimidazole derivatives have been synthesized as Gal-1-mediated anticancer agents. The target compound (**9**) showed significant growth inhibition against breast cancer (MCF-7) cells with an IC_{50} value of 7.01 ± 0.20 μM. The target compound also showed good cytotoxicity in the range of 10.69–14.04 μM against colorectal cancer (HCT-116), breast cancer (MDA-MB-231), prostate cancer (DU-145), and lung cancer (A-549). In addition, *in-vitro* Gal-1 expression in cell supernatant of MCF-7 cells with compound (**9**) was measured in enzymatic GAL-1 ELISA studies and found to show dose-dependent reduction from 10 to 300 μM. The target compound showed Gal-1-mediated apoptosis, which was confirmed by morphological changes in MCF-7-treated cells like blebbing, cell wall deformation, and cell shrinkage, based on the apoptosis studies such as Acridine Orange/Ethidium Bromide (AO/EB) staining, DAPI nucleic acid staining, mitochondrial membrane potential, annexin V/propidium iodide dual staining assay, and dichlorofluorescein (DCF) fluorescence studies. In cell cycle analysis, the target compound selectively arrested MCF-7 cell growth at the G2/M phase and S phase. Further, the binding specificity of target compound toward Gal-1 was confirmed by surface plasmon resonance and fluorescence spectroscopy studies and the specific binding constant value (K_a) of $1.2 \times 10^4\ M^{-1}$ was observed in fluorescence spectroscopy studies, whereas the equilibrium constant (KD) value of 5.76×10^{-4} M was observed in surface plasmon resonance studies. The binding of the target compound to Gal-1 was also confirmed by RP-HPLC studies and found to show 85.44% of binding to Gal-1. The molecular docking studies were also supported based on the strong amino acid interactions such as ARG48, TRP68, and ASP125 with the target compound [29, 30].

Tsung-Chieh Shih et al. reported a novel Gal-1 inhibitor named LLS2 (**10**), which was discovered through the One-Bead-Two-Compound library. The interaction of target gal-1 with LLS2 was confirmed by LC-MS/MS analytical and pull-down assay. The binding complex of LLS2 with Gal-1 selectively decreases membrane-specific H-Ras, and K-Ras pathways, lead to involve in the apoptosis process. The LLS2 exhibited a synergistic effect in combination with paclitaxel against many of the human cancer cell lines such as pancreatic cancer, ovarian cancer, and breast cancer cells *in vitro*. The combination of paclitaxel with LLS2 efficiently reduces the growth of ovarian cancer xenografts in athymic mice *in vivo* (**Figure 2**).

The same group recently published a more potent Gal-1 inhibitor LLS3 (**11**), it impairs castration-resistant prostate cancer progression and invasion. LLS3 targets Gal-1 as an allosteric inhibitor, and reduces Gal-1 binding affinity toward its binding partners and also causes suppression of Akt, and AR signaling pathways. LLS3 showed *in vivo* efficacy in both androgen receptor-positive and negative xenograft models. In addition to potentiating the anticancer effect of docetaxel to cause suppression of tumors, it also efficiently suppresses the progression of prostate cancer cells *in vivo* [31, 32].

3.2 Tubulin protein inhibitors

Tubulin is one of the members of a small family of globular proteins. Several isoforms are present out of which α- and β-tubulins are the most common members of tubulin. The cellular protein tubulin is an important protein for replication.

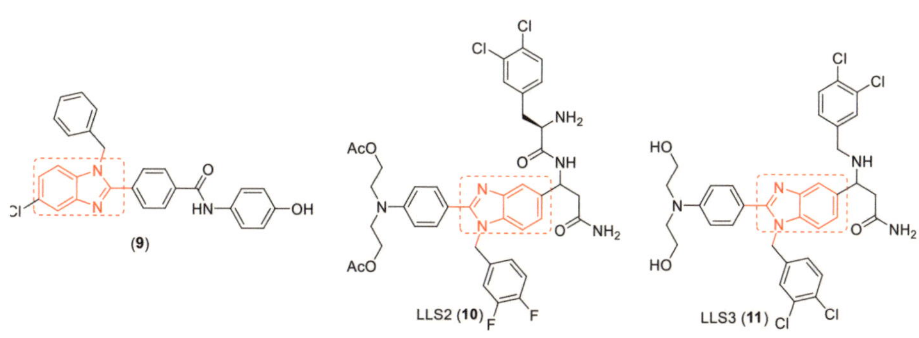

Figure 2.
The novel benzimidazole derivatives as Gal-1 mediated anticancer agents.

Microtubules are hallowing filaments and composed of head and tail polar fashion arrangements of α- and β-tubulins as the constituent subunits. Microtubules contain 13 active protofilaments aligned parallel with the whole axis of the microtubule cylinder. This may provide continuous transport of cellular materials by motor proteins (dynein and kinesin) over distant places. Microtubules also form an integral part of the cytoskeleton and are responsible for the maintenance of cell shape, and motility and intracellular transport of the vesicles, mitochondria, and other components [33, 34]. Moreover, cell division involves the duplication of DNA and the segregation of the replicated chromosomes into two daughter nuclei. The segregation of these chromosomes is mitotic phase is brought by the microtubules. In the formation of the microtubule, the plus (+) end is terminated by β-tubulin whereas the minus (−) end is terminated by α-tubulin. They are always either in a state of polymerization or depolymerization. Microtubules have the ability to shorten or lengthen in a scholastic fashion through loss or addition of α/β-tubulin heterodimers from ends of microtubules. This property is referred to as "dynamic instability" [35, 36]. Microtubules are blessed with a property to grow continuously as long as the free tubulin amount is above a critical level. The critical concentration at the minus end is somewhat higher than at the plus end and the minus end tends to stop growing first. Even above the critical tubulin concentration, its end may suddenly stop growing and begin to shrink. The change from growth to shrinkage has been termed as "catastrophe." After some time, a shrinking microtubule end may "pause" and/or begin to grow again; the latter process is known as "rescue." During mitotic cell division, the chromosomes are segregated by the mitotic spindle, which is formed from tubulin microtubules. Therefore, tubulin dynamics have a distinct role in cell division. Some of the drugs affect the microtubulin dynamics and thus cause either polymerization or depolymerization and thereby alter cellular replication. So at the mechanistic level, tubulin is one of the most attractive and challenging approaches for designing new anticancer compounds.

Zhang et al. have synthesized a series of 1,2-diarylbenzimidazole derivatives and reported as potential anticancer agents. Among all, the target molecule (**12**) has been found to show significant cytotoxicity against human cancer cells such as A549, HepG2, HeLa, and MCF-7 cells in the range of GI_{50} = 0.71–2.41 μM and also found to show normal cytotoxicity toward normal cells. The apoptosis process by the target compound was confirmed by morphological changes on HepG2 and HeLa-treated cells like cell wall deformation, blebbing, and cell shrinkage, based on apoptosis studies such as mitochondrial membrane potential, annexin V/propidium iodide dual staining assay, and dichlorofluorescein (DCF) fluorescence studies. In cell cycle analysis, the target compound selectively arrested tumor growth at the G2/M phase. Further, the target compound showed significant inhibition of microtubule polymerization with an IC_{50} value of 8.47 μM. The molecular docking simulation studies were performed to confirm the binding of the target compound with microtubule protein and found that the target compound has made strong interactions with protein [37].

Miao et al. reported a novel series of 2-aryl-benzimidazole-based dehydroabietic acid derivatives as potential cytotoxic agents via targeting tubulin polymerization. The synthesized molecules were characterized by elemental and analytical techniques. The target compound (**13**) showed significant growth inhibition against hepatocarcinoma cancer (SMMC-7721) cells with an IC_{50} value of 0.08 ± 0.01 μM. The target compound also showed good cytotoxicity in the range of 0.04–0.07 μM against breast cancer (MDA-MB-231), cervical cancer (HeLa), and colon cancer (CT-26). The apoptosis studies such as ROS levels measurements, loss of mitochondrial membrane potential, and cell cycle analysis were performed to confirm the induction of apoptosis in hepatocarcinoma cancer (SMMC-7721) cells. In cell cycle analysis, the target compound selectively arrested tumor growth at the G2/M phase. Further, the target compound showed significant inhibition of microtubule polymerization with an IC_{50} of 5 μM. The molecular docking studies supported the selectivity of the target compound to tubulin protein based on strong electronic interactions between the target compounds and tubulin [38].

Wang et al. reported a new series of benzimidazole containing benzsulfamide-pyrazole ring derivatives as potential tubulin polymerization inhibitors. The target compound (**14**) showed significant growth inhibition against lung cancer (A549) cells with an IC_{50} value of 0.15 ± 0.05 μM and also showed good growth inhibition against Hela, HepG2, and MCF-7 cell lines in the range of 0.17–0.33 μM concentration. Further, the target compound showed significant inhibition of microtubule polymerization with an IC_{50} value of 1.52 μM. In cell cycle analysis, the target compound selectively arrested A549 cell growth at the G2/M phase. The target compound showed A549 cell apoptosis based on the studies of annexin V/ propidium iodide dual staining assay and cell cycle analysis. The molecular docking studies were also supported based on the strong amino acid interactions such as LYS 352, LYS 254, ASN 258, and CYS 241 with the target compound [39] (**Figure 3**).

Baig et al. have reported a series of imidazo [2,1-b] thiazole-benzimidazole derivatives as antiproliferative agents via tubulin polymerization inhibition. The target molecule (**15**) has shown significant cytotoxicity against human lung (A549) cancer with an IC_{50} value of 1.08 μM. It also showed good cytotoxicity toward DU-145 (prostate), MCF-7 (breast cancer), A549 (lung cancer), and HeLa (cervical cancer) in the range of 1.65–7.55 μM. In cell cycle analysis, the target compound selectively arrested A549 cell growth at the G2/M phase. The target compound showed apoptosis, which was confirmed by morphological changes in A549-treated cells like blebbing, cell wall deformation, and cell shrinkage, based on the apoptosis studies such as Hoechst staining, mitochondrial membrane potential, annexin V/ propidium iodide dual staining assay. Further, the target compound exhibits a significant inhibition of microtubule assembly with an IC_{50} of 1.68 μM. The computational studies revealed that the target compound can easily be occupied in the colchicine binding site of the protein [40].

3.3 Carbonic anhydrase inhibitors

The human carbonic anhydrases (hCAs) are an α-family of carbonic anhydrases class and exist in 16 different isoforms [41]. Based on their location in the body,

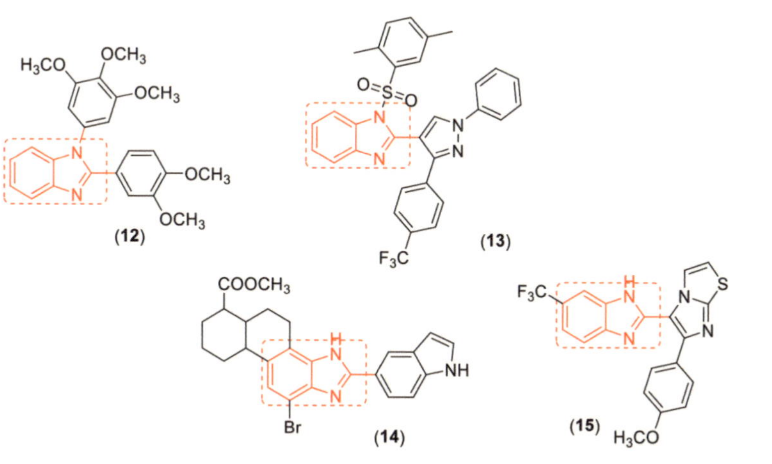

Figure 3.
The target benzimidazole derivatives as selective anticancer agents via targeting tubulin polymerization.

they are classified into cytosolic hCAs such as CA I, CA II, CA III, CA VII, and CA XIII; transmembrane hCAs such as CA IV, CA IX, CA XII, CA XIV, and CA XV; mitochondrial-bound hCAs such as CA Va and Vb; secretory hCAs such as CA VI; and catalytically inactive isoforms like CA VIII, CA X, and CA XI, which are considered as CA-related proteins (CARPs) [42]. Among all, the hCA isoforms IX and XII are overexpressed in many of cancer types as these are tumor-associated transmembrane bound enzymes, mainly hypoxic tumors, which are regarded as emerging potential targets for various tumor types [43]. The overexpression of hCA isoforms IX and XII further contributes to the tumor progression, angiogenesis, metastasis, and proliferation of a variety of tumor cells [44]. In order to exhibit potential cytotoxicity without adverse effects, an anticancer agent should selectively inhibit tumor-associated hCAs IX and XII over other hCAs. Therefore, current cancer research focuses on the development of various heterocycles that selectively target tumor-linked hCA isoforms IX and XII for effective treatment strategies in cancer therapy [45]. Another hCA isoform II is also found to overexpress in some forms of cancer and other conditions like edema, glaucoma, and epilepsy.

Recently, a new series of 2-substituted-benzimidazole-6-sulfonamides have been reported as anticancer potentials by testing against four physiologically relevant hCAs such as CA I, CA II, CA IX, and CA XII. The analysis of hCA inhibition results showed that the new series of benzimidazole-based sulfonamide derivatives exhibited selective inhibition toward tumor-associated isoforms such as CA IX and CA XII. The target molecule (**16**) of this series had shown a promising inhibition at low μM range against hCA IX and XII isoform, with an inhibitory constant (K_i) value of 2.2 and 22.3 μM. Another potent compound (**17**) also exhibited good inhibition at low μM range against hCA IX and XII, with an inhibitory constant (K_i) value of 5.9 and 7.9 μM respectively. Hence, it is concluded that these benzimidazole derivatives might be potential anticancer agents exhibiting a novel mechanism through inhibition of hCA isoforms IX and XII [46]. Asta Zubriene et al. have reported a series of novel benzenesulfonamides with benzimidazole derivatives as selective human carbonic anhydrase I, II, VII, XII, and XIII inhibitors. The target molecules were synthesized from the precursor benzimidazole derivative with different phenacyl bromides. The target molecules (**18, 19**) were evaluated against five physiological relevant hCA isoforms (hCA, EC 4.2.1.1) CA I, CA II, CA VII, CA XII, and CA XIII. The target compound exhibited a promising inhibitory action at a lower nanomolar level against selected hCAs with an inhibitory constant (Ki) value range of 1.67–66.7 μM. Another target molecule has shown significant inhibition at lower nanomolar level against selected hCAs with an inhibitory constant (Ki) value range of 2.86–62.5 μM [47] (**Figure 4**).

3.4 Epidermal growth factor receptor (EGFR) inhibitors

The Epidermal Growth Factor Receptor is a subfamily transmembrane glycoprotein (ErbB-1) of ErbB class of tyrosine kinase receptors and, other subfamilies include HER2/neu (ErbB-2), Her 3 (ErbB-3) and, Her 4 (ErbB-4) [48]. The internal ligands like EGF and TGFα facilitate the growth-promoting signal to cells by interacting with EGFR receptors and regulate epithelial tissue development and homeostasis [49, 50]. In cancer, especially epithelial malignancies, due to overproduction of EGFR ligands in the tumor micro environment causes continual activation (or) mutations of EGFR receptors, result in enhances epithelial tumor growth, metastasis and invasion [51, 52].

In a recent study, a new series of benzimidazole-based triazole and thiadiazole derivatives were synthesized and evaluated as selective EGFR inhibitors. The single-crystal X-ray crystallographic analysis has been performed to confirm the molecular

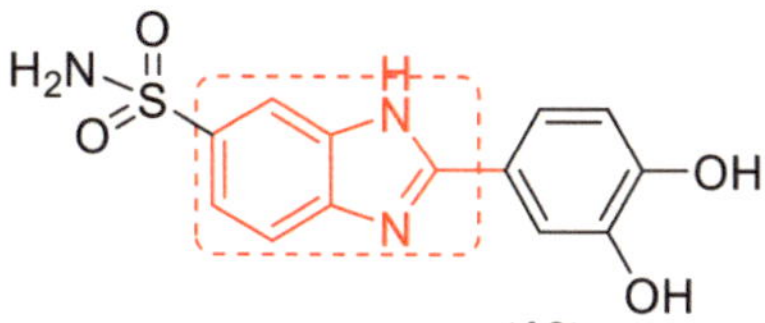

(16)

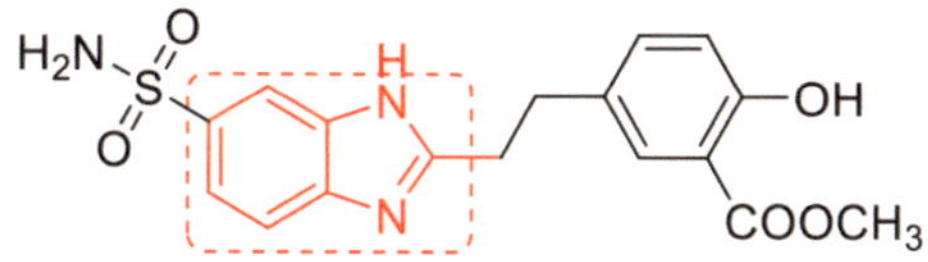

(17)

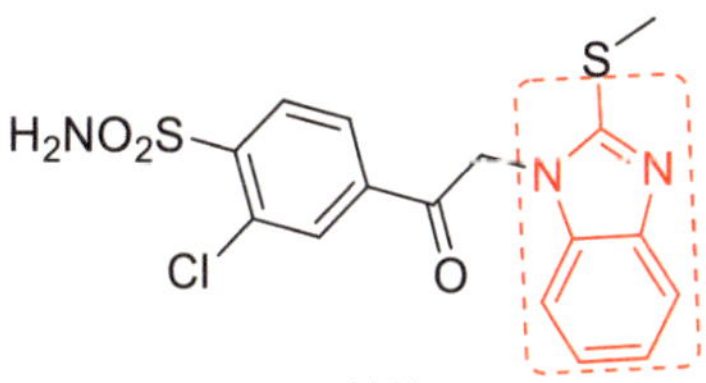

(18)

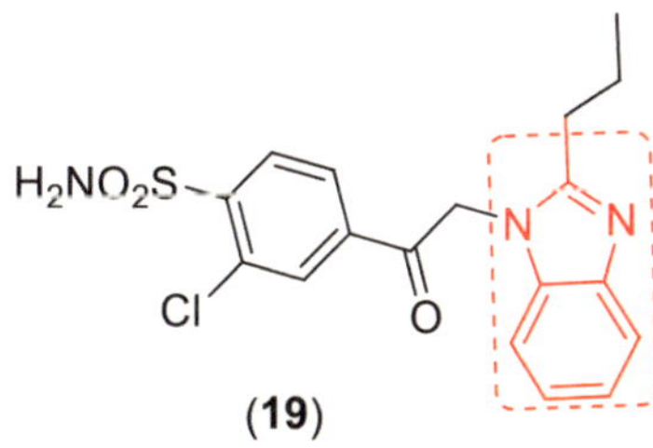

(19)

Figure 4.
The benzimidazole derivatives as human carbonic anhydrase enzyme mediated anticancer agents.

structure of the target compound. The synthesized compounds were evaluated for their EGFR kinase inhibitory potencies with erlotinib as the reference standard and, most of the compounds showed promising activities. The cell inhibition studies were also performed and the target compound (**20**) exhibited a significant inhibition and exhibited EGFR kinase inhibitory activity (over ≥30%) against MCF7 cells. The molecular docking studies indicated that the target compound showed two-hydrogen bonding interactions with residues of LYS721 and THR830 at the binding site of EGFR tyrosine kinase [53]. Akhtar et al. reported the benzimidazole-oxadiazole hybrids as selective EGFR and erbB2 receptor inhibitors. In *in vitro* cell inhibition studies, the target compound (**21**) exhibited a significant inhibition with an IC_{50} of 5.0 μM against breast cancer (MCF-7) cells. The target compound was found to show significant inhibition of EGFR and erbB2 receptor at 0.081 and 0.098 μM respectively. Most of the synthesized compounds exhibited a good cytotoxic activity against selected human cancer cell lines. In cell cycle analysis, the target compound selectively arrested MCF-7 cell growth at the G2/M phase. The computational and 3D-QSAR studies indicated that the target compound exhibited strong interactions with Asp831, Met769, and Thr830 of the EGFR enzyme [54].

Akhtar et al. have synthesized benzimidazole-based pyrazole derivatives through a one-pot multicomponent reaction and evaluated them for their potential anticancer activities. The synthesized compounds were screened against selected human cancer cell lines such as MCF-7, MDA-MB231, A549, HepG2, and HaCaT. The evaluation of EGFR inhibitory activities was performed for all the synthesized compounds. The target compound (**22**) exhibited promising cytotoxicity against the lung (A549) cancer cell lines with an IC_{50} value of 2.2 mM and the EGFR receptor inhibition value with an IC_{50} of 0.97 mM. In cell cycle analysis, the target compound selectively arrested A549 cell growth at the G2/M phase. In addition, it suppressed the growth of lung cancer cells by inducing apoptosis. In molecular docking studies, the target compound showed strong electronic interactions with Met769, Thr830, Lys721, and Phe699 of the active pocket of the EGFR receptor [55]. Yuan et al. have synthesized a library of 6-amide-2-aryl benzoxazole/benzimidazole derivatives and evaluated them for their selective inhibitory activities against VEGFR-2. The library of compounds exhibited selective anticancer activity against the liver hepatocellular carcinoma (HepG2), and human umbilical vein endothelial cells (HUVECs) over the lung cancer (A549) and breast (MDA-MB-231) cancer

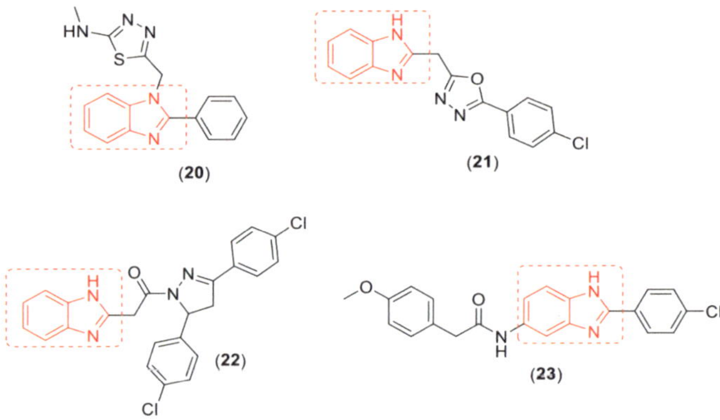

Figure 5.
The benzimidazole derivatives as selective anticancer agents via targeting EGFR.

cell lines. The target compound exhibited a significant growth inhibition against HepG2 and HUVEC with IC_{50} values of 1.47 and 2.57 mM, respectively. The target compound (**23**) showed anti-angiogenesis ability (79% inhibition at 10 nM/eggs) by chick chorioallantoic membrane (CAM) assay and exhibited excellent VEGFR-2 kinase inhibition with an IC_{50} of 0.051 mM. The computational analysis showed that the target compound made strong interactions with the active site of VEGFR-2 kinase. It is concluded that the 6-amide-2-arylbenzoxazole/benzimidazole derivatives are essential inhibitors of VEGFR-2 kinase for the treatment of anti-angiogenesis [56] (**Figure 5**).

4. Miscellaneous agents

Wu et al. synthesized a series of novel benzimidazole-2-substituted phenyl or pyridine propyl ketene derivatives and two representative compounds (**24**) and (**25**) showed significant inhibitory activity against colorectal (HCT116), breast (MCF-7), and liver (HepG2) cell lines, and effective inhibition of tumor growth in BALB/c mice with colon carcinoma HCT116 cells [57]. Reddy et al. reported a series of pyrazole-containing benzimidazole hybrids and evaluated them for their potential anti-proliferative activity against lung (A549), breast (MCF-7), and cervical (HeLa) cell lines. The compounds (**26**) and (**27**) showed potent growth inhibition against all the cell lines tested, with IC_{50} values in the range of 0.83–1.81 μM [58]. Gowda et al. synthesized a series of novel 1-(4-methoxyphenethyl)-1H-benzimidazole-5-carboxylic acid derivatives and the compound (**28**) induced maximum cell death in leukemic cells (K562 and CEM cell lines), through inducing apoptosis via S/G2 cell cycle arrest; down regulation of CDK2, Cyclin B1 and PCNA; cleavage of PARP; and elevated levels of DNA strand breaks [59]. Akhtar et al. reported a series of benzimidazole-linked oxadizole hybrids and the compounds were screened for their anticancer and *in vitro* EGFR and erbB2 receptor inhibition assay. Two of the compounds (**29**) and (**30**) displayed promising activity. The compound 70a showed EGFR inhibition and induced apoptosis by G2/M cell cycle arrest [54] (**Figure 6**).

5. Synthetic strategies

The first benzimidazole (2,5-dimethylbenzimidazole) (**3**) or 2,6-dimethylbenzimidazole (**4**) was prepared in 1872 by Hoebrecker through reduction of

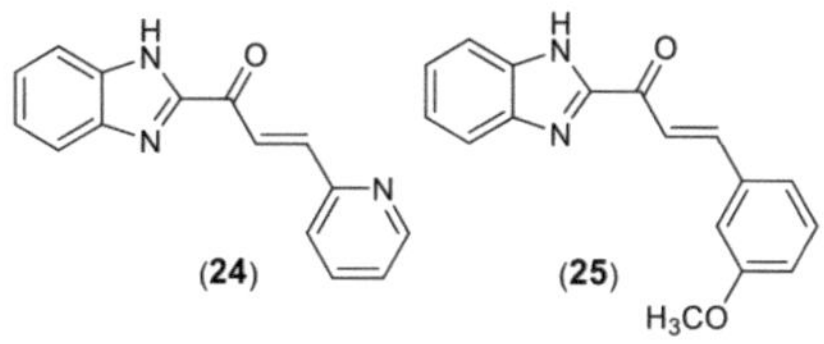

(24) (25)

Antiproliferative activity, IC_{50} in μM		
HCT116	MCF7	HepG2
0.062±1.45	3.64 ± 0.39	0.34 ± 0.68
0.03 ± 0.02	9.80± 1.35	0.75 ± 5.41

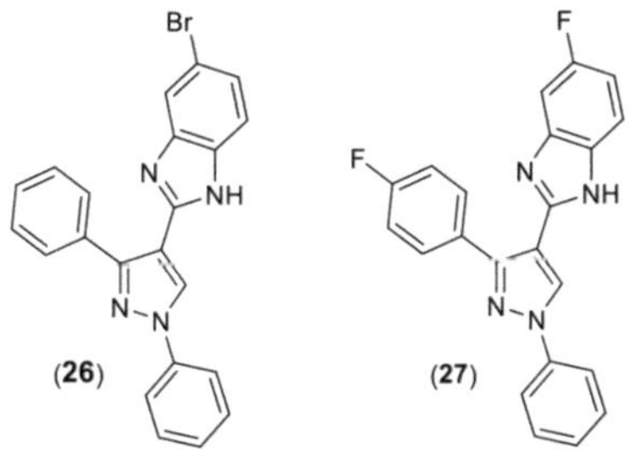

(26) (27)

Antiproliferative activity, IC_{50} in μM		
A549	MCF-7	HeLa
1.81±0.4	0.83±0.3	1.76±0.7
1.13±0.2	0.95±0.3	1.57±0.3

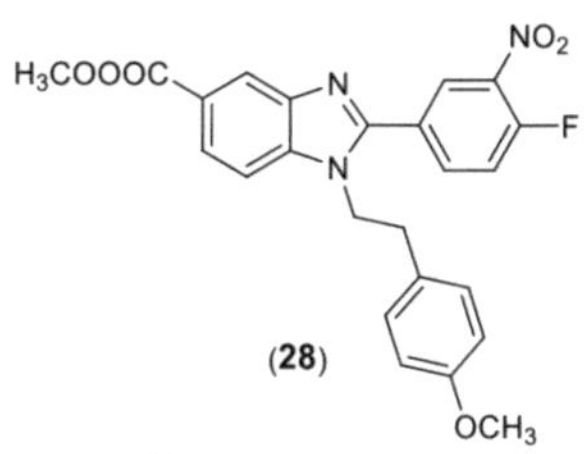

(28)

Antiproliferative activity, IC_{50} in μM	
K562	CEM
3	4

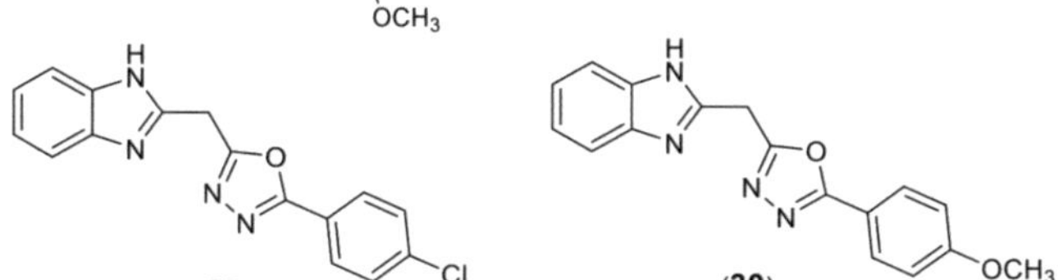

(29) (30)

Antiproliferative activity, IC_{50} in μM						
EGFR	erbB2	MCF7	HaCaT	MDA-MB231	HepG2	A549
0.081	0.61	5.0	9.51	4.51	2.51	5.2
0.098	0.91	2.5	3.8	0.131	15.6	13.2

Figure 6.
The novel benzimidazole derivatives as potential anticancer agents.

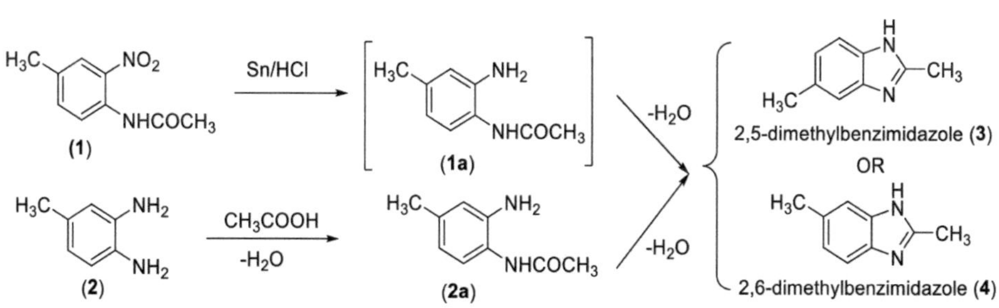

Figure 7.
General syntheses of benzimidazoles from aniline derivatives.

2-nitro-4-methylacetanilide [60] (**1**) (**Figure 7**). Several years later, the synthesis of benzimidazole was reported by refluxing 3,4-diamino toluene (**2**) with acetic acid [61]. Many synthetic ways toward the construction of benzimidazole ring started from commercially available benzene derivatives containing nitrogen functionalities, especially ortho derivatives. Hence, a number of methods have been reported for the synthesis of bioactive benzimidazoles and their derivatives. The majority of these involve the condensation of *O*-phenylene diamines (**5**) and its derivatives with carboxylic acids (**6**), esters, alcohols, or aldehydes [62].

Synthesis of benzimidazoles in the presence of various catalysts involves the condensation of *O*-phenylene diamines with *ortho* esters in the presence of Lewis acids like $ZrCl_4$, $SnCl_4$, $TiCl_4$, $HFCl_4$, etc. The most commonly used method for synthesis of benzimidazoles (7) is Phillip's method, which involves the condensation of *O*-phenylene diamines (**5**) with carboxylic acids (**6**) or its derivatives by heating the reagents in the presence of concentrated hydrochloric acid [62] (**Figure 8**).

The benzimidazole derivatives (**14**) were synthesized under mild conditions with inherently low cost by many researchers using (**8**), (**9**), (**10**), (**11**), (**12**), and (**13**), as reactants (**Figure 9**). Suheyla et al. demonstrated the synthesis of benzimidazoles by condensation of O-phenylene diamine with an appropriate aldehyde (**8**) in the presence of sodium metabisulfite. They proposed the reaction that depends on forming the bisulfite adduct of the aryl aldehyde to prepare benzimidazole.

Hanan et al. have reported one-pot conversion of aromatic and heteroaromatic 2-nitroamines (**9**) into bicyclic 2*H*-benzimidazoles employs formic acid, iron

Figure 8.
Phillip's condensation for the synthesis of benzimidazoles.

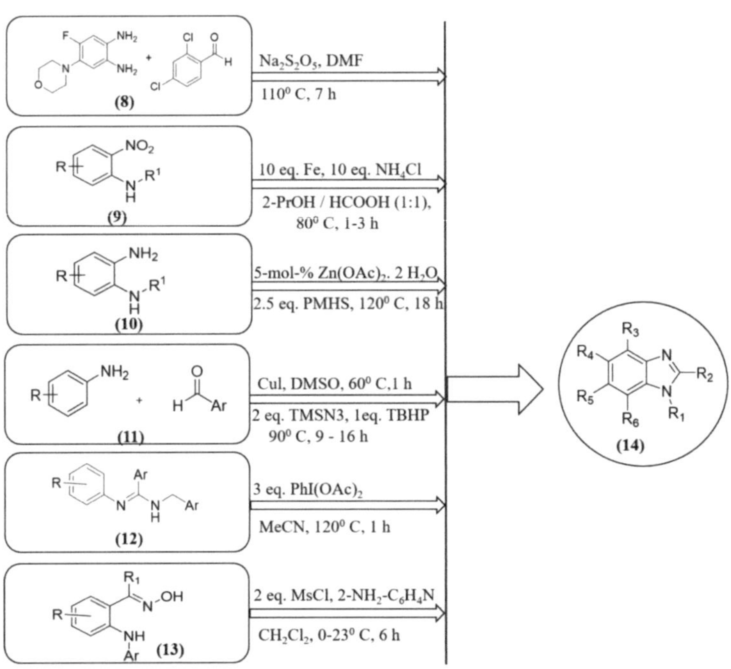

Figure 9.
Synthetic strategies of benzimidazoles.

powder, and NH_4Cl as an additive to reduce the nitro group and effect the imidazole cyclization with high-yields [63]. Nale et al. developed a method for the synthesis of benzimidazole derivatives in the presence of zinc catalysts from N-substituted formamides and various o-phenylenediamines [64] (**10**). Mahesh et al. developed a method of one-pot, multicomponent reaction, which enables the transformation of commercial aryl amines, aldehydes, and azides (**11**) into various benzimidazoles *via* an efficient copper-catalyzed amination of N-aryl imines [64]. Lin et al. developed a method for solvent/oxidant-switchable synthesis of multisubstituted benzimidazoles *via* metal-free selective oxidative annulation of arylamidines [65] (**12**). Wray et al. synthesized various *N*-aryl-1*H*-indazoles and benzimidazoles from common arylamino oximes (**13**) in good to excellent yields [66].

6. Conclusion

There are numerous benzimidazole derivatives for various cancer types involving unique types of mechanism. Although it is a widely used pharmacophore, still very few target-specific benzimidazoles are available. Therefore, researchers across the world need to develop new benzimidazole derivatives that are more target specific and help in the cancer treatment to overcome non-selective toxicity and adverse effects. This chapter mainly focused on target-based benzimidazole derivatives and synthetic strategies. Hence, it would give more ideas to young medicinal researchers to develop target-specific benzimidazole derivatives as potential cytotoxic agents.

Conflict of interest

Authors declare "no conflict of interest."

Author details

Nerella Sridhar Goud*, Pardeep Kumar and Rose Dawn Bharath
Department of Neuroimaging and Interventional Radiology (NI & IR), Radiochemistry Facility, National Institute of Mental Health and Neuro Sciences (NIMHANS), India

*Address all correspondence to: sridhar.ku23@gmail.com

References

[1] Bray F, Ferlay J, Soerjomataram I, Siegel RL, Torre LA, Jemal A. Global cancer statistics 2018: GLOBOCAN estimates of incidence and mortality worldwide for 36 cancers in 185 countries. CA: A Cancer Journal for Clinicians. 2018;**68**(6):394-424

[2] DeSantis CE, Ma J, Gaudet MM, Newman LA, Miller KD, Goding Sauer A, et al. Breast cancer statistics, 2019. CA: A Cancer Journal for Clinicians. 2019;**69**(6):438-451

[3] Siegel RL, Miller KD, Jemal A. Cancer statistics, 2015: Cancer statistics, 2015. CA: A Cancer Journal for Clinicians. 2015;**65**(1):5-29

[4] Siegel RL, Miller KD, Jemal A. Cancer statistics, 2019. CA: A Cancer Journal for Clinicians. 2019;**69**(1):7-34

[5] Arruebo M, Vilaboa N, Sáez-Gutierrez B, Lambea J, Tres A, Valladares M, et al. Assessment of the evolution of cancer treatment therapies. Cancers. 2011;**3**(3):3279-3330

[6] Mokhtari RB, Homayouni TS, Baluch N, Morgatskaya E, Kumar S, Das B, et al. Combination therapy in combating cancer. Oncotarget. 2017;**8**(23):38022-38043

[7] Padma VV. An overview of targeted cancer therapy. Biomedicine. 2015;**5**(4):19

[8] Housman G, Byler S, Heerboth S, Lapinska K, Longacre M, Snyder N, et al. Drug resistance in cancer: An overview. Cancers. 2014;**6**(3):1769-1792

[9] Nussbaumer S, Bonnabry P, Veuthey J-L, Fleury-Souverain S. Analysis of anticancer drugs: A review. Talanta. 2011;**85**(5):2265-2289

[10] Vitaku E, Smith DT, Njardarson JT. Analysis of the structural diversity, substitution patterns, and frequency of nitrogen Heterocycles among U.S. FDA approved pharmaceuticals: Miniperspective. Journal of Medicinal Chemistry. 2014;**57**(24):10257-10274

[11] Akhtar MJ, Shahar Yar M, Sharma VK, Khan AA, Ali Z, Haider MR, et al. Recent progress of Benzimidazole hybrids for anticancer potential. Current Medicinal Chemistry. 2019;**26**

[12] Kamanna K. Synthesis and pharmacological profile of Benzimidazoles. In: Marinescu M, editor. Chemistry and Applications of Benzimidazole and its Derivatives. IntechOpen; 2019

[13] Ganie AM, Dar AM, Khan FA, Dar BA. Benzimidazole derivatives as potential antimicrobial and antiulcer agents: A mini review. Mini-Reviews in Medicinal Chemistry. 2019;**19**(16):1292-1297

[14] Ries UJ, Mihm G, Narr B, Hasselbach KM, Wittneben H, Entzeroth M, et al. 6-substituted Benzimidazoles as new nonpeptide angiotensin II receptor antagonists: Synthesis, biological activity, and structure-activity relationships. Journal of Medicinal Chemistry. 1993;**36**(25):4040-4051

[15] Sethi P, Bansal Y, Bansal G. Synthesis and PASS-assisted evaluation of Coumarin–Benzimidazole derivatives as potential anti-inflammatory and anthelmintic agents. Medicinal Chemistry Research. 2018;**27**(1):61-71

[16] Akhtar W, Khan MF, Verma G, Shaquiquzzaman M, Rizvi MA, Mehdi SH, et al. Therapeutic evolution of Benzimidazole derivatives in the last Quinquennial period. European Journal of Medicinal Chemistry. 2017;**126**:705-753

[17] Strand DS, Kim D, Peura DA. 25 years of proton pump inhibitors: A comprehensive review. Gut and Liver. 2017;**11**(1):27-37

[18] Abongwa M, Martin RJ, Robertson AP. A brief review on the mode of action of Antinematodal drugs. Acta Veterinaria. 2017;**67**(2):137-152

[19] Carini DJ, Duncia JV, Aldrich PE, Chiu AT, Johnson AL, Pierce ME, et al. Nonpeptide angiotensin II receptor antagonists: The discovery of a series of N-(Biphenylylmethyl)Imidazoles as potent, orally active Antihypertensives. Journal of Medicinal Chemistry. 1991;**34**(8):2525-2547

[20] Cheson BD, Leoni L. Bendamustine: Mechanism of action and clinical data. Clinical Advances in Hematology & Oncology. 2011;**9**(8 Suppl 19):1-11

[21] Hasinoff BB, Wu X, Nitiss JL, Kanagasabai R, Yalowich JC. The anticancer multi-kinase inhibitor Dovitinib also targets topoisomerase I and topoisomerase II. Biochemical Pharmacology. 2012;**84**(12):1617-1626

[22] Tamm I. Inhibition of influenza virus multiplication by N-glycosides of Benzimidazoles. Journal of Experimental Medicine. 1954;**99**(3):227-250

[23] Slayden RA. Identification of cell cycle regulators in mycobacterium tuberculosis by inhibition of septum formation and global transcriptional analysis. Microbiology. 2006;**152**(6):1789-1797

[24] Kumar K, Awasthi D, Lee S-Y, Zanardi I, Ruzsicska B, Knudson S, et al. Novel Trisubstituted Benzimidazoles, targeting Mtb FtsZ, as a new class of antitubercular agents. Journal of Medicinal Chemistry. 2011;**54**(1):374-381

[25] Chou F-C, Chen H-Y, Kuo C-C, Sytwu H-K. Role of Galectins in tumors and in clinical immunotherapy. International Journal of Molecular Sciences. 2018;**19**(2):430

[26] Rabinovich GA. Galectin-1 as a potential cancer target. British Journal of Cancer. 2005;**92**(7):1188-1192

[27] Astorgues-Xerri L, Riveiro ME, Tijeras-Raballand A, Serova M, Neuzillet C, Albert S, et al. Unraveling Galectin-1 as a novel therapeutic target for cancer. Cancer Treatment Reviews. 2014;**40**(2):307-319

[28] Cousin J, Cloninger M. The role of Galectin-1 in cancer progression, and synthetic multivalent Systems for the Study of Galectin-1. International Journal of Molecular Sciences. 2016;**17**(9):1566

[29] Goud NS, Ghouse SM, Vishnu J, Komal D, Talla V, Alvala R, et al. Synthesis of 1-benzyl-1H-Benzimidazoles as Galectin-1 mediated anticancer agents. Bioorganic Chemistry. 2019;**89**:103016

[30] Goud NS, Soukya PSL, Ghouse M, Komal D, Alvala R, Alvala M. Human Galectin-1 and its inhibitors: Privileged target for cancer and HIV. Mini-Reviews in Medicinal Chemistry. 2019;**19**(16):1369-1378

[31] Shih T-C, Liu R, Fung G, Bhardwaj G, Ghosh PM, Lam KS. A novel Galectin-1 inhibitor discovered through one-bead two-compound library potentiates the antitumor effects of paclitaxel in vivo. Molecular Cancer Therapeutics. 2017;**16**(7):1212-1223

[32] Shih T-C, Liu R, Wu C-T, Li X, Xiao W, Deng X, et al. Targeting Galectin-1 impairs castration-resistant prostate cancer progression and invasion. Clinical Cancer Research. 2018;**24**(17):4319-4331

[33] Congdon LM, Pourpak A, Escalante AM, Dorr RT, Landowski TH.

Proteasomal inhibition stabilizes topoisomerase IIα protein and reverses resistance to the topoisomerase II poison Ethonafide (AMP-53, 6-Ethoxyazonafide). Biochemical Pharmacology. 2008;**75**(4):883-890

[34] Piao W-H, Wong R, Bai X-F, Huang J, Campagnolo DI, Dorr RT, et al. Therapeutic effect of Anthracene-based anticancer agent Ethonafide in an animal model of multiple sclerosis. The Journal of Immunology. 2007;**179**(11):7415-7423

[35] Lin B, Chen Z, Xu Y, Zhang H, Liu J, Qian X. 7-b, a novel Amonafide analogue, cause growth inhibition and apoptosis in Raji cells via a ROS-mediated mitochondrial pathway. Leukemia Research. 2011;**35**(5):646-656

[36] Qian X, Li Z, Yang Q. Highly efficient antitumor agents of Heterocycles containing sulfur atom: Linear and angular Thiazonaphthalimides against human lung cancer cell in vitro. Bioorganic & Medicinal Chemistry. 2007;**15**(21):6846-6851

[37] Zhang Y-L, Yang R, Xia L-Y, Man R-J, Chu Y-C, Jiang A-Q, et al. Synthesis, anticancer activity and molecular docking studies on 1,2-Diarylbenzimidazole analogues as anti-tubulin agents. Bioorganic Chemistry. 2019;**92**:103219

[38] Miao T-T, Tao X-B, Li D-D, Chen H, Jin X-Y, Geng Y, et al. Synthesis and biological evaluation of 2-aryl-Benzimidazole derivatives of Dehydroabietic acid as novel tubulin polymerization inhibitors. RSC Advances. 2018;**8**(31):17511-17526

[39] Wang Y-T, Shi T-Q, Zhu H-L, Liu C-H. Synthesis, biological evaluation and molecular docking of Benzimidazole grafted Benzsulfamide-containing Pyrazole ring derivatives as novel tubulin polymerization inhibitors. Bioorganic & Medicinal Chemistry. 2019;**27**(3):502-515

[40] Baig MF, Nayak VL, Budaganaboyina P, Mullagiri K, Sunkari S, Gour J, et al. Synthesis and biological evaluation of Imidazo[2,1-b] Thiazole-Benzimidazole conjugates as microtubule-targeting agents. Bioorganic Chemistry. 2018;**77**:515-526

[41] Supuran CT. Carbonic anhydrases: Novel therapeutic applications for inhibitors and activators. Nature Reviews Drug Discovery. 2008;**7**(2):168-181

[42] Supuran C. Carbonic anhydrases an overview. Current Pharmaceutical Design. 2008;**14**(7):603-614

[43] Supuran CT, Winum J-Y. Carbonic anhydrase IX inhibitors in cancer therapy: An update. Future Medicinal Chemistry. 2015;**7**(11):1407-1414

[44] Supuran CT, Alterio V, Di Fiore A, D'Ambrosio K, Carta F, Monti SM, et al. Inhibition of carbonic anhydrase IX targets primary tumors, metastases, and cancer stem cells: Three for the Price of one. Medicinal Research Reviews. 2018;**38**(6):1799-1836

[45] Peng X-M, Damu GL, Zhou C. Current developments of Coumarin compounds in medicinal chemistry. Current Pharmaceutical Design. 2013;**19**(21):3884-3930

[46] Milite C, Amendola G, Nocentini A, Bua S, Cipriano A, Barresi E, et al. Novel 2-substituted-Benzimidazole-6-sulfonamides as carbonic anhydrase inhibitors: Synthesis, biological evaluation against isoforms I, II, IX and XII and molecular docking studies. Journal of Enzyme Inhibition and Medicinal Chemistry. 2019;**34**(1):1697-1710

[47] Zubrienė A, Čapkauskaitė E, Gylytė J, Kišonaitė M, Tumkevičius S,

Matulis D. Benzenesulfonamides with Benzimidazole moieties as inhibitors of carbonic anhydrases I, II, VII, XII and XIII. Journal of Enzyme Inhibition and Medicinal Chemistry. 2014;**29**(1):124-131

[48] van der Geer P, Hunter T, Lindberg RA. Receptor protein-tyrosine kinases and their signal transduction pathways. Annual Review of Cell Biology. 1994;**10**(1):251-337

[49] Schneider MR, Wolf E. The epidermal growth factor receptor ligands at a glance. Journal of Cellular Physiology. 2009;**218**(3):460-466

[50] Seshacharyulu P, Ponnusamy MP, Haridas D, Jain M, Ganti AK, Batra SK. Targeting the EGFR signaling pathway in cancer therapy. Expert Opinion on Therapeutic Targets. 2012;**16**(1):15-31

[51] Sigismund S, Avanzato D, Lanzetti L. Emerging functions of the EGFR in cancer. Molecular Oncology. 2018;**12**(1):3-20

[52] Sasaki T, Hiroki K, Yamashita Y. The role of epidermal growth factor receptor in cancer metastasis and microenvironment. BioMed Research International. 2013;**2013**:1-8

[53] Celik İ, Ayhan-Kılcıgil G, Guven B, Kara Z, Gurkan-Alp AS, Karayel A, et al. Design, synthesis and docking studies of Benzimidazole derivatives as potential EGFR inhibitors. European Journal of Medicinal Chemistry. 2019;**173**:240-249

[54] Akhtar MJ, Siddiqui AA, Khan AA, Ali Z, Dewangan RP, Pasha S, et al. Design, synthesis, docking and QSAR study of substituted Benzimidazole linked Oxadiazole as cytotoxic agents, EGFR and ErbB2 receptor inhibitors. European Journal of Medicinal Chemistry. 2017;**126**:853-869

[55] Akhtar MJ, Khan AA, Ali Z, Dewangan RP, Rafi M, Hassan MQ, et al. Synthesis of stable Benzimidazole derivatives bearing Pyrazole as anticancer and EGFR receptor inhibitors. Bioorganic Chemistry. 2018;**78**:158-169

[56] Yuan X, Yang Q, Liu T, Li K, Liu Y, Zhu C, et al. Design, synthesis and in vitro evaluation of 6-Amide-2-aryl Benzoxazole/Benzimidazole derivatives against tumor cells by inhibiting VEGFR-2 kinase. European Journal of Medicinal Chemistry. 2019;**179**:147-165

[57] Wu L, Jiang Z, Shen J, Yi H, Zhan Y, Sha M, et al. Design, synthesis and biological evaluation of novel Benzimidazole-2-substituted phenyl or pyridine propyl ketene derivatives as antitumour agents. European Journal of Medicinal Chemistry. 2016;**114**:328-336

[58] Reddy TS, Kulhari H, Reddy VG, Bansal V, Kamal A, Shukla R. Design, synthesis and biological evaluation of 1,3-Diphenyl-1 H-Pyrazole derivatives containing Benzimidazole skeleton as potential anticancer and apoptosis inducing agents. European Journal of Medicinal Chemistry. 2015;**101**:790-805

[59] Gowda NRT, Kavitha CV, Chiruvella KK, Joy O, Rangappa KS, Raghavan SC. Synthesis and biological evaluation of novel 1-(4-Methoxyphenethyl)-1H-Benzimidazole-5-carboxylic acid derivatives and their precursors as Antileukemic agents. Bioorganic & Medicinal Chemistry Letters. 2009;**19**(16):4594-4600

[60] Bansal Y, Silakari O. The therapeutic journey of Benzimidazoles: A review. Bioorganic & Medicinal Chemistry. 2012;**20**(21):6208-6236

[61] Alaqeel SI. Synthetic approaches to Benzimidazoles from O-Phenylenediamine: A literature review. Journal of Saudi Chemical Society. 2017;**21**(2):229-237

[62] Hanan E, Chan B, Estrada A, Shore D, Lyssikatos J. Mild and general

one-pot reduction and cyclization of aromatic and Heteroaromatic 2-Nitroamines to bicyclic 2H-Imidazoles. Synlett. 2010;**2010**(18):2759-2764

[63] Nale D, Bhanage B. N-substituted Formamides as C1-sources for the synthesis of Benzimidazole and Benzothiazole derivatives by using zinc catalysts. Synlett. 2015;**26**(20):2835-2842

[64] Mahesh D, Sadhu P, Punniyamurthy T. Copper(I)-catalyzed Regioselective Amination of N -aryl imines using TMSN $_3$ and TBHP: A route to substituted Benzimidazoles. The Journal of Organic Chemistry. 2015;**80**(3):1644-1650

[65] Lin J-P, Zhang F-H, Long Y-Q. Solvent/oxidant-switchable synthesis of multisubstituted Quinazolines and Benzimidazoles via metal-free selective oxidative annulation of Arylamidines. Organic Letters. 2014;**16**(11):2822-2825

[66] Wray BC, Stambuli JP. Synthesis of N -Arylindazoles and Benzimidazoles from a common intermediate. Organic Letters. 2010;**12**(20):4576-4579

Chapter 5

Thiazole Moiety: An Interesting Scaffold for Developing New Antitumoral Compounds

Sandra Ramos-Inza, Carlos Aydillo, Carmen Sanmartín and Daniel Plano

Abstract

Currently, cancer is one of the major health problems of the human population and prominent cause of death. Thiazole ring has demonstrated many pharmacological activities including anticancer. This scaffold has been found alone or incorporated into the diversity of therapeutic active agents such as tiazofurin, dasatinib, and bleomycin, which are well-known antineoplastic drugs. Recently, most of the compounds isolated from natural sources containing thiazole moiety exhibit notable cytotoxicities and present antitumor potential. In this context, several structural changes have been made in the original structure, such as the incorporation of different substituents or the fusion with other carbo- and heterocycles, in order to increase the antitumoral potency. Related to mechanism of action of these derivatives, some of them act through kinase modulation, polymerization inhibition of microtubule, pro-matrix metalloproteinase activation, signal transducer activation of transcription 3, histone deacetylase inhibition, etc.

Keywords: cancer, thiazole, MDM2 inhibitors, mechanism of action, metal complexes, peptides, sulfur

1. Introduction

Cancer is a generic term, which encompasses a wide group of diseases characterized essentially by an uncontrolled growth and propagation of cells with errors in the division mechanisms known as cell cycle. Cancer constitutes a major public health problem worldwide, since it is the second leading cause of death globally, with 9.6 million deaths estimated in 2018 [1]. Due to the limitations and side effects associated with available cancer treatments nowadays, it is an urgent challenge for medicinal researchers to develop more safe and selective anticancer drugs.

Among the design strategies in drug discovery, special attention has been paid to molecules containing sulfur heterocycles in their structures. Several studies have been carried out with plenty of sulfur heterocycles, including thiophene, thione, benzothiophene, and thiazine, towards different pathologies.

Thiazole ring is present in several anticancer drugs, such as bleomycin, sulfathiazole, thiazofurine, and dasatinib, and its derivatives present excellent pharmacological profiles, making this skeleton an ideal candidate to develop more potent and

IntechOpen

safer drugs, especially in cancer. Herein, an extensive revision of the most relevant research published in the past 5 years is gathered.

2. Thiazole rings decorated with different fragments

2.1 Thiazole derivatives with *in vitro* efficacy

Aminothiazoles: Aminothiazoles have been widely used in drug discovery research due to its biological properties. Commercial drugs, such as famotidine, sudoxicam, or cefdinir, contain an aminothiazole core in their structures (**Figure 1**) [2].

Aminothiazole scaffold can be modified by derivatization of the amino group at position 2 of the thiazole ring. Rostom et al. [3] reported a study based on structural modifications including azomethine, *N*-formyl, *N*-acyl, sulfonamide, ureido, and thioureido functionalities. Nine derivatives were evaluated by the NCI *in vitro* screening panel assay, displaying most of them a promising antitumor activity against particular cell lines.

Sun et al. [4] synthesized a series of *N*,4-diaryl-1,3-thiazole-2-amines containing three aromatic rings with an amino linker. Compound **1** (**Figure 2**) was the most cytotoxic agent with IC_{50} values at the submicromolar level. A further biological evaluation showed that this compound inhibited polymerization and disrupted tubulin microtubule dynamics in a similar way to the natural product combretastatin A-4, besides effectively inducing SGC-7901 cell cycle arrest at the G2/M phase.

In other study, a series of tri-substituted aminothiazoles were designed by Lu et al. [5] in order to obtain new antitumoral agents. Compound **2** (**Figure 2**) displayed a EC_{50} value of 0.11 μM in hepatocellular carcinoma along with a selectivity towards nontumoral cells greater than 450 times.

A dysregulation of sirtuin 2 (Sirt2) plays an important role in the pathogenesis of cancer, among other diseases. Schiedel et al. [6] designed a series of novel aminothiazole derivatives with the aim of establishing a well-defined SAR model of sirtuin ligands. These thiazole-bearing compounds behaved as selective human sirtuins (hSirt2) inhibitors.

Chalcones: Chalcones are naturally biarylpropenones, which are classified as a subgroup of flavonoids with a broad spectrum of biological activities, including antimicrobial, anti-inflammatory, and anticancer properties [7].

A series of 4-amino-5-cinnamoylthiazoles as chalcone-like structures were synthesized and evaluated as antitumor agents, showing most of them significant cytotoxic activity against MCF-7, HepG2, and SW480 cell lines [8]. The most promising analog, compound **3** (**Figure 2**), revealed that it could prevent the proliferation of HepG2 cells by blocking cell cycle at the G2 phase and by inducing apoptosis.

Coumarins: Another strategy of design is the incorporation of a coumarin moiety in molecules containing thiazole. Many coumarin-bearing compounds are reported to have significant therapeutic potential, including anticancer activity

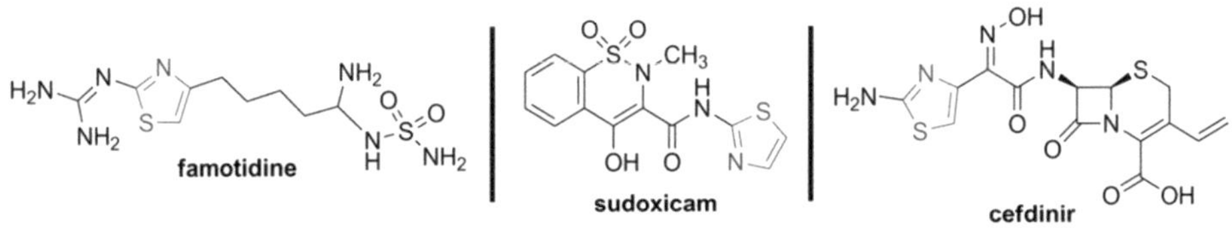

Figure 1.
Some aminothiazole as commercial drugs.

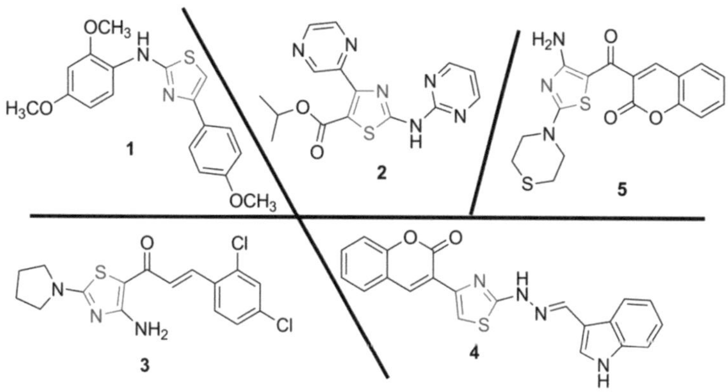

Figure 2.
Thiazole derivatives as potential antitumor agents.

through different mechanisms. Jashari et al. [9] reported the synthesis of new derivatives by combining this coumarin core with different heterocycles. The results showed that the compounds containing thiazoles in the structure had the most promising activity against the cancer cell lines tested.

This strategy can be complementary with the inclusion of other structures with recognized biological activities. A series of indole-incorporated thiazolylcoumarins were synthesized and evaluated against a wide range of tumor cell lines [10]. Among the tested compounds, structure **4** (**Figure 2**) exhibited a broad spectrum of growth inhibition activity with average GI_{50} values of 1.18–2.44 μM against nine cell lines.

Ayati et al. [11] also reported the synthesis of a series of new coumarin-containing compounds developed from the chalcone-like cinnamoylthiazoles mentioned above. Biological evaluations on the most cytotoxic compound **5** (**Figure 2**) against MCF-7 cells revealed the induction of apoptosis and blockage of the cell cycle distribution at the G1-phase.

2.2 Thiazole derivatives with *in vivo* efficacy

For the past 5 years, few examples of scaffolds bearing a thiazole ring have been reported with potent efficacy in xenograft models of various types of cancer. **Figure 3** encompasses the most relevant examples gathered in the literature that are going to be discussed herein.

Attending to their structure, the thiazole analogs can be grouped as follows.

Diaminothiazoles: In 2015, several diaminothiazole derivatives were evaluated *in vitro* against wild-type and resistant colon, breast, and uterine cancer cells lines. All of them showed potent activity in all cell lines with IC_{50} values in the nanomolar range. Among them, DAT1 (4-amino-5-benzoyl-2-(4-methoxyphenylamino)thiazole) (**Figure 3**) also demonstrated *in vivo* tumor growth inhibition of around 60% in a taxol-resistant colon cancer model at a dose of 20 mg/kg [12]. More recently, DAT1 has also demonstrated its capacity to induce apoptosis both *in vitro* and *in vivo* against colon cancer models with mutated p53 through ERK-mediated upregulation of death receptor 5 (DR5) [13]. All these findings have placed DAT1 as a candidate to be tested in clinical trials.

(Thiazole-2-yl)hydrazones: Di Martile et al. reported that a novel pCAF and GCN5 histone deacetylase inhibitor, named CPTH6 (3-methylcyclopentylidene-[4-(4′-chlorophenyl)thiazol-2-yl] hydrazone) (**Figure 3**), was able to reduce tumor growth in a spheroid patient-derived lung cancer stem cells (LCSCs) xenograft model

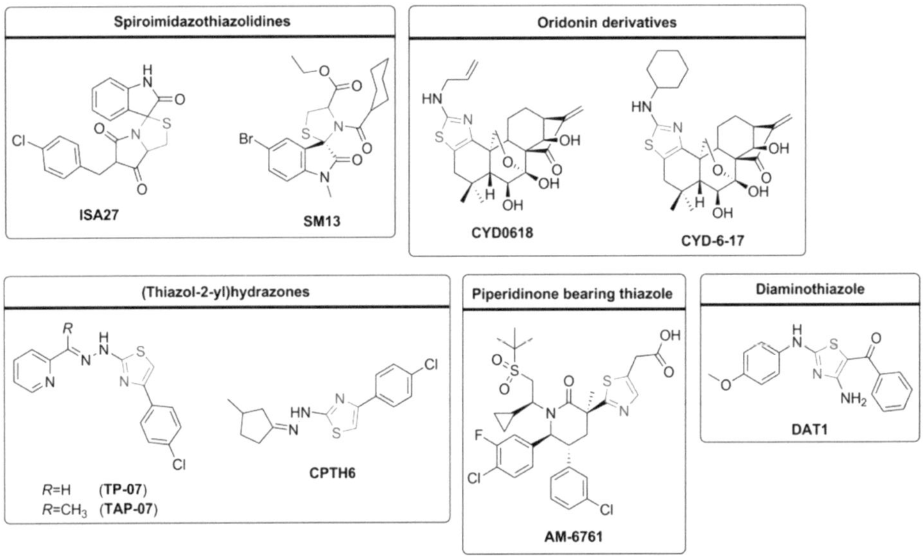

Figure 3.
Representative scaffolds containing thiazole ring with proven in vivo efficacy towards several cancer xenograft models.

accompanied by apoptosis induction and inhibition of α-tubulin acetylation [14]. Likewise, two 2-pyridyl-2,3-thiazole derivatives, TP-07 and TAP-07 (**Figure 3**), possess cytotoxic activity towards several cancer cell lines without antiproliferative effects to normal cells (IC_{50} > 30 μM) along with *in vivo* efficacy against a hepatocellular xenograft cancer model [15]. Thus, both compounds achieved 47% and 73% tumor mass reduction, respectively [15].

Spiroimidazothiazolidines: This class of compounds has demonstrated to be potent inhibitors of the Murine Double Minute-2 (MDM2)-p53 interaction, which ultimately leads to induction of apoptosis. Two analogs withstand in this class of compounds: a) ISA27 (**Figure 3**), which not only presented tumor growth inhibition *in vivo* alone in a glioblastoma xenograft model but also a synergistic effect with temozolomide, a first-line treatment drug against brain cancers [16], and b) SM13 (**Figure 3**), an analog that reduced tumor growth in a human thyroid cancer xenograft model in the absence of p53 transcriptional activity [17].

Piperidinone analogs: Based on previous morpholine and piperidone MDM2 inhibitors, Gonzalez et al. introduced a thiazole ring decorated with a carboxylic acid over the piperidone scaffold. The resulting hit compound, termed AM-6761 (**Figure 3**), maintained the MDM2 inhibition efficacy and presented an ED_{50} value of 11 mg/kg in SJSA-1 osteosarcoma xenograft model [18].

Oridonin derivatives: Oridonin is a complex ent-kaurane diterpenoid isolated from the traditional Chinese herb *Isodon rubescens*, with well-known cytotoxic activity against various human cancers. In 2013, Ding et al. designed a series of novel nitrogen-enriched oridonin derivatives with thiazole-fused A-ring. The hit compound, CYD0618 (**Figure 3**), induced a threefold shrinkage of the tumor volume in a triple-negative breast cancer MDA-MB-231 xenograft model at a dose of 5 mg/kg, showing much higher efficacy than parent oridonin [19]. Later, Zhou et al. reported another oridonin analog, CYD-6-17 (**Figure 3**), which significantly inhibited renal cell carcinoma tumor growth *in vivo* by targeting 3-phosphoinositide-dependent protein kinase 1 (PDPK1) and its downstream pathways [20].

3. Fused thiazole rings decorated with different fragments

Benzothiazoles: In the last few years, benzothiazoles have attracted considerable interest due to their broad spectrum of pharmacological activities, such as antitubercular, antimicrobial, analgesic, and antitumor properties [21].

This moiety can be functionalized with several structural modifications. Novel methylsulfonyl benzothiazoles were synthetized and evaluated against HeLA cell line, with compounds **6** and **7** (**Figure 4**) showing GI_{50} values of 0.1 μM or below [22].

Xie et al. [23] reported a new series of benzothiazole derivatives, with *in vitro* efficacy against HCT116, MCF-7, U87 MG, and A549 cell lines. Compound **8** (**Figure 4**) was proved to retain the antiproliferative activity and the inhibitory activity against PI3K (phosphoinositide 3-kinase) and mTORC1 (mammalian target of rapamycin), which are abnormally active in many tumor cells.

Benzothiazole derivatives bearing pyrimidine moiety were synthetized and evaluated for anticancer activity against MCF-7, A549, and A375 cancer cell lines, with significant antitumor activity. A further study of the most promising compounds indicated an effect on the expression of proteins that cause abnormal cell proliferation, such as ERK1/2, NF-kB, and survivin [24].

This moiety can also be used in the design of new molecules with a chalcone-like structure, as it has been mentioned before. Imidazole bearing benzothiazoles were synthetized by Sultana et al. [25] and evaluated against several cancer cell lines. Compounds **9** and **10** (**Figure 4**) exhibited good cytotoxicity against human breast cancer (MDA MB-231) with IC_{50} values of 1.3 and 1.2 μM, respectively. These compounds were revealed to induce cell cycle arrest in G2/M phase and to inhibit microtubule assembly.

Imidazoles: Imidazole-based compounds have achieved great progress in medicinal chemistry, since they have showed anticancer, antifungal, antibacterial, and antiparasitic activities, among others [26]. Their use as heterocycles merged with thiazole has attracted great attention in the last years [27, 28] due to its therapeutic properties.

A series of imidazo[2,1-*b*]thiazole derivatives were evaluated against different tumor cell lines, showing that compounds **11** and **12** (**Figure 4**) had a significant cytotoxic activity against A549 with IC_{50} values of 0.92 and 0.78 μM, respectively. These derivatives had proven to induce cell cycle arrest in G2/M phase and apoptosis in this cell line [29]. Ali et al. [30] synthesized a series of imidazo[2,1-*b*]thiazoles decorated with pyrazoles that turned out to be promising leads to further develop.

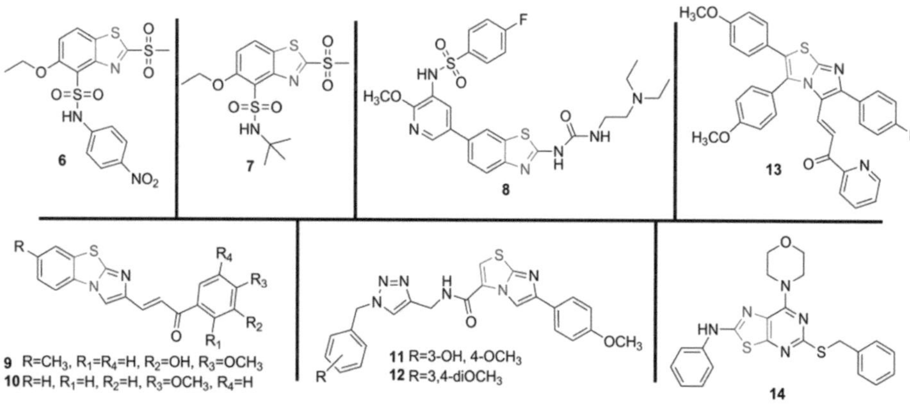

Figure 4.
Some thiazole-fused compounds with antitumor activity.

Due to the pharmacological properties of the imidazo[2,1-*b*]thiazole derivatives and coumarin compounds already mentioned, it has been reported a design that embodied both the active pharmacophores in a single molecule in order to evaluate their synergic activity against a series of tumor cell lines [31], showing some of them prominent cytotoxic activity.

Kamal et al. [32] also designed a novel series of imidazole merged with thiazole as chalcone-like derivatives and evaluated their cytotoxic activity against MCF-7, A549, HeLA, DU-145, and HT-29 cell lines. Among the compounds tested, structure **13** (**Figure 4**) with a pyridyl ring was the most active. This compound also disrupted microtubule dynamics, induced cell cycle arrest in G2/M phase and ultimately trigger apoptosis.

Pyrimidines: Compounds with fused rings can also be formed by merging other heterocyclic moieties with thiazole core. Li et al. [33] reported a novel series of thiazolo[5,4-*d*]pyrimidine derivatives, which were evaluated against three cancer cell lines. Compound **14** (**Figure 4**) showed the most potent antiproliferative activity with good selectivity when compared to normal cells (IC_{50} values of 1.03 μM against MGC803 and 38.95 μM against GES-1). Biological studies indicated that this compound could inhibit the cell colony formation and migration by inducing apoptosis on MGC803 cells.

A series of thiazolo[3,2-*a*]pyrimidines were synthetized and evaluated in the NCI-60 cell lines panel assay, achieving significant cytotoxicity against some of the cell lines tested [34].

4. Miscellaneous structures bearing thiazole ring

Diazepines: Heterocyclic compounds 1,4-diazepines are considered an interesting moiety in drug research due to their broad range of pharmacological activities, including antibacterial, anti-HIV, anticonvulsant, and anticancer [35]. Ramírez et al. designed a series of novel thiazole-based compounds by fusing this structure with pyrimidine, which has also showed biological properties. The results indicated that some compounds showed promising antitumoral activity, with GI_{50} values below 2 μM against NCI's *in vitro* cell line screening [35].

Pyrazoline: Another heterocyclic structure used in combination with thiazole moiety is the pyrazoline ring. New thiazolyl-pyrazoline derivatives were synthetized, and their cytotoxicity was evaluated against A549 human lung adenocarcinoma and NIH/3 T3 mouse embryonic fibroblast cells, presenting in some cases similar IC_{50} values to cisplatin [36].

Curcumins: Bayomi et al. [37] synthetized and evaluated a series of new curcumin analogs bearing thiazole as antitumoral and antioxidant agents, showing similar behavior than that of cisplatin and ascorbic acid, respectively.

Thiazolines: Thiazolines are the reduced form of thiazole and also have attracted interest in drug research due to its biological activity. Altintop et al. [38] evaluated a series of new thiazoline-based derivatives bearing a hydrazone moiety. The results showed that some of the compounds were potent inhibitors of DNA synthesis against C6 tumor cells.

5. Thiazole and metal complexes

There is a great variability of transition metals that in combination with different ligands have been reported as antitumoral agents acting through different mechanisms. The literature revealed the considerable interest in the thiazole pharmacophore alone [39], fused to other rings [40], or incorporated into different structures [41] for cancer therapy.

On the other hand, among the most effective and well-studied class of chemotherapeutic agents are the platinum-based drugs, which comprise cisplatin, carboplatin, and oxaliplatin. Given the clinical success of the platinum-based drugs, extensive research efforts have been made to develop alternative metal ions, that is, ruthenium, copper, zinc, and nickel, with antitumor activity.

Copper complexes: Copper complexes have attracted a vast interest due to their bioavailability, increased uptake in cancerous tissues, role in angiogenesis and photophysical properties, among others. The most common types of copper complexes are those incorporating 1,10-phenanthroline (phen) ligands. Planarity of the intercalative ligand is crucial in the binding of these complexes with DNA. The complexes containing nonplanar ligands favored groove binding [42].

Besides, Shakir et al. [43] have reported several Cu (II) complexes derived from benzothiazole and thiazole, which showed greater antioxidant and anticancer activities than the corresponding free ligands in various cell lines.

Studies carried out with several Cu (II) complexes with 2,2,6′,2′′-terpyridines revealed that these complexes are able to promote the generation of reactive oxygen species (ROS) in the presence of mild reducing agents. This feature has been exploited to oxidatively break the DNA strands, hence inhibiting the proliferation of tumor cells. In this context, the replacement of two pyridine rings by two thiazole nuclei (compound **15** in **Figure 5**) also achieved efficient DNA cleavage in several tumor cell lines [44]. Later, Czerwinska et al. corroborated an increase in the antiproliferative effect of these complexes against ovarian carcinoma cells by apoptosis [45].

In addition, the copper complexes have been recognized as promising drugs for metastatic tumors. For example, copper complexes of pyrrolidine dithiocarbamate ($Cu(PDTC)_2$) possessed potent anticancer activity on cisplatin-resistant neuroblastoma cells. Additionally, two copper thiosemicarbazone complexes showed similar effect on cisplatin-resistant neuroblastoma cells and prostate cancer. Xie et al. [46] reported the synthesis and antitumoral activity of two copper complexes of (4*R*)-2-thioxo-4-thiazolidinecarboxylic acid (TTDC) and 3-rhodaninepropionic acid (RDPA) against prostate cancer, presenting both of them variable potency, likely

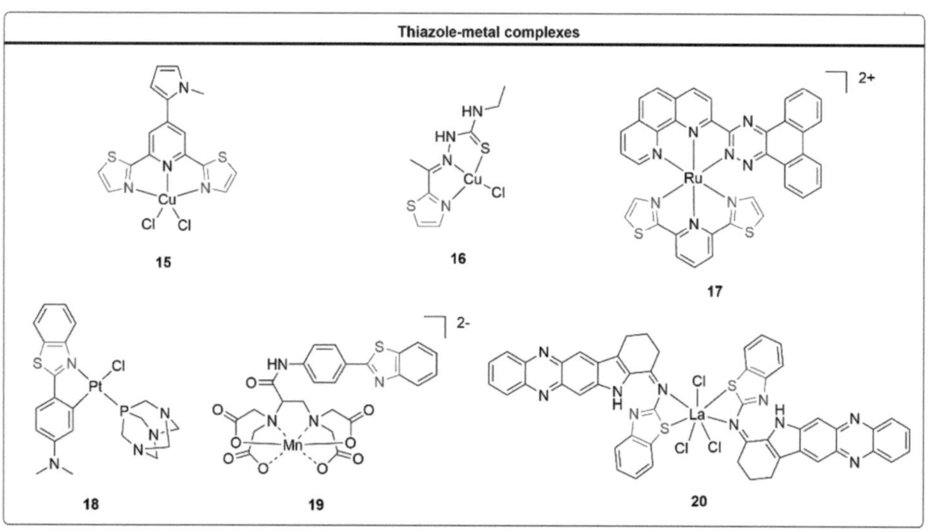

Figure 5.
Some thiazole-metal complexes with antitumor activity.

related to different functional groups on TTDC and RDPA ligands. Owing to the presence of sulfur and amino groups in CuTTDC and CuRDPA, these complexes had emerged as ligands to attach to delivery vehicles, such as peptides or monoclonal antibodies for targeted delivery.

It is notably that a number of copper (II) complexes have been shown to present antitumor activity, through inhibition of human topoisomerase IIα. Recently, Sandhaus et al. [47] have identified a new complex (compound **16** in **Figure 5**) with potent antiproliferative activity towards colon cancer cell lines (HTC-116, Caco-2, and HT-29) and aggressive breast cancer cell lines (HCC 1500, HCC 70, HCC 1806, and HCC 1395).

Ruthenium complexes: Currently, ruthenium complexes are found to be a promising alternative for platinum because of favorable properties as anticancer drugs. Among the ligands, 2,6-di(thiazol-2-yl)pyridine combined with phenantrolines have demonstrated to act as DNA intercalative agents along with topoisomerase I and IIα inhibitors (compound **17** in **Figure 5**) [48]. The assays with other ligands, such as 1,3-thiazolidine-2-thione, with 1,4-bis(diphenylphosphino)butane or 2,2′-bipyridine, displayed strong cytotoxicity against breast and prostate cancer cell lines [49].

Platinum and palladium complexes: Platinum and palladium have similar chemical properties and modes of coordination, but the palladium compounds are more labile from a thermodynamic and a kinetic point of view with relation to platinum derivatives.

Rubino and co-workers [50] have reported two new mono-Pt(II) and binuclear chloro-bridged Pd(II) complexes with 2,2′-dithiobis(benzothiazole) as ligand. Only platinum derivative has emerged as an effective inductor of apoptotic death on HepG2 and MCF-7 cells and caused cell cycle arrest at G0/G1 phase while palladium was inactive. On the other hand, the inclusion of 2-(4-substituted)benzothiazoles (compound **18** in **Figure 5**) as ligands resulted in potent cytotoxic agents through tubulin polymerization in A549 and HeLa cell lines [51].

In addition, thiazolidinone-derived complexes, specifically with (*Z*)-2-((*E*)-(1-(pyridin-2-yl)ethylidene)hydrazono)thiazolidin-4-one, were markedly cytotoxic to MCF-7, HepG2, and NCI-H460 and presented better profile than cisplatin [52]. Other relevant strategy is the combination with scaffold with proven anticancer activity. In this context, the coumarin-thiazole analogs complexed with platinum or palladium showed that the Pd complex had higher antitumor effects than its Pt analogs in several cancer cell lines [53].

Other metal complexes and applications: Manganese is a metal that plays a critical role in cell development, and it is required for mitochondrial function. As novelty, Islam et al. [54] have described a new Mn-EDTA complex (compound **19** in **Figure 5**) incorporating a benzothiazole that has been investigated as potential agents for diagnosis of liver cancer by magnetic resonance.

Cobalt (II) complexes are one of the most studied, and they have been reported as cytotoxic agents *in vitro* against breast cancer cell lines [55]. However, the cobalt (III) complexes are less known, although a new Co(III)sulfathiazole complex have been reported as cytotoxic compound without genotoxic effects [56].

In recent years, lanthanum (III) complexes are emerging as promising agents due to their more physiological activities and lower toxicities after coordination with ligands. The main mechanism of action associated is the interaction with DNA by intercalation mode. Likewise, these compounds are useful as clinical biomarkers for early diagnosis of the presence of prostate cancer. One of the most relevant lanthanum (III) derivative is 2,3-dihydro-1*H*-indolo[2,3-*b*]phenazin-4-(5*H*)-ylidene)benzothiazole-2-amine (compound **20** in **Figure 5**) that showed excellent anticancer activity in PC-3 cells [57].

Another relevant option is the gold(I) compounds that can act as prodrugs. Thus, 2-mercapto thiazoline as ligand by reaction with K[Au(CN)$_2$] resulted in the nitrogen-coordinated complex [NCAu(N-mtz)]. On the other hand, reaction with [(Ph$_3$P)AuCl] yielded the sulfur-coordinated complex [(Ph$_3$P)Au(S-mtz)]. Both of them inhibited the growth of tumorigenic cell lines such as the human ovarian carcinoma (A2780), human colon carcinoma (HCT116), and human breast adenocarcinoma (MCF7) [58].

Finally, another strategy to design new complexes as antitumoral agents is the combination of anti-inflammatory derivatives with metals. In this approach, the 1,2-benzothiazines nuclei, which are present in meloxicam and piroxicam, were complexed with ruthenium and osmium to obtain new derivatives with potent activity against cancer cell lines [59].

6. Peptidic thiazoles

Thiazoles and thiazolines are quite common motifs present in peptides isolated from natural sources, many of them known for having biological activity, typically antibacterial. These peptides are biosynthesized from nonribosomal peptide synthase (NRPS) or ribosomally produced and post-translationally modified. Both processes involve cyclodehydrations of cysteine residues to yield thiazolines and subsequent dehydrogenations to give thiazoles [60]. In this context, marine organism (cyanobacteria, fungi, sponges, tunicates, ascidians, etc.) provide an endless source of new structures with biological potential, cancer included [61]. Many isolated thiazole-containing peptides from nature have anticancer properties *per se*, but more efforts are continuously needed by scientific community to enhance and modulate its anticancer activity through structure modifications. Recent developments in this area are included here and listed by its cyclic/acyclic nature.

6.1 Linear peptides

Cyanobacteria-derived bisebromoamide was isolated and tested against HeLa S3 cells, showing a very low IC_{50} [62]. It was also shown to induce apoptosis through ERK and mTOR inhibition in renal carcinoma cell lines [63]. A modification of central thiazoline of bisebromoamide by a thiazole and alanine scanning [64] provided new analogs, getting insights in the structural dependence of the cytotoxicity. Four analogs showed nanomolar cytotoxicity activity against human colon tumor cell line HCT116.

P-glycoprotein (P-gp, multidrug resistance protein 1) is overexpressed in patients suffering from chemotherapy resistance. In this sense, cyclic and acyclic (*S*)-valine-derived thiazole peptide dimers, trimers, and tetramers were found to be potent P-gp efflux transport inhibitors [65]. Based on this hit, further derivatization led to peptidomimetic TTT-28 (**Figure 6**), which was found to be a potent P-gp transport inhibitor and superior to parent compound in reversal of resistance to placitaxel in SW620/Ad300 and HEK/ABCB1 cell lines [66]. *In vivo* study [67] showed TTT-28 enhanced intratumoral concentration of placitaxel, inhibiting the growth of ABCB1 overexpressing tumors. Additional extensive derivatizations of TTT-28 in terminal groups and central thiazole building block side chain helped to understand the drug/substrate interactions with P-gp [68]. Modifications on these sites led to divergent effects in ATPase efflux pump, from initial stimulation in TTT-28 to inhibition.

Polyamides based on 2 and 3 repeating units of 2-aminothiazole-4-carboxylic acid were synthetized [69] and proved to bind selectively to c-*MYC* quadruplex

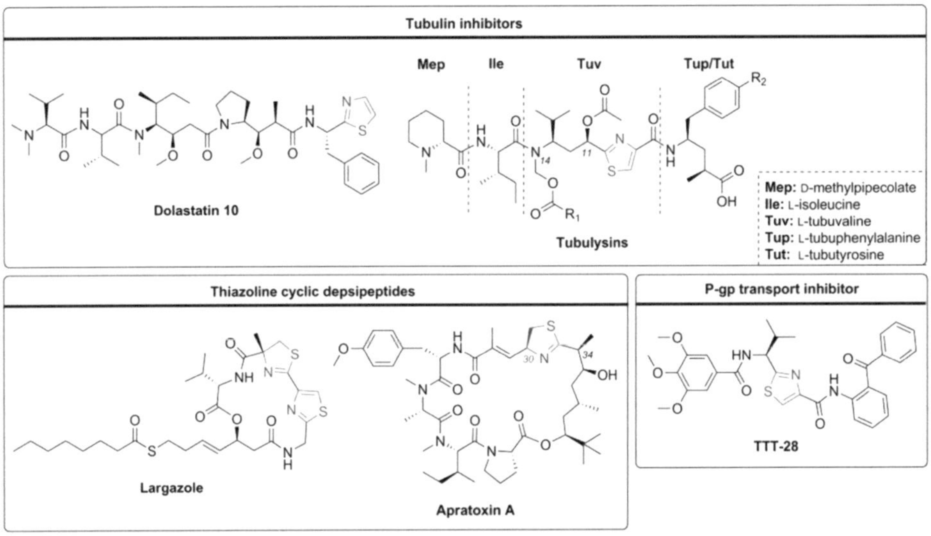

Figure 6.
Thiazole/thiazoline containing peptides and peptidomimetics with anticancer activity.

over other G-quadruplex and duplex DNA and therefore inhibiting c-*MYC* oncogene transcription. Antiproliferative activity of the tripeptide was found in HeLa cells, caused by apoptotic pathway.

Thiazole scaffold is also present in short peptides known for inhibiting tubulin polymerization. Dolastatin 10 was firstly isolated from *Dolabella auricularia* and is composed of five unnatural amino acids with a thiazole ring in C-terminal (**Figure 6**). It was demonstrated as very potent in cell proliferation assays (IC_{50} < 5.0 nM), but due to its high toxicity at maximum tolerated dose, new analogs have been developed. *N*-Terminal modified dolastatin analog (PF-06380101) bearing a quaternary amino acid was found to have improved potency and suitable ADME properties for antibody-drug conjugates [70]. Modified dolastatins at thiazole moiety by addition of new functionalities as alcohols, amines, and thiols have also been reported [71]. These analogs also showed low IC_{50} for several cancerous cell lines.

Another thiazole-containing peptides targeting to tubulin polymerization are the tubulysins (**Figure 6**), isolated first from myxobacteria. Great number of modifications have been attempted to date, and numerous SAR studies have shed light into tubulysin mode of action (for a review, see ref. [72]).

In this context, a pretubulysin (tubulysin biosynthetic precursors) lacking of C^{11} acetate and bearing a methyl group at N^{14} showed efficacy against various *in vivo* metastatic bladder, breast, and lung cancer models [73]. New tubulysin derivative KEMTUB10 with a N^{14}-benzyl-Tuv and 4-fluorophenyl moiety in Tup exhibited activity in the picomolar range in the main breast cancer cell lines [74]. It blocks cells in G2/M phase of the cell cycle and stimulates apoptosis. In line with these results, attachment of alkyl groups at mentioned Tuv N^{14} as benzyl, 4-fluorobenzyl, and cyclopropylmethyl in tubulysins also led to superpotent cytotoxic activity [75]. More Tuv modifications have been reported, like the incorporation of tetrahydropyranyl ring by Diels-Alder reaction for conformational restriction of tubulysin [76], but rigidification seemed to affect negatively to polymerization inhibition. Systematic derivatization by substitution of each subunit of tubulysin by diverse moieties, including three-dimensional structural motifs such as cubane and [1.1.1]-bicyclopentane, was reported [77]. A profound structure-activity study indicated that thiazole in Tuv unit cannot be substituted by 3D motifs but can be replaced by aromatics such as pyridine without significant loss of activity.

One objective for researchers working with tubulysins is the improvement of their therapeutic efficacy by the targeted cancer therapy as antibody-drug conjugation (ADC) or small molecule drug conjugates (SMDC), acting the tubulysins as payloads. This represents a very powerful tool, which is already being applied to all class of tubulin inhibitors [78]. Tubulysin warheads are therefore being used in ADC; one of them (AZ13599185) is in phase I clinical trials targeting HER2 receptors, involved in breast cancer development [79]. Following this trend, the modifications of tubulysins for an easier linking to conjugates is a new goal. New derivatizations at *C*-terminal Tup showed broad tolerance with no loss of activity, enabling more opportunities to conjugate to biomolecules and receptor ligands [80]. Another issue that arose during ADC conjugation of tubulysins to trastuzumab is the metabolism of C^{11} acetate *in vivo*, inactivating the payload [81, 82]. The problem was solved replacing the acetate ester by a more inert functionality to esterases like carbamates, retaining the activity. Tubulysin warheads have also been applied in a SMDC strategy in conjugation with folic acid to address folate receptor (FR), expressed in many cancers [83]. The EC1456 conjugate was tested in mice bearing FR-positive xenografts leading to curation of 100%. Results against human vintafolide-resistant xenografts were also positive.

6.2 Cyclic peptides

Cyclic depsipeptide largazole was discovered from cyanobacteria *Symploca* sp. [84], and its distinctive structural feature is the presence of a thiazole fused linearly to a 4-methylthiazoline and a labile thioester (**Figure 6**). Largazole possesses great activity as inhibitor of class I histone deacetylase (HDAC) metalloenzymes [85], a promising target for chemotherapy. Many largazole structure-activity relationship studies have been reported. Among multiple sites of modification performed, thiazole-thiazoline fragment located in the macrocycle seems to be the most promising to achieve more potent and selective analogues. Substitution of thiazole by pyridyl residues and depsipeptide framework alteration to peptide isostere led to equipotent largazole analogues but with improved selectivity for different HDACs [86]. The replacement of thiazole and thiazoline by bipyridyl fragments led to derivatives with a similar activity of largazole, but with an improved selectivity for class I HDAC [87].

Largazole inhibition of HDAC is actually attained by largazole thiol derived from thioester hydrolysis. The thiol forms a thiolate-Zn^{2+} complex [88], a critical binding for the activity, since substitution by other poorer Zn-binding groups correlated to less HDAC inhibition and lower cytotoxicity [89]. The octanoyl side chain on the thioester allows good cell permeability. Thus, modifications can be made for a better membrane permeability and thiol liberation inside the cell. In this sense, controlled release of largazole thiol from an isobutylene-caged largazole thiol derivative, which possesses a high permeability, has been achieved by UV light photoactivation of a thiol-ene triggering reaction [90].

Cyclic thiazole- and oxazoline-containing octapeptides and patellamides, isolated from marine tunicates, have also been an object of modification. It has been shown that changes in the position of thiazoles and oxazolines in ascidiacyclamide can influence their cytotoxic activity, obtaining inactive derivatives and 10 times more active compounds depending on the conformations attained [91].

Apratoxin- and thiazoline-containing depsipeptides are known potent cytotoxins isolated from marine cyanobacteria. Apratoxins are known for being potent anticancer agents and co-translational translocation inhibitors. Different derivatives have been synthetized, involving thiazoline stereogenic configuration change at C^{30} and substitution in C^{34} [92] (**Figure 6**). A new derivative, apratoxin S10 with

(*R*)-C^{30} thiazoline and the addition of a methyl group at C^{34}, shows potent *in vitro* angiogenesis and vascularized tumor cell growth inhibition [93]. It showed antipancreatic cancer activity, including in orthotopic pancreatic patient-derived xenograft mouse model [94]. Other derivatizations consisting of thiazoline substitution by piperidinecarboxylic acid moiety have been developed [95]. Apratoxins M16 showed comparable cytotoxicity to apratoxin A in HCT116 cancerous cells.

7. Conclusion

Incorporation of thiazole ring into different molecules have demonstrated to be a novel and promising approach to design more potent and safer antitumor drugs. The results of this chapter might help to enlighten other researchers to better design bioactive molecules. This thiazole ring can be incorporated as part of mono- or fused-cycles, metal complexes, or as a part of peptides. In many of these cases, the deletion of thiazole ring entails the loss of the bioactivity pointing out the importance of this ring for the anticancer activity. Thus, we consider this class of compounds and excellent starting point to achieve future drug candidates to treat cancer.

Author details

Sandra Ramos-Inza[1,2], Carlos Aydillo[1,2], Carmen Sanmartín[1,2] and Daniel Plano[1,2*]

1 Faculty of Pharmacy and Nutrition, Department of Pharmaceutical Technology and Chemistry, University of Navarra, Pamplona, Spain

2 Instituto de Investigación Sanitaria de Navarra (IdiSNA), Pamplona, Spain

*Address all correspondence to: dplano@alumni.unav.es

References

[1] Bray F, Ferlay J, Soerjomataram I, Siegel RL, Torre LA, Jemal A. Global cancer statistics 2018: GLOBOCAN estimates of incidence and mortality worldwide for 36 cancers in 185 countries. CA: A Cancer Journal for Clinicians. 2018;**68**:394-424. DOI: 10.3322/caac.21492

[2] Das D, Sikdar P, Bairagi M. Recent developments of 2-aminothiazoles in medicinal chemistry. European Journal of Medicinal Chemistry. 2016;**109**:89-98. DOI: 10.1016/j.ejmech.2015.12.022

[3] Rostom SA, Faidallah HM, Radwan MF, Badr MH. Bifunctional ethyl 2-amino-4-methylthiazole-5-carboxylate derivatives: Synthesis and in vitro biological evaluation as antimicrobial and anticancer agents. European Journal of Medicinal Chemistry. 2014;**76**:170-181. DOI: 10.1016/j.ejmech.2014.02.027

[4] Sun M, Xu Q, Xu J, Wu Y, Wang Y, Zuo D, et al. Synthesis and bioevaluation of N,4-diaryl-1, 3-thiazole-2-amines as tubulin inhibitors with potent antiproliferative activity. PLoS One. 2017;**12**:e0174006. DOI: 10.1371/journal.pone.0174006

[5] Lu H, Rogowskyj J, Yu W, Venkatesh A, Khan N, Nakagawa S, et al. Novel substituted aminothiazoles as potent and selective anti-hepatocellular carcinoma agents. Bioorganic & Medicinal Chemistry Letters. 2016;**26**:5819-5824. DOI: 10.1016/j.bmcl.2016.10.015

[6] Schiedel M, Rumpf T, Karaman B, Lehotzky A, Olah J, Gerhardt S, et al. Aminothiazoles as potent and selective Sirt2 inhibitors: A structure-activity relationship study. Journal of Medicinal Chemistry. 2016;**59**:1599-1612. DOI: 10.1021/acs.jmedchem.5b01517

[7] Singh P, Anand A, Kumar V. Recent developments in biological activities of chalcones: A mini review. European Journal of Medicinal Chemistry. 2014;**85**:758-777. DOI: 10.1016/j.ejmech.2014.08.033

[8] Ayati A, Esmaeili R, Moghimi S, Oghabi Bakhshaiesh T, Eslami SZ, Majidzadeh AK, et al. Synthesis and biological evaluation of 4-amino-5-cinnamoylthiazoles as chalcone-like anticancer agents. European Journal of Medicinal Chemistry. 2018;**145**:404-412. DOI: 10.1016/j.ejmech.2018.01.015

[9] Jashari A, Imeri F, Ballazhi L, Shabani A, Mikhova B, Drager G, et al. Synthesis and cellular characterization of novel isoxazolo- and thiazolohydrazinylidene-chroman-2, 4-diones on cancer and non-cancer cell growth and death. Bioorganic & Medicinal Chemistry. 2014;**22**: 2655-2661. DOI: 10.1016/j.bmc.2014.03.026

[10] Gali R, Banothu J, Gondru R, Bavantula R, Velivela Y, Crooks PA. One-pot multicomponent synthesis of indole incorporated thiazolylcoumarins and their antibacterial, anticancer and DNA cleavage studies. Bioorganic & Medicinal Chemistry Letters. 2015;**25**:106-112. DOI: 10.1016/j.bmcl.2014.10.100

[11] Ayati A, Oghabi Bakhshaiesh T, Moghimi S, Esmaeili R, Majidzadeh AK, Safavi M, et al. Synthesis and biological evaluation of new coumarins bearing 2,4-diaminothiazole-5-carbonyl moiety. European Journal of Medicinal Chemistry. 2018;**155**:483-491. DOI: 10.1016/j.ejmech.2018.06.015

[12] Vasudevan S, Thomas SA, Sivakumar KC, Komalam RJ, Sreerekha KV, Rajasekharan KN, et al. Diaminothiazoles evade multidrug resistance in cancer cells and xenograft tumour models and develop transient

specific resistance: Understanding the basis of broad-spectrum versus specific resistance. Carcinogenesis. 2015;**36**: 883-893. DOI: 10.1093/carcin/bgv072

[13] Thamkachy R, Kumar R, Rajasekharan KN, Sengupta S. ERK mediated upregulation of death receptor 5 overcomes the lack of p53 functionality in the diaminothiazole DAT1 induced apoptosis in colon cancer models: Efficiency of DAT1 in Ras-Raf mutated cells. Molecular Cancer. 2016;**15**:22. DOI: 10.1186/s12943-016-0505-7

[14] Di Martile M, Desideri M, De Luca T, Gabellini C, Buglioni S, Eramo A, et al. Histone acetyltransferase inhibitor CPTH6 preferentially targets lung cancer stem-like cells. Oncotarget. 2016;**7**:11332-11348. DOI: 10.18632/oncotarget.7238

[15] Dos Santos Silva TD, Bomfim LM, da Cruz Rodrigues ACB, Dias RB, Sales CBS, Rocha CAG, et al. Anti-liver cancer activity in vitro and in vivo induced by 2-pyridyl 2,3-thiazole derivatives. Toxicology and Applied Pharmacology. 2017;**329**:212-223. DOI: 10.1016/j.taap.2017.06.003

[16] Costa B, Bendinelli S, Gabelloni P, Da Pozzo E, Daniele S, Scatena F, et al. Human glioblastoma multiforme: p53 reactivation by a novel MDM2 inhibitor. PLoS One. 2013;**8**:e72281. DOI: 10.1371/journal.pone.0072281

[17] Sorriento D, Del Giudice C, Bertamino A, Ciccarelli M, Gomez-Monterrey I, Campiglia P, et al. New small molecules, ISA27 and SM13, inhibit tumour growth inducing mitochondrial effects of p53. British Journal of Cancer. 2015;**112**:77-85. DOI: 10.1038/bjc.2014.577

[18] Gonzalez AZ, Li Z, Beck HP, Canon J, Chen A, Chow D, et al. Novel inhibitors of the MDM2-p53 interaction featuring hydrogen bond acceptors as carboxylic acid isosteres. Journal of Medicinal Chemistry. 2014;**57**: 2963-2988. DOI: 10.1021/jm401911v

[19] Ding C, Zhang Y, Chen H, Yang Z, Wild C, Chu L, et al. Novel nitrogen-enriched oridonin analogues with thiazole-fused A-ring: Protecting group-free synthesis, enhanced anticancer profile, and improved aqueous solubility. Journal of Medicinal Chemistry. 2013;**56**:5048-5058. DOI: 10.1021/jm400367n

[20] Zhou J, Yun EJ, Chen W, Ding Y, Wu K, Wang B, et al. Targeting 3-phosphoinositide-dependent protein kinase 1 associated with drug-resistant renal cell carcinoma using new oridonin analogs. Cell Death & Disease. 2017;**8**:e2701. DOI: 10.1038/cddis.2017.121

[21] Gill RK, Rawal RK, Bariwal J. Recent advances in the chemistry and biology of benzothiazoles. Archiv der Pharmazie (Weinheim). 2015;**348**: 155-178. DOI: 10.1002/ardp.201400340

[22] Lad NP, Manohar Y, Mascarenhas M, Pandit YB, Kulkarni MR, Sharma R, et al. Methylsulfonyl benzothiazoles (MSBT) derivatives: Search for new potential antimicrobial and anticancer agents. Bioorganic & Medicinal Chemistry Letters. 2017;**27**:1319-1324. DOI: 10.1016/j.bmcl.2016.08.032

[23] Xie XX, Li H, Wang J, Mao S, Xin MH, Lu SM, et al. Synthesis and anticancer effects evaluation of 1-alkyl-3-(6-(2-methoxy-3-sulfonylaminopyridin-5-yl)benzo[d]thiazol-2-yl)urea as anticancer agents with low toxicity. Bioorganic & Medicinal Chemistry. 2015;**23**: 6477-6485. DOI: 10.1016/j.bmc.2015.08.013

[24] Kumbhare RM, Dadmal TL, Ramaiah MJ, Kishore KS, Pushpa Valli

SN, Tiwari SK, et al. Synthesis and anticancer evaluation of novel triazole linked N-(pyrimidin-2-yl)benzo[d] thiazol-2-amine derivatives as inhibitors of cell survival proteins and inducers of apoptosis in MCF-7 breast cancer cells. Bioorganic & Medicinal Chemistry Letters. 2015;**25**:654-658. DOI: 10.1016/j.bmcl.2014.11.083

[25] Sultana F, Reddy Bonam S, Reddy VG, Nayak VL, Akunuri R, Rani Routhu S, et al. Synthesis of benzo[d] imidazo[2,1-b]thiazole-chalcone conjugates as microtubule targeting and apoptosis inducing agents. Bioorganic Chemistry. 2018;**76**:1-12. DOI: 10.1016/j.bioorg.2017.10.019

[26] Zhang L, Peng XM, Damu GL, Geng RX, Zhou CH. Comprehensive review in current developments of imidazole-based medicinal chemistry. Medicinal Research Reviews. 2014;**34**:340-437. DOI: 10.1002/med.21290

[27] Abdel-Maksoud MS, Kim MR, El-Gamal MI, Gamal El-Din MM, Tae J, Choi HS, et al. Design, synthesis, in vitro antiproliferative evaluation, and kinase inhibitory effects of a new series of imidazo[2,1-b]thiazole derivatives. European Journal of Medicinal Chemistry. 2015;**95**:453-463. DOI: 10.1016/j.ejmech.2015.03.065

[28] Romagnoli R, Baraldi PG, Prencipe F, Balzarini J, Liekens S, Estevez F. Design, synthesis and antiproliferative activity of novel heterobivalent hybrids based on imidazo[2,1-b][1,3,4]thiadiazole and imidazo[2,1-b][1,3]thiazole scaffolds. European Journal of Medicinal Chemistry. 2015;**101**:205-217. DOI: 10.1016/j.ejmech.2015.06.042

[29] Shaik SP, Nayak VL, Sultana F, Rao AVS, Shaik AB, Babu KS, et al. Design and synthesis of imidazo[2,1-b] thiazole linked triazole conjugates: Microtubule-destabilizing agents. European Journal of Medicinal Chemistry. 2017;**126**:36-51. DOI: 10.1016/j.ejmech.2016.09.060

[30] Ali AR, El-Bendary ER, Ghaly MA, Shehata IA. Synthesis, in vitro anticancer evaluation and in silico studies of novel imidazo[2,1-b] thiazole derivatives bearing pyrazole moieties. European Journal of Medicinal Chemistry. 2014;**75**:492-500. DOI: 10.1016/j.ejmech.2013.12.010

[31] Gali R, Banothu J, Porika M, Velpula R, Bavantula R, Abbagani S. Synthesis and in vitro cytotoxic activity of novel coumarinylimidazo[2,1-b] thiazole derivatives. RSC Advances. 2014;**4**:53812-53819. DOI: 10.1039/C4RA11428K

[32] Kamal A, Balakrishna M, Nayak VL, Shaik TB, Faazil S, Nimbarte VD. Design and synthesis of imidazo[2,1-b] thiazole-chalcone conjugates: Microtubule-destabilizing agents. ChemMedChem. 2014;**9**:2766-2780. DOI: 10.1002/cmdc.201402310

[33] Li ZH, Zhang J, Liu XQ, Geng PF, Ma JL, Wang B, et al. Identification of thiazolo[5,4-d]pyrimidine derivatives as potent antiproliferative agents through the drug repurposing strategy. European Journal of Medicinal Chemistry. 2017;**135**:204-212. DOI: 10.1016/j.ejmech.2017.04.056

[34] Hassan GS, El-Messery SM, Abbas A. Synthesis and anticancer activity of new thiazolo[3,2-a]pyrimidines: DNA binding and molecular modeling study. Bioorganic Chemistry. 2017;**74**:41-52. DOI: 10.1016/j.bioorg.2017.07.008

[35] Ramirez J, Svetaz L, Quiroga J, Abonia R, Raimondi M, Zacchino S, et al. Synthesis of novel thiazole-based 8,9-dihydro-7H-pyrimido[4,5-b][1,4] diazepines as potential antitumor and antifungal agents. European Journal of

Medicinal Chemistry. 2015;**92**:866-875. DOI: 10.1016/j.ejmech.2015.01.053

[36] Altintop MD, Ozdemir A, Turan-Zitouni G, Ilgin S, Atli O, Demirel R, et al. A novel series of thiazolyl-pyrazoline derivatives: Synthesis and evaluation of antifungal activity, cytotoxicity and genotoxicity. European Journal of Medicinal Chemistry. 2015;**92**:342-352. DOI: 10.1016/j.ejmech.2014.12.055

[37] Bayomi SM, El-Kashef HA, El-Ashmawy MB, Nasr MN, El-Sherbeny MA, Abdel-Aziz NI, et al. Synthesis and biological evaluation of new curcumin analogues as antioxidant and antitumor agents: Molecular modeling study. European Journal of Medicinal Chemistry. 2015;**101**:584-594. DOI: 10.1016/j.ejmech.2015.07.014

[38] Altintop MD, Kaplancikli ZA, Ciftci GA, Demirel R. Synthesis and biological evaluation of thiazoline derivatives as new antimicrobial and anticancer agents. European Journal of Medicinal Chemistry. 2014;**74**:264-277. DOI: 10.1016/j.ejmech.2013.12.060

[39] Asati V, Mahapatra DK, Bharti SK. Thiazolidine-2,4-diones as multi-targeted scaffold in medicinal chemistry: Potential anticancer agents. European Journal of Medicinal Chemistry. 2014;**87**:814-833. DOI: 10.1016/j.ejmech.2014.10.025

[40] Kamal A, Syed MA, Mohammed SM. Therapeutic potential of benzothiazoles: A patent review (2010 - 2014). Expert Opinion on Therapeutic Patents. 2015;**25**:335-349. DOI: 10.1517/13543776.2014.999764

[41] Sovic I, Jambon S, Kraljevic Pavelic S, Markova-Car E, Ilic N, Depauw S, et al. Synthesis, antitumor activity and DNA binding features of benzothiazolyl and benzimidazolyl substituted isoindolines. Bioorganic & Medicinal Chemistry. 2018;**26**:1950-1960. DOI: 10.1016/j.bmc.2018.02.045

[42] Perontsis S, Hatzidimitriou AG, Begou OA, Papadopoulos AN, Psomas G. Characterization and biological properties of copper(II)-ketoprofen complexes. Journal of Inorganic Biochemistry. 2016;**162**:22-30. DOI: 10.1016/j.jinorgbio.2016.06.001

[43] Shakir M, Hanif S, Sherwani MA, Mohammad O, Azam M, Al-Resayes SI. Pharmacophore hybrid approach of new modulated bis-diimine Cu(II)/Zn(II) complexes based on 5-chloro Isatin Schiff base derivatives: Synthesis, spectral studies and comparative biological assessment. Journal of Photochemistry and Photobiology. B. 2016;**157**:39-56. DOI: 10.1016/j.jphotobiol.2016.01.019

[44] Li L, Du K, Wang Y, Jia H, Hou X, Chao H, et al. Self-activating nuclease and anticancer activities of copper(II) complexes with aryl-modified 2,6-di(thiazol-2-yl)pyridine. Dalton Transactions. 2013;**42**:11576-11588. DOI: 10.1039/c3dt50395j

[45] Czerwinska K, Machura B, Kula S, Krompiec S, Erfurt K, Roma-Rodrigues C, et al. Copper(ii) complexes of functionalized 2,2′:6′,2″-terpyridines and 2,6-di(thiazol-2-yl)pyridine: Structure, spectroscopy, cytotoxicity and catalytic activity. Dalton Transactions. 2017;**46**:9591-9604. DOI: 10.1039/c7dt01244f

[46] Xie F, Peng F. Anticancer activity of copper complex of (4R)-(−)-2-Thioxo-4-thiazolidinecarboxylic acid and 3-Rhodaninepropionic acid on prostate and breast cancer cells by fluorescent microscopic imaging. Journal of Fluorescence. 2018;**28**:89-96. DOI: 10.1007/s10895-017-2177-0

[47] Sandhaus S, Taylor R, Edwards T, Huddleston A, Wooten Y, Venkatraman R, et al. A novel copper(II) complex identified as a potent drug against colorectal and breast cancer cells and as a poison inhibitor for human topoisomerase IIalpha. Inorganic Chemistry Communications. 2016;**64**:45-49. DOI: 10.1016/j.inoche.2015.12.013

[48] Du K, Liang J, Wang Y, Kou J, Qian C, Ji L, et al. Dual inhibition of topoisomerases I and IIalpha by ruthenium(II) complexes containing asymmetric tridentate ligands. Dalton Transactions. 2014;**43**:17303-17316. DOI: 10.1039/c4dt02142h

[49] Correa RS, da Silva MM, Graminha AE, Meira CS, Santos JA, Moreira DR, et al. Ruthenium(II) complexes of 1,3-thiazolidine-2-thione: Cytotoxicity against tumor cells and anti-Trypanosoma cruzi activity enhanced upon combination with benznidazole. Journal of Inorganic Biochemistry. 2016;**156**:153-163. DOI: 10.1016/j.jinorgbio.2015.12.024

[50] Rubino S, Busa R, Attanzio A, Alduina R, Di Stefano V, Girasolo MA, et al. Synthesis, properties, antitumor and antibacterial activity of new Pt(II) and Pd(II) complexes with 2,2′-dithiobis(benzothiazole) ligand. Bioorganic & Medicinal Chemistry. 2017;**25**:2378-2386. DOI: 10.1016/j.bmc.2017.02.067

[51] Lalinde E, Lara R, Lopez IP, Moreno MT, Alfaro-Arnedo E, Pichel JG, et al. Benzothiazole-based cycloplatinated chromophores: Synthetic, optical, and biological studies. Chemistry. 2018;**24**:2440-2456. DOI: 10.1002/chem.201705267

[52] Song XQ, Liu YH, Shao J, Zhang ZL, Xie CZ, Qiao X, et al. Rapid induction of apoptosis in tumor cells treated with a new platinum(II) complex based on amino-thiazolidinone. European Journal of Medicinal Chemistry. 2018;**157**: 188-197. DOI: 10.1016/j.ejmech.2018.07.075

[53] Sahin O, Ozdemir UO, Seferoglu N, Genc ZK, Kaya K, Aydiner B, et al. New platinum (II) and palladium (II) complexes of coumarin-thiazole Schiff base with a fluorescent chemosensor properties: Synthesis, spectroscopic characterization, X-ray structure determination, in vitro anticancer activity on various human carcinoma cell lines and computational studies. Journal of Photochemistry and Photobiology. B. 2018;**178**:428-439. DOI: 10.1016/j.jphotobiol.2017.11.030

[54] Islam MK, Kim S, Kim HK, Park S, Lee GH, Kang HJ, et al. Manganese complex of ethylenediaminetetraacetic acid (EDTA)-Benzothiazole aniline (BTA) conjugate as a potential liver-targeting MRI contrast agent. Journal of Medicinal Chemistry. 2017;**60**:2993-3001. DOI: 10.1021/acs.jmedchem.6b01799

[55] Vellaiswamy G, Ramaswamy S. Co(II) complexes of4-((3-ethoxy-2-hydroxybenzylidene)amino)-N-(thiazol-2-yl)benzenesulphonamide and 4-((pyridin-2-ylmethylene)amino)-N-(thiazol-2-tl)benzenesulfonamide: Synthesis, fluorescence properties and anticancer activity. Journal of Fluorescence. 2017;**27**:1559-1565. DOI: 10.1007/s10895-017-2096-0

[56] Pontoriero A, Mosconi N, Monti L, Bellu S, Williams PAM, Raimondi M, et al. Synthesis, characterization and biological studies of a cobalt(III) complex of sulfathiazole. Chemico-Biological Interactions. 2017;**278**: 152-161. DOI: 10.1016/j.cbi.2017.10.004

[57] Neelima, Poonia K, Siddiqui S, Arshad M, Kumar D. In vitro anticancer activities of Schiff base and its lanthanum complex.

Spectrochimica Acta. Part A, Molecular and Biomolecular Spectroscopy. 2016;**155**:146-154. DOI: 10.1016/j.saa.2015.10.015

[58] Abbehausen C, Manzano CM, Corbi PP, Farrell NP. Effects of coordination mode of 2-mercaptothiazoline on reactivity of Au(I) compounds with thiols and sulfur-containing proteins. Journal of Inorganic Biochemistry. 2016;**165**:136-145. DOI: 10.1016/j.jinorgbio.2016.05.011

[59] Aman F, Hanif M, Kubanik M, Ashraf A, Sohnel T, Jamieson SM, et al. Anti-inflammatory oxicams as multi-donor ligand systems: pH- and solvent-dependent coordination modes of meloxicam and piroxicam to Ru and Os. Chemistry. 2017;**23**:4893-4902. DOI: 10.1002/chem.201700263

[60] Walsh CT, Acker MG, Bowers AA. Thiazolyl peptide antibiotic biosynthesis: A cascade of post-translational modifications on ribosomal nascent proteins. The Journal of Biological Chemistry. 2010;**285**:27525-27531. DOI: 10.1074/jbc.R110.135970

[61] Kang H, Choi M-C, Seo C, Park Y. Therapeutic properties and biological benefits of marine-derived anticancer peptides. International Journal of Molecular Sciences. 2018;**19**:919. DOI: 10.3390/ijms19030919

[62] Teruya T, Sasaki H, Fukazawa H, Suenaga K. Bisebromoamide, a potent cytotoxic peptide from the marine *Cyanobacterium lyngbya* sp.: Isolation, stereostructure, and biological activity. Organic Letters. 2009;**11**:5062-5065. DOI: 10.1021/ol9020546

[63] Suzuki K, Mizuno R, Suenaga K, Teruya T, Tanaka N, Kosaka T, et al. Bisebromoamide, an extract from Lyngbya species, induces apoptosis through ERK and mTOR inhibitions in renal cancer cells. Cancer Medicine. 2013;**2**:32-39. DOI: 10.1002/cam4.53

[64] Johnston HJ, Boys SK, Makda A, Carragher NO, Hulme AN. Naturally inspired peptide leads: Alanine scanning reveals an actin-targeting thiazole analogue of bisebromoamide. Chembiochem. 2016;**17**:1621-1627. DOI: 10.1002/cbic.201600257

[65] Singh S, Prasad NR, Kapoor K, Chufan EE, Patel BA, Ambudkar SV, et al. Design, synthesis, and biological evaluation of (S)-Valine Thiazole-derived cyclic and noncyclic peptidomimetic oligomers as modulators of human P-glycoprotein (ABCB1). Chembiochem. 2014;**15**:157-169. DOI: 10.1002/cbic.201300565

[66] Singh S, Prasad NR, Chufan EE, Patel BA, Wang YJ, Chen ZS, et al. Design and synthesis of human ABCB1 (P-glycoprotein) inhibitors by peptide coupling of diverse chemical scaffolds on carboxyl and amino termini of (S)-valine-derived thiazole amino acid. Journal of Medicinal Chemistry. 2014;**57**:4058-4072. DOI: 10.1021/jm401966m

[67] Wang Y-J, Patel BA, Anreddy N, Zhang Y-K, Zhang G-N, Alqahtani S, et al. Thiazole-valine peptidomimetic (TTT-28) antagonizes multidrug resistance in vitro and in vivo by selectively inhibiting the efflux activity of ABCB1. Scientific Reports. 2017;7:42106. DOI: 10.1038/srep42106

[68] Patel BA, Abel B, Barbuti AM, Velagapudi UK, Chen Z-S, Ambudkar SV, et al. Comprehensive synthesis of amino acid-derived Thiazole Peptidomimetic analogues to understand the enigmatic drug/substrate-binding site of P-glycoprotein. Journal of Medicinal Chemistry. 2018:61. DOI: 834-864. DOI: 10.1021/acs.jmedchem.7b01340

[69] Dutta D, Debnath M, Müller D, Paul R, Das T, Bessi I, et al. Cell penetrating thiazole peptides inhibit c-MYC expression via site-specific targeting of c-MYC G-quadruplex. Nucleic Acids Research. 2018;**46**:5355-5365. DOI: 10.1093/nar/gky385

[70] Maderna A, Doroski M, Subramanyam C, Porte A, Leverett CA, Vetelino BC, et al. Discovery of cytotoxic Dolastatin 10 analogues with N-terminal modifications. Journal of Medicinal Chemistry. 2014;**57**: 10527-10543. DOI: 10.1021/jm501649k

[71] Yokosaka S, Izawa A, Sakai C, Sakurada E, Morita Y, Nishio Y. Synthesis and evaluation of novel dolastatin 10 derivatives for versatile conjugations. Bioorganic & Medicinal Chemistry. 2018;**26**:1643-1652. DOI: 10.1016/j.bmc.2018.02.011

[72] Murray BC, Peterson MT, Fecik RA. Chemistry and biology of tubulysins: Antimitotic tetrapeptides with activity against drug resistant cancers. Natural Product Reports. 2015;**32**:654-662. DOI: 10.1039/c4np00036f

[73] Braig S, Wiedmann RM, Liebl J, Singer M, Kubisch R, Schreiner L, et al. Pretubulysin: A new option for the treatment of metastatic cancer. Cell Death & Disease. 2014;**5**:e1001. DOI: 10.1038/cddis.2013.510

[74] Lamidi OF, Sani M, Lazzari P, Zanda M, Fleming IN. The tubulysin analogue KEMTUB10 induces apoptosis in breast cancer cells via p53, Bim and Bcl-2. Journal of Cancer Research and Clinical Oncology. 2015;**141**: 1575-1583. DOI: 10.1007/s00432-015-1921-6

[75] Sani M, Lazzari P, Folini M, Spiga M, Zuco V, De Cesare M, et al. Synthesis and Superpotent anticancer activity of tubulysins carrying non-hydrolysable N-substituents on tubuvaline. Chemistry - A European Journal. 2017;**23**:5842-5850. DOI: 10.1002/chem.201700874

[76] Park Y, Bae SY, Hah J-M, Lee SK, Ryu J-S. Synthesis of stereochemically diverse cyclic analogs of tubulysins. Bioorganic & Medicinal Chemistry. 2015;**23**:6827-6843. DOI: 10.1016/j.bmc.2015.10.003

[77] Nicolaou KC, Yin J, Mandal D, Erande RD, Klahn P, Jin M, et al. Total synthesis and biological evaluation of natural and designed tubulysins. Journal of the American Chemical Society. 2016;**138**:1698-1708. DOI: 10.1021/jacs.5b12557

[78] Chen H, Lin Z, Arnst K, Miller D, Li W. Tubulin inhibitor-based antibody-drug conjugates for cancer therapy. Molecules. 2017;**22**:1281. DOI: 10.3390/molecules22081281

[79] Li JY, Perry SR, Muniz-Medina V, Wang X, Wetzel LK, Rebelatto MC, et al. A Biparatopic HER2-targeting antibody-drug conjugate induces tumor regression in primary models refractory to or ineligible for HER2-targeted therapy. Cancer Cell. 2016;**29**:117-129. DOI: 10.1016/j.ccell.2015.12.008

[80] Colombo R, Wang Z, Han J, Balachandran R, Daghestani HN, Camarco DP, et al. Total synthesis and biological evaluation of tubulysin analogues. The Journal of Organic Chemistry. 2016;**81**:10302-10320. DOI: 10.1021/acs.joc.6b01314

[81] Leverett CA, Sukuru SCK, Vetelino BC, Musto S, Parris K, Pandit J, et al. Design, synthesis, and cytotoxic evaluation of novel tubulysin analogues as ADC payloads. ACS Medicinal Chemistry Letters. 2016;**7**:999-1004. DOI: 10.1021/acsmedchemlett.6b00274

[82] Tumey N, Leverett CA, Vetelino B, Li F, Rago B, Han X, et al. Optimization of tubulysin antibody-drug conjugates: A case study in addressing ADC metabolism. ACS Medicinal Chemistry Letters. 2016;**7**:977-982. DOI: 10.1021/acsmedchemlett.6b00195

[83] Reddy JA, Dorton R, Bloomfield A, Nelson M, Dircksen C, Vetzel M, et al. Pre-clinical evaluation of EC1456, a folate-tubulysin anti-cancer therapeutic. Scientific Reports. 2018;**8**:8943-8943. DOI: 10.1038/s41598-018-27320-5

[84] Ying Y, Taori K, Kim H, Hong J, Luesch H. Total synthesis and molecular target of largazole, a histone deacetylase inhibitor. Journal of the American Chemical Society. 2008;**130**:8455-8459. DOI: 10.1021/ja8013727

[85] Poli G, Di Fabio R, Ferrante L, Summa V, Botta M. Largazole analogues as histone deacetylase inhibitors and anticancer agents: An overview of structure-activity relationships. ChemMedChem. 2017;**12**:1917-1926. DOI: 10.1002/cmdc.201700563

[86] Clausen DJ, Smith WB, Haines BE, Wiest O, Bradner JE, Williams RM. Modular synthesis and biological activity of pyridyl-based analogs of the potent class I histone deacetylase inhibitor Largazole. Bioorganic & Medicinal Chemistry. 2015;**23**:5061-5074. DOI: 10.1016/j.bmc.2015.03.063

[87] Almaliti J, Al-Hamashi AA, Negmeldin AT, Hanigan CL, Perera L, Pflum MKH, et al. Largazole analogues embodying radical changes in the depsipeptide ring: Development of a more selective and highly potent analogue. Journal of Medicinal Chemistry. 2016;**59**:10642-10660. DOI: 10.1021/acs.jmedchem.6b01271

[88] Cole KE, Dowling DP, Boone MA, Phillips AJ, Christianson DW. Structural basis of the antiproliferative activity of largazole, a depsipeptide inhibitor of the histone deacetylases. Journal of the American Chemical Society. 2011;**133**:12474-12477. DOI: 10.1021/ja205972n

[89] Kim B, Ratnayake R, Lee H, Shi G, Zeller SL, Li C, et al. Synthesis and biological evaluation of largazole zinc-binding group analogs. Bioorganic & Medicinal Chemistry. 2017;**25**:3077-3086. DOI: 10.1016/j.bmc.2017.03.071

[90] Sun S, Oliveira B, Jiménez-Osés G, Bernardes GJL. Radical-mediated thiol-ene strategy for photoactivation of thiol-containing drugs in cancer cells. Angewandte Chemie, International Edition. 2018;**57**:15832-15835. DOI: 10.1002/anie.201811338

[91] Asano A, Yamada T, Taniguchi T, Sasaki M, Yoza K, Doi M. Ascidiacyclamides containing oxazoline and thiazole motifs assume square conformations and show high cytotoxicity. Journal of Peptide Science. 2018;**24**:e3120. DOI: 10.1002/psc.3120

[92] Chen Q-Y, Liu Y, Cai W, Luesch H. Improved total synthesis and biological evaluation of potent apratoxin S4 based anticancer agents with differential stability and further enhanced activity. Journal of Medicinal Chemistry. 2014;**57**:3011-3029. DOI: 10.1021/jm4019965

[93] Cai W, Chen Q-Y, Dang LH, Luesch H. Apratoxin S10, a dual inhibitor of angiogenesis and cancer cell growth to treat highly vascularized tumors. ACS Medicinal Chemistry Letters. 2017;**8**:1007-1012. DOI: 10.1021/acsmedchemlett.7b00192

[94] Cai W, Ratnayake R, Gerber MH, Q-y C, Yu Y, Derendorf H, et al. Development of apratoxin S10 (Apra S10) as an anti-pancreatic cancer agent and its preliminary evaluation in an orthotopic patient-derived

xenograft (PDX) model. Investigational New Drugs. 2018;**10**. DOI: 10.1007/s10637-018-0647-0

[95] Onda Y, Masuda Y, Yoshida M, Doi T. Conformation-based design and synthesis of apratoxin a mimetics modified at the α,β-unsaturated thiazoline moiety. Journal of Medicinal Chemistry. **2017**(60):6751-6765. DOI: 10.1021/acs.jmedchem.7b00833

Chapter 6

One-Pot-Condensation Reaction of Heterocyclic Amine, 1,3-Diketone and Aldehydes Using *In Situ* Generated Superoxide Ion: A Rapid Synthesis of Structurally Diverse Drug-Like Complex Heterocycles

Sundaram Singh and Savita Kumari

Abstract

A novel, convenient one-pot multicomponent synthesis of tetraheterocyclicbenzimidazolo/benzothiazolo quinazolin-1-one derivatives has been reported in the presence of tetraethylammonium superoxide under non-aqueous condition. The superoxide induced three-component reaction of various aromatic aldehydes, 2-aminobenzimadazole/2-aminobenzothiazole and dimedone/1,3- cyclohexanedione produced tetraheterocyclicbenzimidazolo/benzothiazolo quinazolin-1-one derivatives at room temperature under the mild reaction conditions. The tetraethylammonium superoxide has been generated by phase transfer reaction of potassium superoxide and tetraethylammonium bromide in dry DMF at room temperature. The present study extended the applicability of tetraethylammonium bromide as a phase transfer catalyst for the efficient use of superoxide ion in multi-component synthesis of structurally diverse drug-like complex heterocycles (quinazolines).

Keywords: superoxide ion, multicomponent reaction, Tetraethylammonium bromide, phase transfer catalyst, KO_2

1. Introduction

The importance of oxygen in sustaining life is unquestionable but the aerobic life-style is fraught with danger. However, some recent reports about oxygen toxicity have caused much concern among the whole scientific community. The oxygen toxicity is due to various reactive oxygen species (ROS) such as hydroxyl radical ($HO^{\bullet}$), superoxide anion radical $O_2^{\bullet -}$, and perhydroxyl radical. Hypochlorous acid (HOCl), hydrogen peroxide (H_2O_2), singlet oxygen and ozone are also included in this category, although they are not free radicals but can lead to free radical reaction. Out of all the reactive oxygen species, superoxide anion radical is probably the most important ROS, which has come to the forefront of current chemical and biochemical research for the two reasons [1–4]. First superoxide ion as a biochemical species which causes many diseases such as cancer, ageing, inflammation, heart attack and lung injury, etc. More recently, it has been implicated to play a key role in both aging and cancer. Second superoxide ion as a novel reagent. Further from its elementary reactivity pattern, this anion radical has been recognized as a multi-potent reagent, which acts as a base, nucleophile, oxidant and reductant. In view of these two points, superoxide research has become an area of interdisciplinary investigation [5–13].

Multi-component reactions (MCRs), in which multiple reactions are combined into one synthetic operation, have been used extensively to form carbon-carbon bonds in synthetic chemistry. Such reactions offer greater possibilities for molecular diversity per step with minimum reaction time, labor, cost, and waste production. The rapid assembly of molecular diversity utilizing MCRs has gained a great deal of attention, most notably for the construction of 'drug-like' libraries [14–20].

Quinazolines are very interesting heterocycles [21–25] as they serve as building blocks in numerous natural and synthetic products [26]. They exhibit a wide spectrum of biological and pharmacological activities such as propyl hydroxylase inhibitor [27], antidiabetics [28], anti-inflammatory [29], antiviral [30], antimicrobial [31], antineoplastic [32] and potent immunosuppressive agents [33]. Moreover, benzimidazolo quinazolines have also been an important class of heterocyclic compounds in drug research, as they are formed from both biodynamic heterosystems, benzimidazole and quinazoline, which have shown significant anticancer activities. Many useful methods, have been reported for synthesis of tetrahydrobenzoimidazo [2,1-b] quinazolin-1(2H)-ones ring system skeletons, such as the condensation of aminoazoles with benzylidene compounds, or three-component condensation of 2-aminobenzothiazole or 2-aminobenzimidazole and an aldehyde with cyclic 1,3-diketone. These reported methodologies produce good results in many cases [34, 35]. However, some of them suffer with certain limitations such as expensive catalysts, low yields of products, long reaction times, tedious procedures for preparations of catalysts, and tedious workup conditions [36–40]. Thus, there is enough room for further investigation in this direction. With a view to investigate the behavior of the superoxide ion in multicomponent organic synthesis, which is of importance in itself and further to assess its synthetic scope, the reaction of this novel reagent was studied.

2. Results and discussion

In continuation of our ongoing program on superoxide research and the synthesis of biologically active compounds, it is our current endeavor to extent the applicability of Et_4NO_2 for the synthesis of tetraheterocyclicBenzimidazolo/benzothiazolo quinazolin-1-one ring systems **4** by a one-pot three-component condensation reaction of various aromatic aldehydes **2** and 1,3-diketones **3** with 2-aminobenzimidazole/2-aminobenzothiazole **1** using tetraethylammonium superoxide under non aqueous conditions (**Figure 1**).

In order to achieve the optimum yield of the product, the effect of various parameters such as effect of solvents (DMF, DMSO, and CH_3CN) and molar proportion of the reactants were investigated in detail using benzaldehyde **2**, dimedone **3** with 2-aminobenzimidazole **1** as a model reaction.

To investigate the effect of solvents, the model reaction was carried out in different aprotic solvents. The results obtained clearly indicate that DMF was the best solvent among all investigated solvents in terms of product yield and the reaction time (**Table 1**).

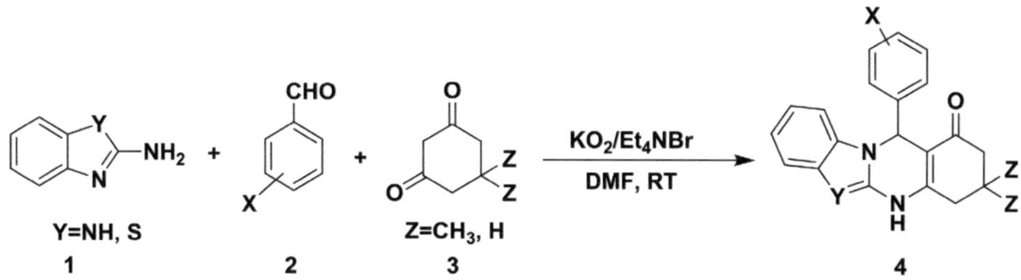

Figure 1.
*One-pot synthesis of compounds **4** by the three-component condensation reaction of various aldehydes and 1,3-diketones with 2-aminobenzimidazole/2-aminobenzothiazole under superoxide ion at room temperature.*

Entry	Solvents	Time	%Yield
1	Dichloromethane	12 h	Trace
2	Acetonitrile	8 h	70
3	Tetrahydrofuran	14 h	42
4	Dimethylsulfoxide	20 h	Trace
5	Dimethylformamide	6 h	88

Table 1.
Effect of solvents on the yield of the product 4a.

Entry	Reactants molar ratio					Product yield* (%)
	Benzaldehydes:dimidone:2-aminobezimidazole:KO_2:Et_4NBr					
1	1.0	1.0	1.0	1.0	1.0	40
2	1.0	1.0	1.0	1.0	0.5	38
3	1.0	1.0	1.0	1.0	0.25	28
4	1.0	1.0	1.0	2.5	1.25	69
5	**1.0**	**1.0**	**1.0**	**4.0**	**2.00**	**88**
6	1.0	1.0	1.0	6.0	3.00	90

**Isolated yield based on aldehyde.*
Optimized condition has been shown by bold letter (entry 5).

Table 2.
Effect of reactants molar ratio on the yield of product 4a.

In order to establish the reactants molar ratio on the yield of product the model reaction was carried out in different concentration of reactants (**Table 2**).

A perusal of the table clearly indicates the profound effect of the concentration of KO_2 and Et_4NBr on the yield of the product **4a**. As regards the ratio of KO_2 and Et_4NBr, it is evident from the entries 1, 2 and 3 that with the diminution of the concentration of Et_4NBr, the yield of product **4a** decreases. But as may be seen only a little difference in the yield of the product in the case of entries 1 and 2, the ratio of KO_2 and Et_4NBr was further kept to be 2:1. Therefore, in subsequent studies, the concentration of KO_2 has been increased manifold but the ratio of KO_2 and Et_4NBr was all along maintained to be 2:1. Furthermore, in case of entries 5 and 6, there is just a 2% increase in the yield of the product and for that 2% increase, the concentration of KO_2 and Et_4NBr have been increased substantially (6 fold and 3 fold respectively). As a result, considering the high cost of KO_2 and Et_4NBr, the entry 5, with the reactants ratio **1:1:1:4:2**, has been selected as the optimum ratio.

The scope and limitations of this reaction were fully illustrated with various *ortho-, meta- and para-substituted* benzaldehydes in the presence of 2-aminobenzimidazole and 2-aminobenzothiazole.

As indicated in **Table 3**, the reaction proceeded efficiently with both electron-withdrawing and electron releasing *ortho-, meta- and para-*substituted benzaldehydes.

The products were identified by their physical and spectral data, which were in full agreement with the reported values.

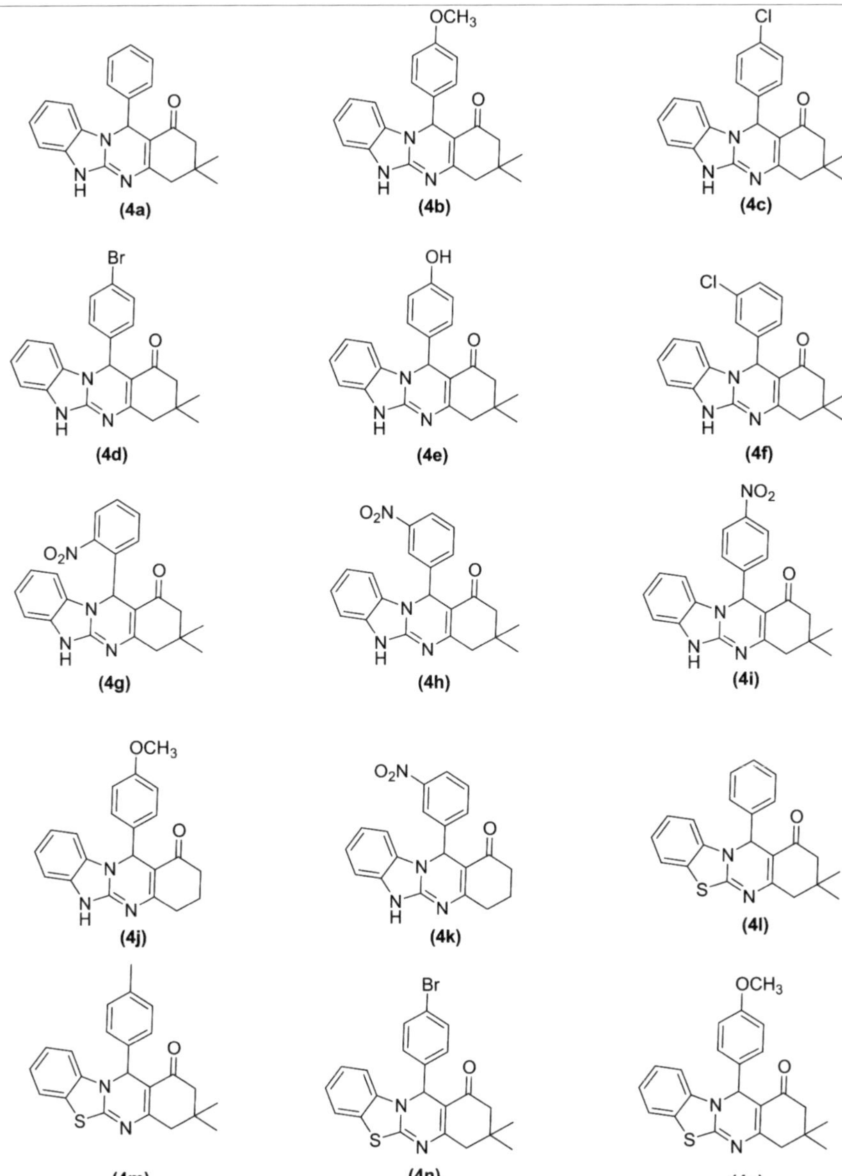

Table 3.
Synthesis of tetraheterocyclicbenzimidazolo/benzothiazolo quinazolin-1-ones.

2.1 Mechanism for the synthesis of tetraheterocyclicbenzimidazolo/ benzothiazolo quinazolin-1-ones

The proposed mechanism for the formation of tetraheterocyclicbenzimidazolo/benzothiazolo quinazolin-1-ones ring system is given in **Figure 2**. The reaction was initiated by the abstraction of proton from 1,3-diketones **3** by tetraethylammonium superoxide which was *in situ* generated by the phase transfer reaction of potassium superoxide with tetraethylammonium bromide. Now, Knoevenagel condensation takes place between benzaldehyde **2** and subsequently, by dehydration, olefin 3-benzylidene-2,4-hexanedione **5** is produced. Then 2-aminobenzimidazole/2-aminobenzothiazole**1** is reacted with compound **5** through a Michael addition to producc a product of type **6** and after cyclisation to afford tetraheterocyclicbenzimidazolo/benzothiazolo quinazolin-1-one ring systems **4**.

Potassium superoxide (1.42 g, 0.02 mol) and tetraethylammonium bromide (2.10 g, 0.01 mol) were weighed under nitrogen atmosphere using an atmosbag and were transferred into a three-necked R. B. flask, dry DMF (20 mL) was added to it and the mixture was agitated magnetically for 15 min to facilitate the formation of tetraethylammoniumsuperoxide. To the stirred reaction mixture, dimedone (0.70 g, 0.005 mol) **3** were added. After 10 min, benzaldehyde (0.53 g, 0.005 mol) **2** and 2-aminobenzimidazole (0.665 g, 0.005 mmol) **1** were introduced, and the stirring was continued 6 h. After the reaction was over as indicated by TLC, mixture was treated with cold brine solution (2 mL) followed by saturated sodium hydrogen carbonate solution (2 mL) to decompose the unreacted KO_2. The mixture was then extracted with dichloromethane (3 × 15 mL) and the combined organic phase was dried over anhydrous Na_2SO_4, filtered, and evaporated to give the products **4a**, which were purified by column chromatography.

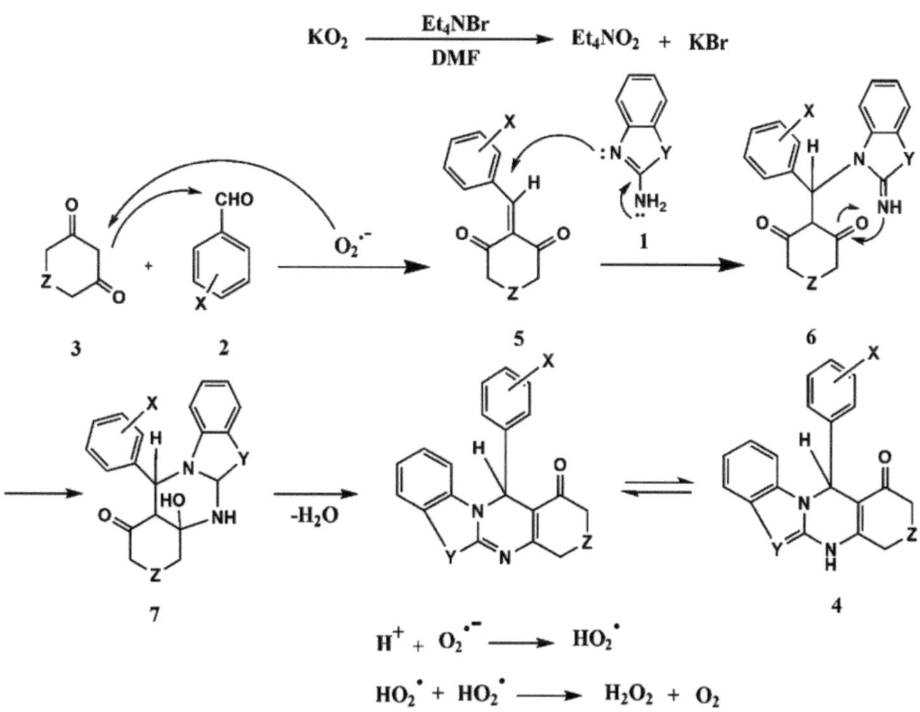

Figure 2.
*Plausible mechanism for the formation of tetraheterocyclicbenzimidazolo/benzothiazolo quinazolin-1-one derivatives (**4a–o**).*

All the products were characterized by IR and ^{1}H NMR (because of low solubility of compounds **4a-o**, ^{13}C NMR was not obtained).

3,3–Dimethyl–12–phenyl–3,4,6,12–tetrahydrobenzo[4,5]imidazo[2,1-b] quinazolin–1(2H)–one (4a): M.p. > 300°C; IR (KBr, υ = cm^{-1}) 3445, 2885, 1640, 1618, 1610, 1565 cm^{-1}; ^{1}H NMR (500 MHz, DMSO–d_6): δ = 11.16 (br. s, 1H, NH), 7.39–7.30 (m, 6H), 7.23 (d, J = 8.0 Hz, 1H), 7.07–7.04 (m, 1H), 6.98–6.95 (m, 1H), 6.44 (s, 1H), 2.26 (d, J = 16.0 Hz, 2H), 2.06 (d, J = 16.0 Hz, 2H), 1.06 (s, 3H), 0.92 (s, 3H). Anal. Calcd for $C_{22}H_{21}N_3O$: C, 76.94; H, 6.16; N, 12.24; O, 4.66. Found: C, 76.90; H, 6.20; N, 12.26; O, 4.64.

12–(4–Methoxyphenyl)–3,3–dimethyl–3,4,6,12–tetrahydrobenzo[4,5] imidazo[2,1-b]quinazolin–1(2H)–one (4b): M.p. > 300°C; IR (KBr, υ = cm^{-1}) 3391, 2850, 1670, 1644, 1610, 1590 cm^{-1};^{1}H NMR (500 MHz, DMSO–d_6): δ = 11.06 (br. s, 1H, NH), 7.36 (d, J = 8.0 Hz, 1H), 7.26–7.24 (m, 3H), 7.04 (t, J = 7.5 Hz, 1H), 6.95 (t, J = 7.5 Hz, 1H), 6.78 (d, J = 8.5 Hz, 2H), 6.36 (s, 1H), 3.65 (s, 3H), 2.64–2.52 (m, 2H), 2.25 (d, J = 16.0 Hz, 1H), 2.05 (d, J = 16.0 Hz, 1H), 1.06 (s, 3H), 0.94 (s, 3H). Anal Calcd for $C_{23}H_{23}N_3O_2$: C, 73.97; H, 6.21; N, 11.25; O, 8.57. Found: C, 73.92; H, 6.26; N, 11.23; O, 8.59.

12–(4–Chlorophenyl)–3,3–dimethyl–3,4,6,12–tetrahydrobenzo[4,5] imidazo[2,1-b] quinazolin–1(2H)–one (4c): M.p. > 300°C; IR (KBr, υ = cm^{-1}) 3440, 2934, 1655, 1650, 1613, 1580 cm^{-1}; ^{1}H NMR (500 MHz, DMSO–d_6): δ = 11.10 (br. s, 1H, NH), 7.36 (d, J = 7.5 Hz, 1H), 7.33 (d, J = 6.5 Hz, 2H), 7.24 (s, 2H), 7.15 (s, 1H), 7.04 (s, 1H), 6.95 (s, 1H), 6.41 (s, 1H), 2.63 (d, J = 16.0 Hz, 2H), 2.26 (d, J = 16.0 Hz, 2H), 1.06 (s, 3H), 0.93 (s, 3H). Anal. Calcd for $C_{22}H_{20}ClN_3O$: C, 69.93; H, 5.34; Cl, 9.38; N, 11.12; O, 4.23. Found: C, 69.90; H, 5.37; Cl, 9.34; N, 11.15; O, 4.24.

12–(4–Bromophenyl)–3,3–dimethyl–3,4,6,12–tetrahydrobenzo[4,5] imidazo[2,1-b] quinazolin–1(2H)–one (4d): M.p. >300°C; IR (KBr, υ = cm^{-1}) 3441, 2956, 1645, 1614, 1590, 1566 cm^{-1}; ^{1}H-NMR (500 MHz, DMSO-d_6): δ = 10.01 (br. s, 1H, NH), 6.99–7.89 (m, Ar–H), 6.43 (s, 1H), 2.59–2.67 (m, 2H), 2.20 (d, J = 16.00 Hz, 1H), 2.00 (d, J = 16.01 Hz, 1H) 1.05 (s, 3H), 0.94 (s, 3H). Anal. Calcd for $C_{22}H_{20}BrN_3O$: C, 62.57; H, 4.77; Br, 18.92; N, 9.95; O, 3.79. Found: C, 62.67; H, 4.86; Br, 18.80; N, 9.83; O, 3.90.

12–(4–Hydroxyphenyl)–3,3–dimethyl–3,4,6,12–tetrahydrobenzo[4,5] imidazo[2,1-b] quinazolin–1(2H)–one (4e): M.p. > 300°C; IR (KBr, υ = cm^{-1}) 3449, 2891, 1642, 1613, 1587, 1566 cm^{-1}; ^{1}H-NMR (500 MHz, DMSO-d_6) δ = 11.02 (br. s, 1H, NH), 9.33 (s, 1H, OH), 6.61–7.36 (m, 8H, Ar–H), 6.18 (s, 1H), 2.51–2.74 (m, 2H), 2.25 (d, J = 9.24 Hz, 1H), 2.05(d, J = 8.94 Hz, 1H), 1.07 (s, 3H), 0.96 (s, 3H), Anal. Calcd for $C_{22}H_{21}N_3O_2$: C, 73.52; H, 5.89; N, 11.69; O, 8.90. Found: C, 73.63; H, 5.97; N, 11.80; O, 8.71.

12–(3–Chlorophenyl)–3,3–dimethyl–3,4,6,12–tetrahydrobenzo[4,5] imidazo[2,1-b] quinazolin–1(2H)–one (4f) M.p. > 300°C; IR (KBr, υ = cm^{-1}) 3400, 2891, 1660, 1652, 1613, 1575 cm^{-1}; ^{1}H NMR (500 MHz, DMSO–d_6): δ = 11.18 (br. s, 1H, NH), 7.46 (s, 1H), 7.39 (d, J = 8.0 Hz, 1H), 7.30–7.21 (m, 5H), 7.06 (s, 1H), 6.98 (s, 1H), 6.46 (s, 1H), 2.58 (d, J = 16.0 Hz, 1H), 2.26 (d, J = 16.0 Hz, 1H), 2.08 (d, J = 16.0 Hz, 1H), 1.06 (s, 3H), 0.93 (s, 3H). Anal. Calcd for $C_{22}H_{20}ClN_3O$: C, 69.93; H, 5.34; Cl, 9.38; N, 11.12; O, 4.23. Found: C, 69.90; H, 5.37; Cl, 9.35; N, 11.14; O, 4.24.

3,3–Dimethyl–12–(2–nitrophenyl)–3,4,6,12–tetrahydrobenzo[4,5] imidazo[2,1-b]quinazolin–1(2H)–one (4 g): M.p. > 300°C; IR (KBr, υ = cm^{-1}) 3398, 2972, 1664, 1645, 1618, 1594 cm^{-1}; ^{1}H NMR (500 MHz, DMSO–d_6): δ = 11.18 (br. s, 1H, NH), 7.46 (s, 1H), 7.31–7.19 (m, 5H), 7.06 (t, J = 7.5 Hz, 1H), 6.98 (t, J = 7.5 Hz, 1H), 6.46 (s, 1H), 2.62 (d, J = 16.0 Hz, 1H), 2.55 (s, 1H), 2.26 (d, J = 16.0 Hz, 1H), 2.08 (d, J = 16.0 Hz, 1H), 1.06 (s, 2H), 0.93 (s, 2H). Anal. Calcd for $C_{22}H_{20}N_4O_3$: C, 68.03; H, 5.19; N, 14.42; O, 12.36. Found: C, 68.07; H, 5.15; N, 14.46; O, 12.32.

3,3-Dimethyl-12-(3-nitrophenyl)-3,4,6,12-tetrahydrobenzo[4,5]imidazo[2,1-b]quinazolin-1(2H)-one (4 h): M.p. > 300°C; IR (KBr, υ = cm^{-1}) 3394, 2970, 1660, 1648, 1615, 1598 cm^{-1}; ^{1}H NMR (500 MHz, DMSO-d_6): δ = 11.26 (br. s, 1H, NH), 8.27 (s, 1H), 8.04 (d, J = 8.0 Hz, 1H), 7.73 (d, J = 7.5 Hz, 1H), 7.56 (t, J = 8.0 Hz, 1H), 7.40 (d, J = 8.0 Hz, 1H), 7.29 (d, J = 8.0 Hz, 1H), 7.07 (t, J = 7.5 Hz, 1H), 6.97 (t, J = 7.5 Hz, 1H), 6.65 (s, 1H), 2.27 (d, J = 16.0 Hz, 2H), 2.07 (d, J = 16.0 Hz, 2H), 1.06 (s, 3H), 0.91 (s, 3H). Anal. Calcd for $C_{22}H_{20}N_4O_3$: C, 68.03; H, 5.19; N, 14.42; O, 12.36. Found: C, 68.08; H, 5.14; N, 14.44; O, 12.34.

3,3-Dimethyl-12-(4-nitrophenyl)-3,4,6,12 tetrahydrobenzo[4,5]imidazo[2,1-b]quinazolin-1(2H)-one (4i): M.p. > 300°C; IR (KBr, υ = cm^{-1}) 3396, 2980, 1662, 1641, 1612, 1594 cm^{-1}; ^{1}H NMR (500 MHz, DMSO-d_6): δ = 11.27 (br. s, 1H, NH), 8.12 (d, J = 8.5 Hz, 2H), 7.61 (d, J = 9.0 Hz, 2H), 7.40 (d, J = 7.5 Hz, 1H), 7.23 (d, J = 8.0 Hz, 1H), 7.07 (t, J = 7.5 Hz, 1H), 6.96 (t, J = 7.5 Hz, 1H), 6.60 (s, 1H), 2.65 (d, J = 16.0 Hz, 1H), 2.54 (d, J = 16.0 Hz, 1H), 2.27 (d, J = 16.0 Hz, 1H), 2.06 (d, J = 16.0 Hz, 1H), 1.06 (s, 3H), 0.91 (s, 3H). Anal. Calcd for $C_{22}H_{20}N_4O_3$: C, 68.03; H, 5.19; N, 14.42; O, 12.36. Found: C, 68.04; H, 5.18; N, 14.40; O, 12.38.

12-(4-Methoxyphenyl)-3,4,6,12-tetrahydrobenzo[4,5]imidazo[2,1-b]quinazolin-1(2H)-one (4j): M.p. = 238–240°C; IR (KBr, υ = cm^{-1}) 3398, 2976, 1666, 1642, 1616, 1575 cm^{-1}; ^{1}H NMR (500 MHz, DMSO-d_6): δ = 11.07 (br. s, 1H, NH), 7.36 (d, J = 8.0 Hz, 1H), 7.26–7.22 (m, 3H), 7.06–7.01 (m, 1H), 6.95 (t, J = 7.2 Hz, 1H), 6.78 (d, J = 8.0 Hz, 2H), 6.37 (s, 1H), 3.65 (s, 3H), 2.68 (d, J = 5.0 Hz, 2H), 2.29 (dd, J = 10.5, 5.0 Hz, 1H), 2.22 (dd, J = 16.0, 5.0 Hz, 1H), 2.02–1.93 (m, 1H), 1.88–1.80 (m, 1H). Anal. Calcd for $C_{21}H_{19}N_3O_2$: C, 73.03; H, 5.54; N, 12.17; O, 9.26. Found: C, 73.01; H, 5.56; N, 12.14; O, 9.29.

12-(3-Nitrophenyl)-3,4,6,12-tetrahydrobenzo[4,5]imidazo[2,1-b]quinazolin-1(2H)-one(4 k): M.p. > 300°C; IR (KBr, υ = cm^{-1}) 3412, 2872, 2855, 1670, 1640, 1617, 1601 cm^{-1};^{1}H NMR (500 MHz, DMSO-d_6): δ = 11.28 (br. s, 1H, NH), 8.26 (s, 1H), 8.03 (d, J = 8.5 Hz, 1H), 7.70 (d, J = 7.5 Hz, 1H), 7.54 (t, J = 8.0 Hz, 1H), 7.40 (d, J = 8.0 Hz, 1H), 7.26 (d, J = 7.5 Hz, 1H), 7.06 (t, J = 7.4 Hz, 1H), 6.96 (t, J = 7.5 Hz, 1H), 6.66 (s, 1H), 2.40–2.18 (m, 2H), 1.93 (dd, J = 16.0, 2H);. Anal. Calcd for $C_{20}H_{16}N_4O_3$: C, 66.66; H, 4.48; N, 15.55; O, 13.32. Found: C, 66.64; H, 4.50; N, 15.53; O, 13.34.

3,3-Dimethyl-12-phenyl-2,3,4,12-tetrahydro-1H-benzo[4,5]thiazolo[2,3-b]quinazolin-1-one (4 l): M.p. = 208–210°C; IR (KBr, υ = cm^{-1}) 3428, 2965, 1680, 1655, 1589, 1516, 1370 cm^{-1}; ^{1}H NMR (500 MHz, DMSO-d_6): δ = 7.79 (d, J = 10.0 Hz, 1H), 7.43 (dd, J = 17.5, 7.7 Hz, 3H), 7.28 (dd, J = 16.0, 3H), 7.20 (dd, J = 16.0, 8.0 Hz, 2H), 6.51 (s, 1H), 2.47–2.36 (m, 2H), 2.24 (d, J = 16.0 Hz, 1H), 2.05 (d, J = 16.0 Hz, 1H), 1.02 (s, 3H), 0.86 (s, 3H). Anal. Calcd for $C_{22}H_{20}N_2OS$: C, 73.30; H, 5.59; N, 7.77; O, 4.44; S, 8.89. Found: C, 73.33; H, 5.56; N, 7.79; O, 4.41; S, 8.88.

3,3-Dimethyl-12-(4-methylphenyl)-2,3,4,12-tetrahydro-1H-benzo[4,5]thiazolo[2,3-b]quinazolin-1-one(4 m):

M.p. = 203–205°C. ^{1}H NMR (500 MHz, DMSO-d_6): δ = 7.49–7.47 (m, 1H), 7.34 (d, J = 8 Hz, 2H), 7.28–7.22 (m, 1H), 7.18–7.15 (m, 2H), 7.06 (d, J = 8 Hz, 2H), 6.47 (s, 1H), 2.49 (s, 2H), 2.28–2.17 (m, 5H), 1.09 (s, 3H), 0.97 (s, 3H). Anal. Calcd for $C_{23}H_{22}N_2OS$: C, 73.77; H, 5.92; N, 7.48; O, 4.27; S, 8.56. Found: C, 73.68; H, 5.71; N, 7.60; O, 4.35; S, 8.70.

3,3-Dimethyl-12-(4-bromophenyl)-2,3,4,12-tetrahydro-1H-benzo[4,5]thiazolo[2,3-b]quinazolin-1-one (4n): M.p. 182–184°C. ^{1}H NMR (500 MHz, DMSO-d_6): δ = 7.47 (d, J = 8 Hz, 1H), 7.37–7.28 (m, 5H), 7.19 (d, J = 8 Hz, 1H), 7.06

(d, *J* = 8 Hz, 1H), 6.45 (s, 1H), 2.47 (s, 2H), 2.29–2.20 (m, 2H), 1.07 (s, 3H), 0.91 (s, 3H), Anal. Calcd for $C_{22}H_{19}BrN_2OS$: C, 60.14; H, 4.36; Br, 18.19; N, 6.38; O, 3.64; S, 7.30. Found: C, 60.35; H, 4.49; Br, 18.37; N, 6.50; O, 3.80; S, 7.45.

3,3–Dimethyl–12–(4-methoxyphenyl)–2,3,4,12–tetrahydro–1H–benzo[4,5] thiazolo[2,3–b]quinazolin–1–one (4o): M.p. 87–88°C. ^{1}H NMR (500 MHz, DMSO–d_6): δ = 7.49–7.46 (m, 2H), 7.38 (d, *J* = 8 Hz, 2H), 7.23–7.07 (m, 2H), 6.74 (d, *J* = 8 Hz, 2H), 6.44 (s, 1H), 3.62 (s, 3H), 2.48 (s, 2H), 2.31–2.17 (m, 2H), 1.06 (s, 3H), 0.93 (s, 3H). Anal. Calcd for $C_{23}H_{22}N_2O_2S$: C, 70.74; H, 5.68; N, 7.17; O, 8.19; S, 8.21. Found: C, 70.89; H, 5.80; N, 7.35; O, 8.39; S, 8.40.

3. Conclusion

In conclusion, the reaction of *in situ* generated $O_2^{\bullet-}$ with imidazoles is able to mimic the *in vivo* biochemical reactions involved and corroborate the role of $O_2^{\bullet-}$ in living cells. Since the investigation has been performed at an ambient temperature in the presence of *in situ* generated $O_2^{\bullet-}$, the results may be easily correlated with those occurring at physiological temperatures in more complex biological counterparts.

A novel synthetic route has been developed for the synthesis of tetraheterocyclic benzimidazolo/benzothiazolo quinazolin-1-one ring systems using tetraethylammonium superoxide under non aqueous condition at room temperature (mild reaction condition) within 6 h. The yield of the products was obtained up to 88% without using any tedious purification process. The applicability of tetraethylammonium bromide as an inexpensive alternative to 18-crown-6 for superoxide ion generation has been extended in present report.

Acknowledgements

The authors are thankful to IIT(BHU),Varanasi for financial support.

Conflict of interest

No conflict of interest.

Author details

Sundaram Singh* and Savita Kumari
Indian Institute of Technology (BHU), Varanasi, UP, India

*Address all correspondence to: sundaram.apc@itbhu.ac.in

References

[1] Capozzi G, Modena G. In: Patai S, editor. The Chemistry of Thiol Group: Part 2. New York: John Wiley & Sons, Inc; 1874. pp. 78-839

[2] Reid EE. Organic Chemistry of Bivalent Sulfur. Vol. I. New York: Chemical Publishing Co. Inc.; 1958. pp. 118-126

[3] Tarbell DS. In: Kharash N, editor. Organic Sulfur Compounds: Chapter 10. Vol. 1. New York: Pergamon Press; 1961. pp. 97-102

[4] Wallace JG. Hydrogen Peroxide in Organic Chemistry, E. I. Wilmington, U.S.A: du Pont de Nemours and Co.; 1960

[5] Halliwell B, Gutteridge JMC. Free Radicals in Biology and Medicine. 4th ed. Oxford, UK: Clarendon Press; 2007

[6] Bahorun T, Soobrattee MA, Luximon-Ramma V, Aruoma OI. Free radicals and antioxidants in cardiovascular health and disease. Internet Journal of Medical Update. 2006;**1**:1-17

[7] Valko M, Izakovic M, Mazur M, Rhodes CJ, Telser J. Role of oxygen radicals in DNA damage and cancer incidence. Molecular and Cellular Biochemistry. 2004;**266**:37-56

[8] Valko M, Leibfritz D, Moncol J, Cronin MT, Mazur M, Telser J. Free radicals and antioxidants in normal physiological functions and human disease. The International Journal of Biochemistry and Cell Biology. 2007;**39**:44-84

[9] Droge W. Free radicals in the physiological control of cell function. Physiological Reviews. 2002;**82**:47-95

[10] Willcox JK, Ash SL, Catignani GL. Antioxidants and prevention of chronic disease. Critical Reviews in Food Science and Nutrition. 2004;**44**:275-295

[11] Pacher P, Beckman JS, Liaudet L. Nitric oxide and peroxynitrite in health and disease. Physiological Reviews. 2007;**87**:315-424

[12] Genestra M. Oxyl radicals, redox-sensitive signalling cascades and antioxidants. Cellular Signalling. 2007;**19**:1807-1819

[13] Halliwell B. Biochemistry of oxidative stress. Biochemical Society Transactions. 2007;**35**:1147-1150

[14] Witt A, Bergman J. Recent developments in the field of quinazoline chemistry. Current Organic Chemistry. 2003;**7**:659-677

[15] Connolly DJ, Cusack D, O'Sullivan TP, Guiry PJ. Synthesis of quinazolinones and quinazolines. Tetrahedron. 2005;**61**:10153-10202

[16] Grasso S, Micale N, Monforte A-M, Monforte P, Polimeni S, Zappala M. Synthesis and in vitro antitumour activity evaluation of 1-aryl-1H, 3H-thiazolo [4, 3-b] quinazolines. European Journal of Medicinal Chemistry. 2000;**35**:1115-1119

[17] Testard A, Picot L, Lozach O, Blairvacq M, Meijer L, Murillo L, et al. Synthesis and evaluation of the antiproliferative activity of novel thiazoloquinazolinone kinases inhibitors. Journal of Enzyme Inhibition and Medicinal Chemistry. 2005;**20**:557-568

[18] Weber L. The application of multi-component reactions in drug discovery. Current Medicinal Chemistry. 2002;**9**:2085-2093

[19] Hulme C, Gore V. Multi-component Reactions: Emerging Chemistry in

Drug Discovery From Xylocain to Crixivan. Current Medicinal Chemistry. 2003;**10**:51-80

[20] Dömling A, Ugi I. Multicomponent reactions with isocyanides. Angewandte Chemie, International Edition. 2000;**39**:3168-3210

[21] Shaabani A, Rahmati A, Naderi S. A novel one-pot three-component reaction: Synthesis of triheterocyclic 4H-pyrimido [2, 1-b] benzazoles ring systems. Bioorganic & Medicinal Chemistry Letters. 2005;**15**:5553-5557

[22] Johne S, Herz W, Grisebach H, Kirby GW, Tamm Ch. The quinazoline alkaloids. InFortschritte der Chemie organischer Naturstoffe/Progress in the Chemistry of Organic Natural Products. Vol. 46. Wien: Springer; 1984. pp. 159-229

[23] Johne S. In: Brossi A, editor. The Alkaloids, Chemistry and Pharmacology. Vol. 29. New York: Academic; 1986. pp. 99-140

[24] Liu JF, Kaselj M, Isome Y, Ye P, Sargent K, Sprague K, et al. Design and synthesis of a quinazolinone natural product-templated library with cytotoxic activity. Journal of Combinatorial Chemistry. 2006;**8**:7-10

[25] Hattori K, Kido Y, Yamamoto H, Ishida J, Kamijo K, Murano K, et al. Rational approaches to discovery of orally active and brain-penetrable quinazolinone inhibitors of poly (ADP-ribose) polymerase. Journal of Medicinal Chemistry. 2004;**47**:4151-4154

[26] Young I, Woodside J. Antioxidants in health and disease. Journal of Clinical Pathology. 2001;**54**:176-186

[27] Chai D, Fitch DM. WO Patent 09039322 A1. March, 2009

[28] Kato F, Kimura H, Omatsu M, Yamamoto K, Miyamoto R. WO Patent 02040485. May 23, 2002

[29] McMaster B. WO Patent 03105857. December 24, 2003

[30] Shigeta S, Mori S, Baba M, Hosoya M, Mochizuki N, Chiba T, et al. Inhibitory effect of pyridobenzoazoles on orthomyxo-and paramyxovirus replication in vitro. Antiviral Chemistry and Chemotherapy. 1992;**3**:171-177

[31] Nofal ZM, Fahmy HH, Mohamed HS. Synthesis, antimicrobial and molluscicidal activities of new benzimidazole derivatives. Archives of Pharmacal Research. 2002;**25**:28-38

[32] Abdel-hafez AA. Benzimidazole condensed ring systems: New synthesis and antineoplastic activity of substituted 3, 4-dihydro-and 1, 2, 3, 4-tetrahydro-benzo [4, 5] imidazo [1, 2-a] pyrinnidine derivatives. Archives of Pharmacal Research. 2007;**30**:678-684

[33] Lunn W, Harper R, Stone R. Benzimidazo [2, 1-b] quinazolin-12-ones. New class of potent immunosuppressive compounds. Journal of Medicinal Chemistry. 1971;**14**:1069-1071

[34] Lipson VV, Desenko SM, Shirobokova MG, Borodina VV. Synthesis of 9-Aryl-6, 6-dimethyl-5, 6, 7, 9-tetrahydro-1, 2, 4-triazolo [5, 1-b] quinazolin-8 (4H) ones. Chemistry of Heterocyclic Compounds. 2003;**39**:1213-1217

[35] Lipson VV, Desenko SM, Shishkina SV, Shirobokova MG, Shishkin OV, Orlov VD. Cyclocondensation of 2-aminobenzimidazole with dimedone and its arylidene derivatives. Chemistry of Heterocyclic Compounds. 2003;**39**:1041-1047

[36] Shaabani A, Farhangi E, Rahmati A. A Rapid Combinatorial

Library Synthesis of Benzazolo [2, 1-b] quinazolinones and Triazolo [2, 1-b] quinazolinones. Iranian Journal of Chemistry and Chemical Engineering (IJCCE). 2013;**32**:3-10

[37] Ali M, Nakisa G. A three-component onepot procedure for the synthesis benzimidazoloquinazolinone derivatives in the presence of chitosan-supported metal nanocomposite as a green and reusable catalyst. In: ECSOC-17; 2013. pp. 1-9

[38] Ahmad S, Elham F, Abbas R. Synthesis of tetrahydrobenzimidazo [1, 2-b] quinazolin-1 (2H)-one and tetrahydro-1, 2, 4-triazolo [5, 1-b] quinazolin-8 (4H)-one ring systems under solvent-free conditions. Combinatorial Chemistry & High Throughput Screening. 2006;**9**:771-776

[39] Atar AB, Jeong YS, Jeong YT. Iron fluoride: The most efficient catalyst for one-pot synthesis of 4H-pyrimido [2, 1-b] benzothiazoles under solvent-free conditions. Tetrahedron. 2014;**70**:5207-5213

[40] Heravi MM, Derikvand F, Ranjbar L. Sulfamic acid–catalyzed, three-component, one-pot synthesis of [1, 2, 4] triazolo/benzimidazolo quinazolinone derivatives. Synthetic Communications®. 2010;**40**:677-685

Chapter 7

The Oxygen-Containing Fused Heterocyclic Compounds

Hillemane Venkatachalam
and Nanjangud Venkatesh Anil Kumar

Abstract

The oxygen-containing heterocycles are an important class of compounds in organic chemistry. These compounds are used as drugs (coumarin and oxazole), solvent (tetrahydrofuran), flavors, and fragrances (lactones). The fusion of aromatic ring to the oxygen-heterocycle will change the electron density; thereby, the physical/chemical/biological properties will alter. Also, the preparation of these fused molecules will require a different strategy/method/reaction condition. The topics covered in this chapter are the general synthetic methods and uses of fused heterocyclic compounds containing oxygen as a heteroatom. The derivatization of the primary scaffold is excluded from this chapter. Some of the fused compounds are coumarin (benzopyrans) and piclozotan (benzoxazepines).

Keywords: heterocycles, oxygen heteroatom, fused molecules, coumarin, a flavonoid

1. Introduction

The oxygen-containing heterocycles are an important class of compounds in organic chemistry mainly because of their natural abundance and diverse biological functions. Natural and semi-synthetic oxygen heterocyclic compounds such as Taxol [1] (anticancer), Digoxin (CHF treatment), Cyclosporine-A (immunosuppressant) and Lovastatin (hypolipidemic) are well known used as promising therapeutic compounds [2]. Kaur et al. reviewed the oxygen heterocycles wherein saturated and unsaturated compounds are considered. They discussed the classification and chemistry of each of those compounds [1]. Reports are available wherein the synthesis of natural products containing oxygen as heteroatom is reviewed by Cossy and Guérinot [2]. Also, Rowlands and Farley chaptered the book on the anion radicals from oxygen-containing heterocycles [3]. None of these reports target the oxygen-containing heterocyclic compounds where fused rings are taken into consideration.

2. Classification of oxygen heterocycles

The oxygen-containing heterocycles can be classified in several ways like the classification based on (a) the number of oxygen atoms, (b) saturation level, (c) aromaticity or (d) abundance. For the clarity of the concept, the classification

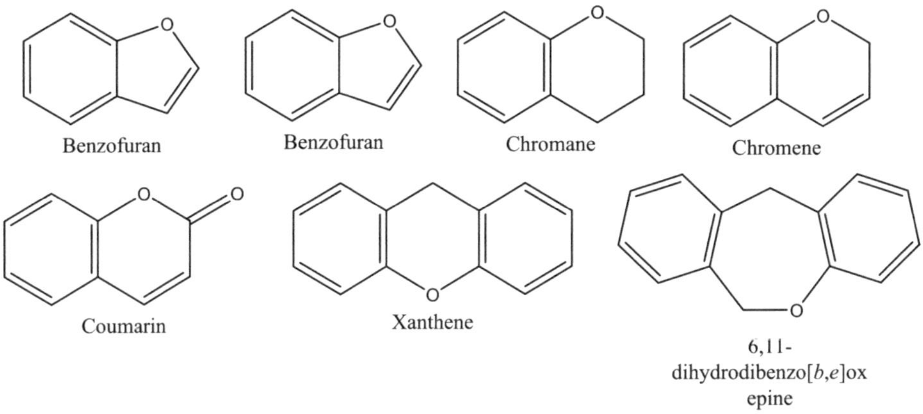

Figure 1.
Mono-oxygen containing fused heterocyclic compounds.

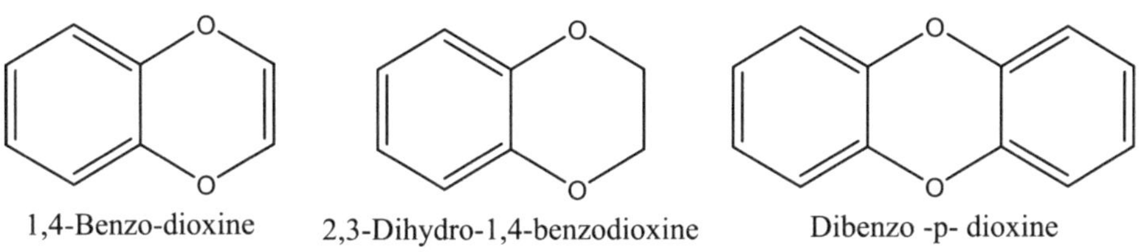

Figure 2.
Di-oxygen containing fused heterocyclic compounds.

based on a number of oxygen atoms is used. The benzene fused furan and pyrans are listed in **Figure 1**, whereas **Figure 2** represents the compounds with two oxygen atoms in the ring system.

3. Synthesis of benzofurans

3.1 Mono-substituted benzofurans

3.1.1 Substitution on furan ring

The ortho-hydroxystilbenes are cyclized in 70–90% yield in a metal-free environment using hypervalent iodine and 1 eq. of (diacetoxyiodo) benzene to get 2-substituted benzofurans [4]. The acids are converted to 2,4,6- trichloro-1,3,5-triazine esters, which are subsequently added to toluene, 2-hydroxybenzyl triphenylphosphonium bromide (1 eq.), and NEt_3 and irradiated at 110°C for two cycles of 30 min to get the 2-substituted benzofurans having a chiral stereocenter adjacent to the heterocycle in 60–80% yield [5]. Pd-catalyzed cyclisation of 2-chloroaryl alkynes using KOH at 100°C for 8 h resulted in the formation of benzofurans [6]. All three reactions are as shown in **Figure 3**.

3.1.2 Substitution on the benzene ring

Also, the substituted 1-allyl-2-allyloxybenzenes cyclizes to give substituted benzofurans by isomerization followed by ring-closure metathesis reaction using 5 mol.% catalyst [7]. The homologous members of benzofurans can be prepared by 0.1 eq. of Ru-catalyzed cycloisomerization of homo- and bis-homopropargylic alcohols in presence of pyridine at 90°C for 1–6 h [8] as shown in **Figure 4**.

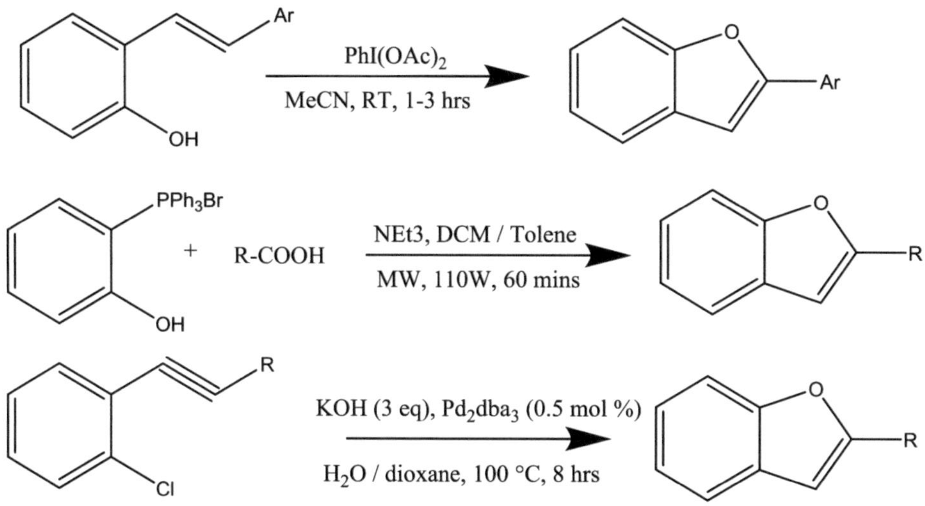

Figure 3.
Preparation of mono-substituted benzofuran.

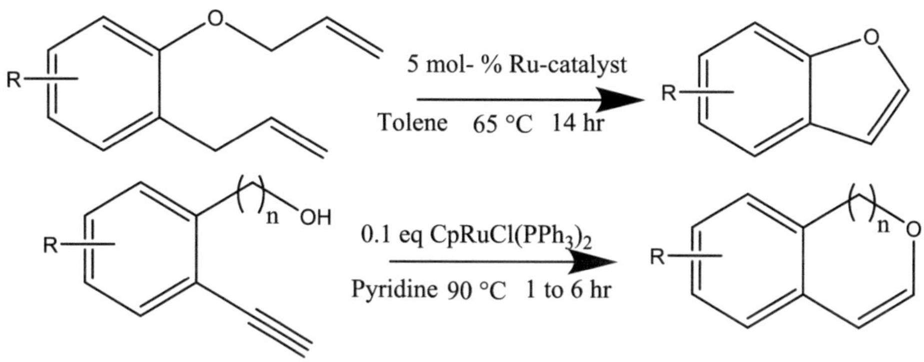

Figure 4.
Benzene ring substituted benzofuran.

3.2 Di-substituted benzofurans

3.2.1 Substitution on furan and benzene ring of benzofuran

The reaction between 1 eq. of 2-(2-hydroxyphenyl)acetonitriles with 2 eq of aryl boronic acids along with $Pd(OAc)_2$ (5 mol.%), bpy (10 mol.%), TFA (10 eq.), with TFA as solvent heated to 80°C for 36 h resulted in benzofuran derivatives with more than 80% yield [9]. Sonogashira cross-coupling reaction of halide with terminal alkynes followed by cyclization of the resulting 2-alkynylphenols in one-pot method results in benzofuran. The method employs the t-BuOH, $PdCl_2$-$(CH_3CN)_2$ (2 mol.%), t-BuOLi (3.6 eq.) and 2-chlorophenol (1 eq.), alkyne (1.5 eq.) were taken in a sealed tube and heated to 110°C for 22 h to get benzofuran [10]. Reaction of N-tosylhydrazones and terminal alkyne in a ligand free environment, using 10 mol.% of CuBr and 3 eq. of Cs_2CO_3 at 100°C for 4 h resulted in the formation of benzofurans with 38–91% yield [11]. Reductive cyclization of 1-(2-hydroxyphenyl)-propargyl alcohols in presence of $Pd(OAc)_2$ (5 mol.%), t-BuNC (1.2 eq.), Cs_2CO_3 (1.2 eq.) and MeCN as solvent gives benzofurans [12]. These reaction schemes are as shown in **Figure 5**.

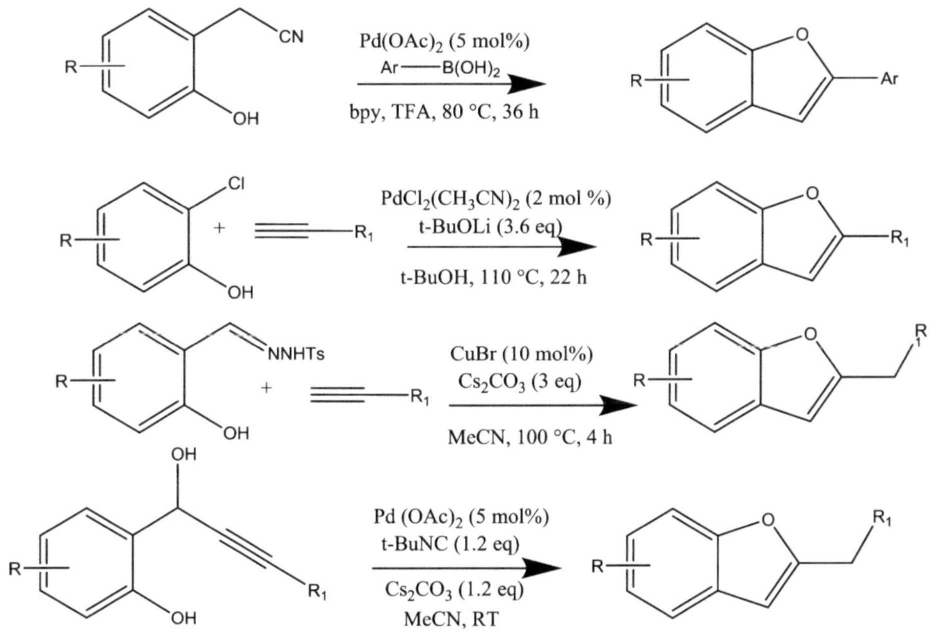

Figure 5.
Di-substituted benzofurans.

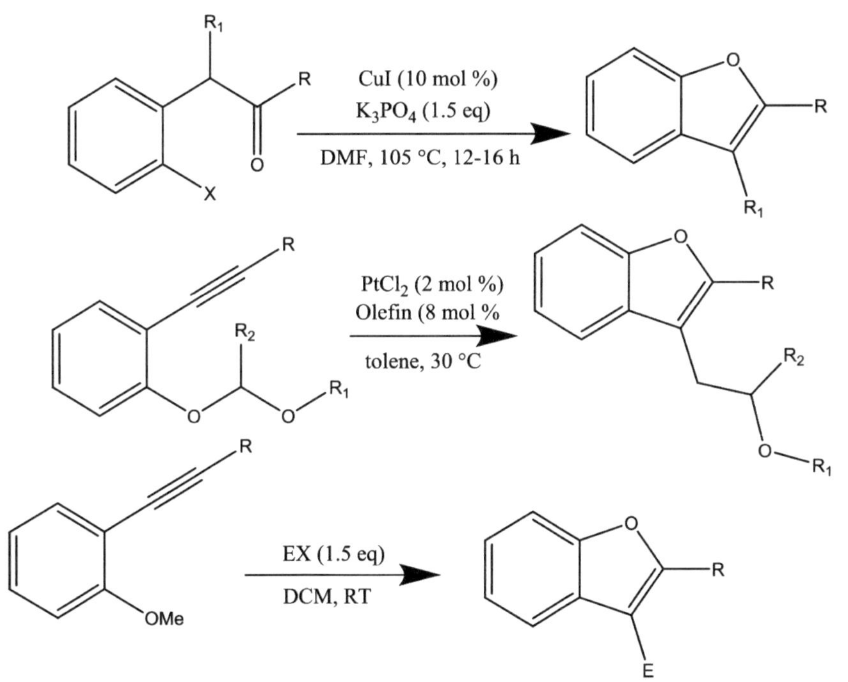

Figure 6.
Preparation of di-substituted benzofuran.

3.2.2 2,3-Substituted benzofuran

Ring closure of 2-haloaromatic ketones (1 eq.) in presence of K_3PO_4 (1.5 eq.), CuI (10 mol.%) and DMF at 105°C for 12–16 h results in di-substituted benzofurans [13]. The good yield coupled with atom economy was achieved, when o-alkynyl phenyl acetals are cyclized using $PtCl_2$ (2 mol.%), olefin (8 mol.%) in tolene at 30°C [14]. Cyclization of o-iodoanisoles and terminal alkynes under mild conditions using an electrophile (EX like PhSeCl or p-$O_2NC_6H_4SCl$) (1.5 eq.), DCM at room temperature for 2–6 h yields 2,3-disubstituted benzofurans [15]. These reaction schemes are as shown in **Figure 6**.

Selective dehydrative C—H alkylation reaction of alkenes with alcohols using $[(C_6H_6)(PCy_3)(CO)RuH]BF_4$, cyclopentene, toluene at 100°C for 6–12 h results in 2,3-substituted benzofurans [16]. O-Arylhydroxylamine hydrochloride (1 eq.) with cyclic or acyclic ketones (1 eq.) in the presence of methanesulfonic acid (2 eq.), THF at 60°C for 2–24 h yields benzofurans in 40–90% yield [17]. Ketones (1 eq.) on treatment with Grignard reagents (3 eq.), benzofurans are formed, in a regioselective manner via [1,2]-aryl migration [18]. These reactions are depicted in **Figure 7**.

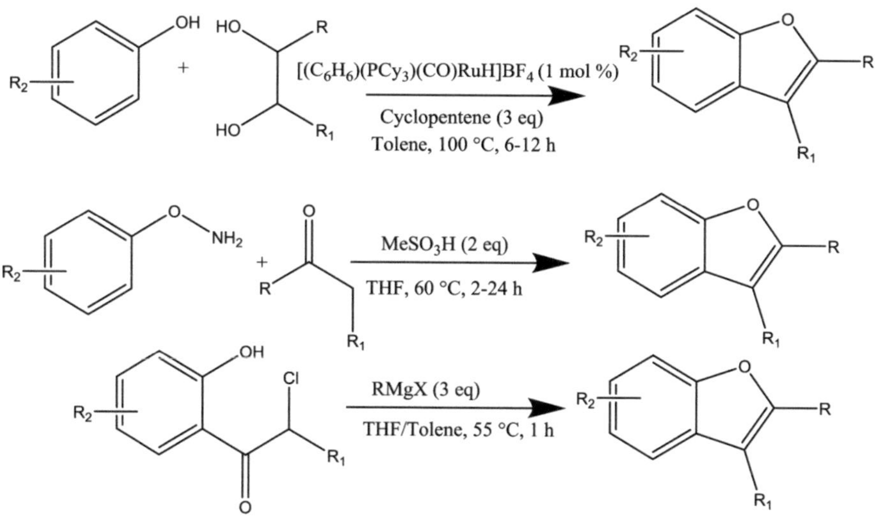

Figure 7.
Substituted benzofurans.

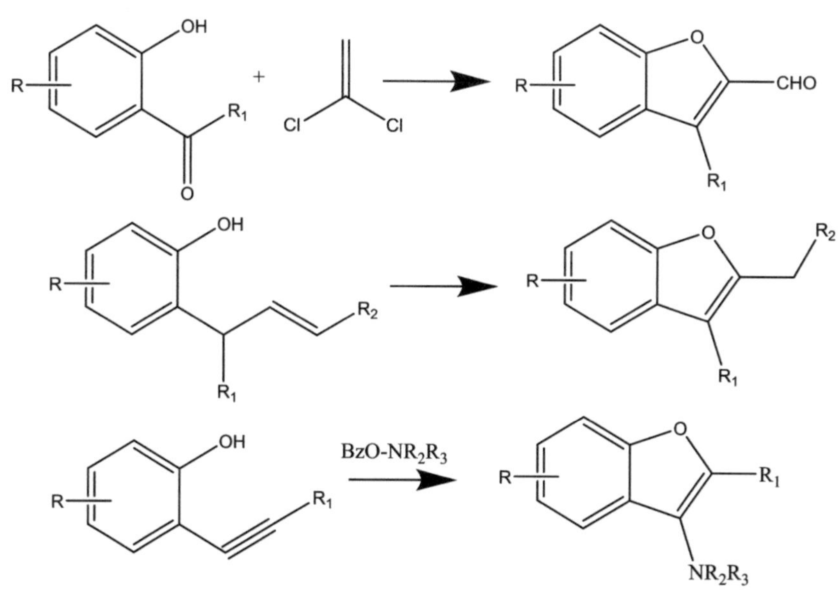

Figure 8.
Substituted benzofuran carbaldehydes.

Base-catalyzed, the condensation of o-hydroxyphenones with 1,1-dichloroethylene, gives substituted benzofuran carbaldehydes [19].

Similarly, o-cinnamyl phenols, on oxidative cyclisation, results in 2-benzyl benzofurans [20], while o-alkylphenols, on annulative reaction, gives 3-aminobenzofurans [21] (**Figure 8**).

4. Synthesis of benzofuranones

Alkenylphenols and phenyl formate reacts to give benzofuranones [22], while phenylacetic acids undergo cyclisation to give benzofuranones [23] (**Figure 9**).

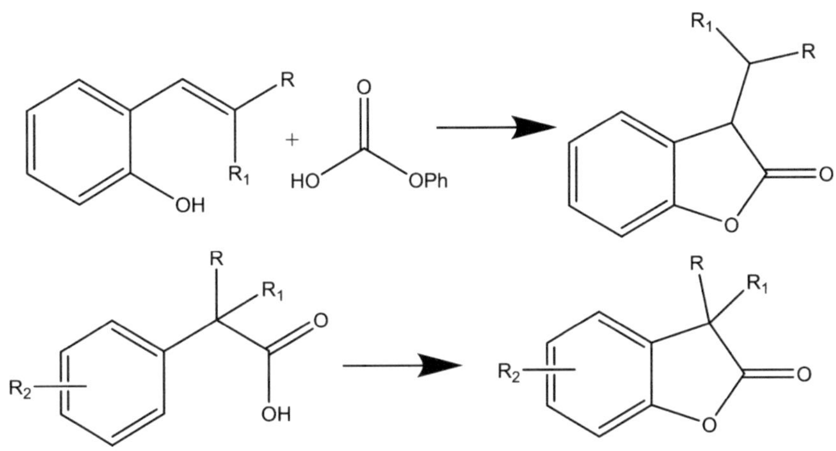

Figure 9.
Synthesis of benzofuranones.

5. Synthesis of dibenzofurans

Iododiaryl ether cyclizes under mild conditions to yield dibenzofurans [24]. Ortho-diazonium slats of diaryl ethers undergo intramolecular cyclisation,

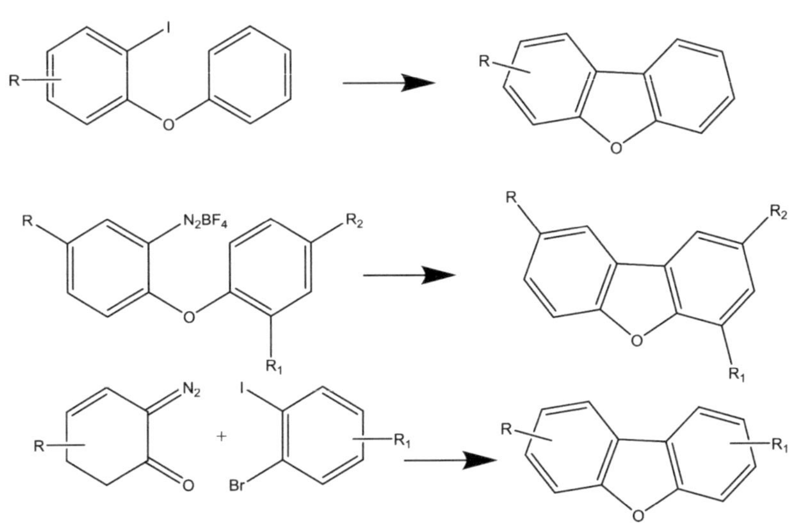

Figure 10.
Mono substituted dibenzofurans.

resulting in dibenzofurans [25]. Cross-coupling of 6-diazo-2-cyclohexenones and ortho-haloiodobenzenes gives substituted dibenzofuran [26] (**Figure 10**).

6. Synthesis of coumarins

Phenols react with beta-keto esters to give coumarins [27]. If phenolic acetates are used, then, acrylates are used [28]. Aromatic alkynoate undergoes cyclisation with aldehydes to form 3-acyl-4-arylcoumarins [29] (**Figure 11**).

Substituted 2-hydroxybenzaldehydes react with phenylacetic acids resulting in substituted 3-aryl coumarins [30]. With dialkyl acetylenedicarboxylate, 2-hydroxybenzaldehydes gives 4-carboxyalkyl-8-formyl coumarins [31]. 2-Hydroxybenzaldehydes (or 2-hydroxybenzaldehydes) cyclizes with aryl acetic acids to give 3-aryl coumarins (or 3-aryl-4-methyl-coumarins) [32] (**Figure 12**).

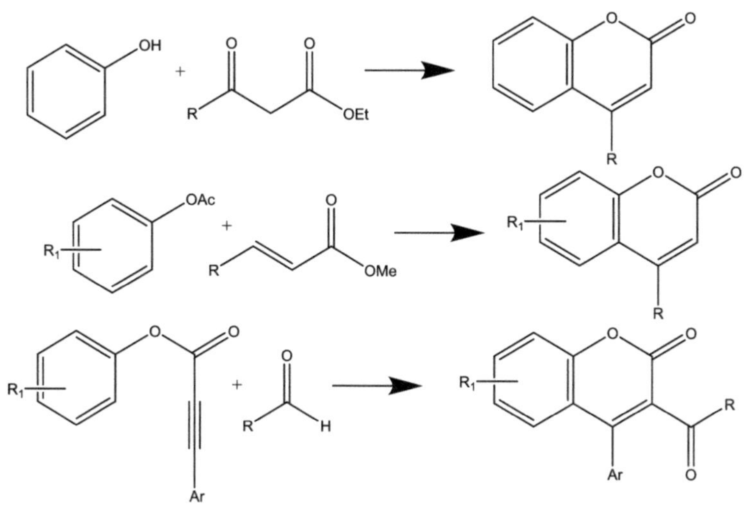

Figure 11.
Coumarin synthesis using phenol derivatives.

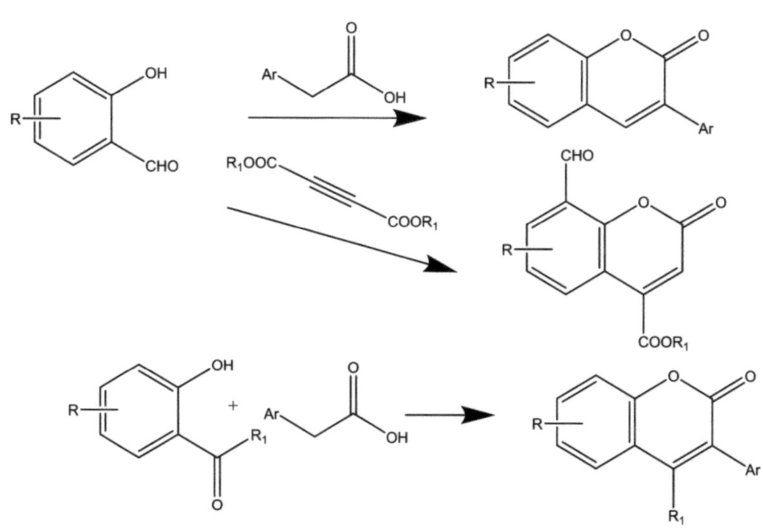

Figure 12.
Coumarin synthesis using benzaldehyde derivatives.

7. Synthesis of isocoumarins

Bromoalkynes reacts with benzoic acid and produces 3-substituted isocoumarins [33]. o-Halobenzoic acids and 1,3-diketones reacts to give 3-substituted isocoumarins [34]. 2-Halobenzoates and ketones react to give the same product [35]. o-Halobenzoic acids add to alkynes resulting in isocoumarin derivatives [36] (**Figure 13**).

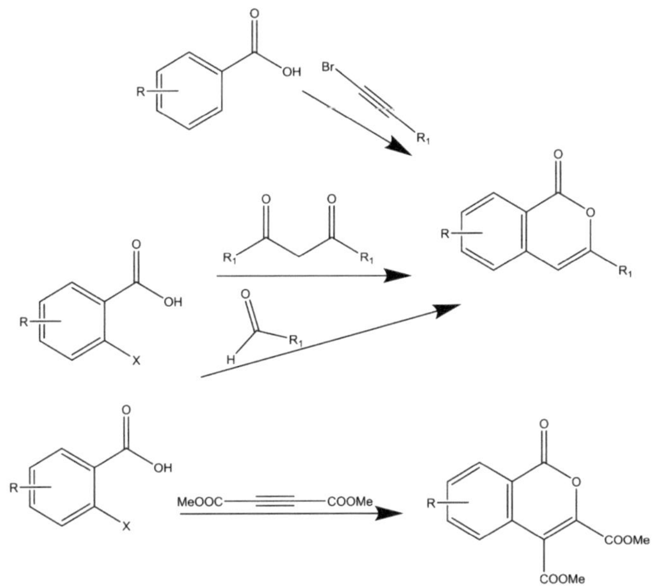

Figure 13.
Synthesis of isocoumarins.

8. Synthesis of flavones and flavonols

8.1 Baker-Venkataraman rearrangement

The chemical reaction between 2-hydroxyacetophenone and acid chloride in the presence of base yields 1,3-diketone which undergo rearrangement and

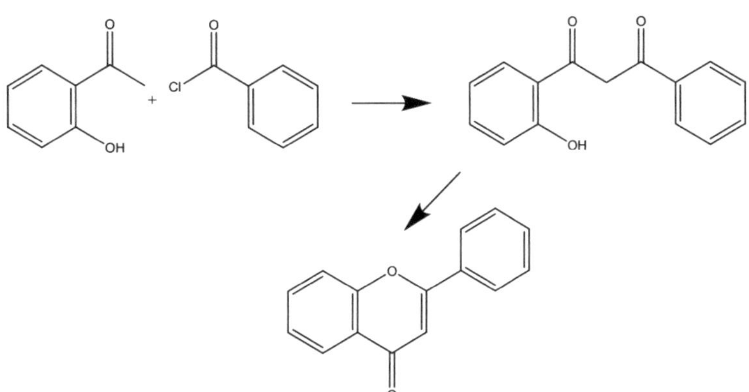

Figure 14.
Synthesis of flavone (from 1,3-diketone).

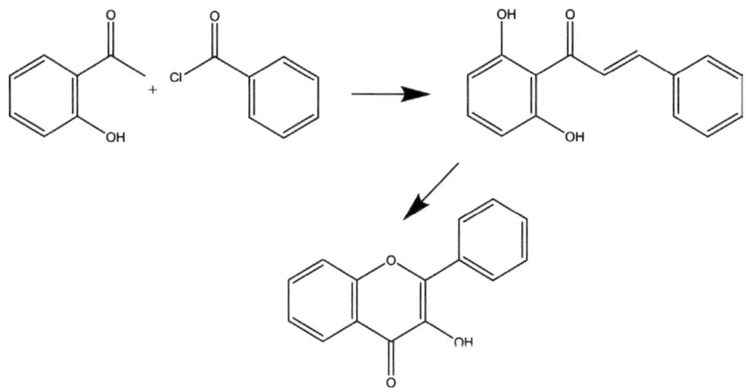

Figure 15.
Synthesis of flavone (from chalcone).

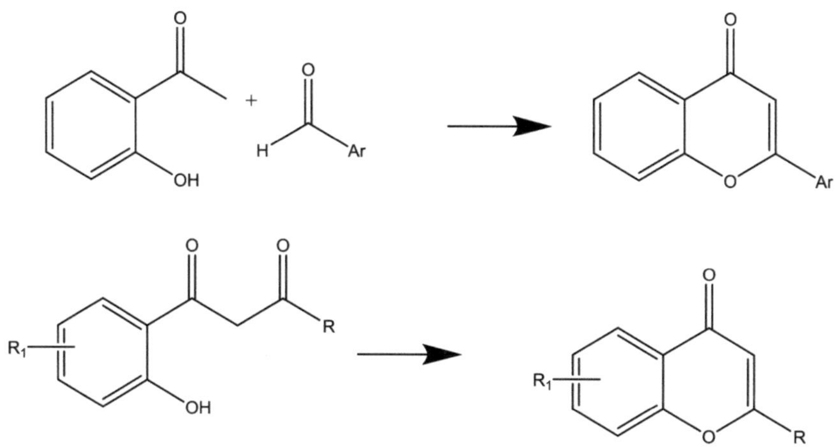

Figure 16.
Preparation of 2-phenyl γ-benzopyrones.

concomitant cyclisation to produce flavone [37]. This reaction is often used to synthesize chromones and flavones (**Figure 14**).

8.2 Algar-Flynn-Oyamada synthesis

In this method, chalcone (1,3-diaryl-2-propen-1-one) is produced by condensing 2-hydroxyacetophenone with an aryl aldehyde in alkaline medium (Claisen-Schmidt condensation) followed by oxidative cyclisation of chalcone to get flavone [38] (**Figure 15**).

A mixture of acetophenone and aromatic aldehyde when exposed to Microwave irradiation, 2-phenyl γ-benzopyrones are obtained [39]. The same compound can be prepared by cyclizing 1,3-diaryl propanediones (**Figure 16**).

9. Conclusions

In this chapter, the synthesis of fused heterocyclic compounds having oxygen as heteroatom is considered. The care is taken not to consider the reactions, where the reactions of the compounds leading to derivatizations are not included.

Benzofurans, benzofuranones, dibenzofurans, coumarins, isocoumarins, chromones, and flavones are the fused heterocyclic compounds considered in this chapter. Also, the reactions are indicative and not the detailed reaction conditions, and appropriate reagents are not included in this chapter.

Conflict of interest

No conflict of interest from both the authors.

Author details

Hillemane Venkatachalam and Nanjangud Venkatesh Anil Kumar*
Department of Chemistry, Manipal Institute of Technology, Manipal Academy of Higher Education, Manipal, India

*Address all correspondence to: nv.anil@manipal.edu

References

[1] Kaur P, Arora R, Gill NS. Review on oxygen heterocycles. Indo American Journal of Pharmaceutical Research. 2013;**3**(11):18

[2] Cossy J, Guérinot A. Natural products containing oxygen heterocycles—Synthetic advances between 1990 and 2015. Advances in Heterocyclic Chemistry. 2016;**119**:107-142

[3] Rowlands CC, Farley RD. 15.3.1 Oxygen Containing Heterocycles, vol. 26E1. 2008. pp. 308-315

[4] Singh FV, Wirth T. Hypervalent iodine mediated oxidative cyclization of o-hydroxystilbenes into benzo- and naphthofurans. Synthesis. 2012;**44**(08):1171-1177

[5] De Luca L, Giacomelli G, Nieddu G. A facile approach to the synthesis of chiral 2-substituted benzofurans. The Journal of Organic Chemistry. 2007;**72**(10):3955-3957

[6] Anderson KW et al. The selective reaction of aryl halides with KOH: Synthesis of phenols, aromatic ethers, and benzofurans. Journal of the American Chemical Society. 2006;**128**(33):10694-10695

[7] van Otterlo WAL et al. An isomerization-ring-closing metathesis strategy for the synthesis of substituted benzofurans. Tetrahedron. 2005;**61**(32):7746-7755

[8] Varela-Fernández A et al. Cycloisomerization of aromatic homo- and bis-homopropargylic alcohols via catalytic Ru vinylidenes: Formation of benzofurans and isochromenes. Organic Letters. 2009;**11**(22):5350-5353

[9] Wang X et al. Palladium-catalyzed reaction of arylboronic acids with aliphatic nitriles: Synthesis of alkyl aryl ketones and 2-arylbenzofurans. Synthesis. 2013;**45**(16):2241-2244

[10] Wang J-R, Manabe K. Hydroxyterphenylphoshine–palladium catalyst for benzo[b]furan synthesis from 2-chlorophenols. Bifunctional ligand strategy for cross-coupling of chloroarenes. The Journal of Organic Chemistry. 2010;**75**(15):5340-5342

[11] Zhou L et al. CuBr-catalyzed coupling of N-tosylhydrazones and terminal alkynes: Synthesis of benzofurans and indoles. Organic Letters. 2011;**13**(5):968-971

[12] Rajesh M et al. Pd-catalyzed isocyanide assisted reductive cyclization of 1-(2-hydroxyphenyl)-propargyl alcohols for 2-alkyl/benzyl benzofurans and their useful oxidative derivatization. The Journal of Organic Chemistry. 2015;**80**(24):12311-12320

[13] Chen C-Y, Dormer PG. Synthesis of benzo[b]furans via CuI-catalyzed ring closure. The Journal of Organic Chemistry. 2005;**70**(17):6964-6967

[14] Nakamura I, Mizushima Y, Yamamoto Y. Synthesis of 2,3-disubstituted benzofurans by platinum–olefin-catalyzed carboalkoxylation of o-alkynylphenyl acetals. Journal of the American Chemical Society. 2005;**127**(43):15022-15023

[15] Yue D, Yao T, Larock RC. Synthesis of 2,3-disubstituted benzo[b]furans by the palladium-catalyzed coupling of o-iodoanisoles and terminal alkynes, followed by electrophilic cyclization. The Journal of Organic Chemistry. 2005;**70**(25):10292-10296

[16] Lee D-H, Kwon K-H, Yi CS. Dehydrative C–H alkylation and alkenylation of phenols with alcohols: Expedient synthesis for substituted

phenols and benzofurans. Journal of the American Chemical Society. 2012;**134**(17):7325-7328

[17] Contiero F et al. Direct preparation of benzofurans from O-arylhydroxylamines. Synlett. 2009;**2009**(18):3003-3006

[18] Pei T et al. Controlled synthesis of 2- and 3-substituted benzo[b]furans. Organic Letters. 2010;**12**(21):4972-4975

[19] Schevenels F, Markó IE. Efficient and connective assembly of highly functionalized benzofurans using o-hydroxyphenones and dichloroethylene. Organic Letters. 2012;**14**(5):1298-1301

[20] Rehan M et al. Synthesis of functionalized benzo[b]furans via oxidative cyclization of o-cinnamyl phenols. The Journal of Organic Chemistry. 2017;**82**(7):3411-3424

[21] Hirano K, Satoh T, Miura M. Copper-catalyzed annulative amination of ortho-alkynylphenols with hydroxylamines: Synthesis of 3-aminobenzofurans by umpolung amination strategy. Organic Letters. 2011;**13**(9):2395-2397

[22] Wang H et al. A palladium-catalyzed regioselective hydroesterification of alkenylphenols to lactones with phenyl formate as CO source. Organic Letters. 2014;**16**(1):186-189

[23] Cheng X-F et al. Pd(II)-catalyzed enantioselective C—H activation/C—O bond formation: Synthesis of chiral benzofuranones. Journal of the American Chemical Society. 2013;**135**(4):1236-1239

[24] Panda N, Mattan I, Nayak DK. Synthesis of dibenzofurans via C—H activation of o-iodo diaryl ethers. The Journal of Organic Chemistry. 2015;**80**(13):6590-6597

[25] Du Z et al. Synthesis of dibenzofurans by palladium-catalysed tandem denitrification/C—H activation. Synlett. 2011;**2011**(20):3023-3025

[26] Zhao H et al. A one-pot synthesis of dibenzofurans from 6-diazo-2-cyclohexenones. Organic Letters. 2015;**17**(23):5744-5747

[27] Potdar MK, Mohile SS, Salunkhe MM. Coumarin syntheses via Pechmann condensation in Lewis acidic chloroaluminate ionic liquid. Tetrahedron Letters. 2001;**42**(52):9285-9287

[28] Gadakh SK, Dey S, Sudalai A. Rh-catalyzed synthesis of coumarin derivatives from phenolic acetates and acrylates via C—H bond activation. The Journal of Organic Chemistry. 2015;**80**(22):11544-11550

[29] Mi X et al. Preparation of 3-acyl-4-arylcoumarins via metal-free tandem oxidative acylation/cyclization between alkynoates with aldehydes. The Journal of Organic Chemistry. 2014;**80**(1):148-155

[30] Sashidhara K et al. Efficient and general synthesis of 3-aryl coumarins using cyanuric chloride. Synlett. 2012;**23**(04):611-621

[31] Majumdar K et al. Aromatic electrophilic substitution vs. intramolecular Wittig reaction: Vinyltriphenylphosphonium salt mediated synthesis of 4-carboxyalkyl-8-formyl coumarins. Synlett. 2011;**2011**(05):694-698

[32] Phakhodee W et al. Ph_3P/I_2-mediated synthesis of 3-aryl-substituted and 3,4-disubstituted coumarins. Synlett. 2017;**28**(07):825-830

[33] Jiang G et al. Palladium-catalyzed sequential nucleophilic addition/oxidative annulation of bromoalkynes with benzoic acids to construct

functionalized isocoumarins. Organic Letters. 2017;**19**(17):4440-4443

[34] Cai S, Wang F, Xi C. Assembly of 3-substituted Isocoumarins via a CuI-catalyzed domino coupling/addition/deacylation process. The Journal of Organic Chemistry. 2012;**77**(5):2331-2336

[35] Casnati A et al. Pd-catalyzed/iodide-promoted α-arylation of ketones for the regioselective synthesis of isocoumarins. The Journal of Organic Chemistry. 2017;**82**(15):8296-8303

[36] Guo X-X. Synthesis of isocoumarin derivatives by copper-catalyzed addition of o-halobenzoic acids to active internal alkynes. The Journal of Organic Chemistry. 2013;**78**(4):1660-1664

[37] Baker W. 322. Molecular rearrangement of some o-acyloxyacetophenones and the mechanism of the production of 3-acylchromones. Journal of the Chemical Society (Resumed). 1933:1381-1389

[38] Algar J, Flynn JP. A new method for the synthesis of flavonols. Proceedings of the Royal Irish Academy. Section B: Biological, Geological, and Chemical Science. 1934;**42**:1-8

[39] Fridén-Saxin M et al. Synthesis of 2-alkyl-substituted chromone derivatives using microwave irradiation. The Journal of Organic Chemistry. 2009;**74**(7):2755-2759

Section 2

Synthetic and Biological Activity Section

Chapter 8

Synthesis and Anticancer Evaluation of Benzenesulfonamide Derivatives

Dattatraya Navnath Pansare and Rohini Narayan Shelke

Abstract

A highly efficient protocol was developed for the synthesis of 3-(indoline-1-carbonyl)-N-(substituted) benzene sulfonamide analogs with excellent yields. The new 3-(indoline-1-carbonyl)-N-(substituted) benzene sulfonamide derivatives (4a-g and 5a-g) were evaluated *in vitro* anticancer activity against a series of different cell lines like A549 (lung cancer cell), HeLa (cervical), MCF-7 (breast cancer cell) and Du-145 (prostate cancer cell) respectively. The results of the anticancer activity data revealed that most of the tested compounds showed IC_{50} values from 1.98 to 9.12 μM in different cell lines. Compounds 4b, 4d, 5d, and 5g were the most potent, with IC_{50} values ranging from 1.98 to 2.72 μM in different cell lines.

Keywords: indoline, sulfonamide, anticancer

1. Introduction

Antibiotic resistant bacteria are rapidly emerging worldwide [1]. The various biological active heterocyclic compounds, the indole derivatives are the key structural feature commonly found in natural products [2, 3] and bioactive molecules, such as tryptophan [4], tryptamine [5], and auxin [6]. Furthermore, it has been reported that sharing of the indole 3-carbon in the formation of spiroindoline derivatives highly enhances biological activity [7]. Moreover, some of the compounds containing benzenesulfonamide moiety also show broad spectrum biological properties such as elastase inhibitors [8], carbonic anhydrase inhibitors [9], clostridium histolyticum collogenase inhibitors [10] as well as herbicides and plant growth regulators [11]. Sulfonamides are common motifs in many drugs and medicinal compounds and play an important role in their bioactivity since the development of sulfa antibiotics in the 1930s [12]. Common drugs such as glibenclamide [13], sultiame [14], and COX-II inhibitors Piroxicam [15], Ampiroxicam [16], and Celecoxib [17] containing a sulfonyl moiety, which displays potential activity across a variety of biological targets. The sulfonamides are organic sulfur compounds which have attracted the attention for their better pharmacological activity [18–20]. It is interesting to note that the sulfonamide containing moiety is known to have some biological and pharmaceutical properties, such as, antitumor, antibacterial, thrombin inhibition, and antifungal activities [21–23].

In view of the above considerations, in continuation of our previous work on triazoles, pyrimidine, thiazoles and thiazolidinones of pharmaceutical interest [24–30] we report here on the synthesis and anticancer activity of new 3-(indoline-1-carbonyl)-N-(substituted) benzene sulfonamide analogs.

2. Results and discussion

2.1 Chemistry

The aim of this work was to design and synthesize a novel series of benzenesulfonamide incorporating biologically active indoline moieties to evaluate their anticancer activity. We have synthesized new derivatives containing sulfonamide linkage in frame work. The synthetic methods adopted for the preparation of the N-(substituted phenyl)-3-(indoline-1-carbonyl)benzenesulfonamide derivatives (**5a-g**) in **Figure 1**. We herein report the synthesis of new substituted sulfonamide derivatives with the aim of investigating their anticancer activity (**Table 2**). The synthetic methods adopted for the preparation of the title compounds (**5a-g**) are presented below. We have tried to develop simplified reaction conditions for all the steps by avoiding costly reagents, tedious purifications and longer reactions times, we have screened peptide coupling condition in **Table 1** to obtain better yield, good purity, shorter reaction time, avoiding costly reagents and mainly reproducibility of yields.

For synthesis of compound **2** was done by using **1** treated with sulfonyl chloride at 0°C in DCM for 30 min and at room temperature for 1 h. The reaction mixture was evaporated under reduced pressure and the obtained gummy material was washed with excess of n-hexane. The material was crystallized using 20% ethyl acetate: n-hexane mixture, no purification was required and the pure compound is obtained as yellow solid. This was used further used for sulfonamide reaction.

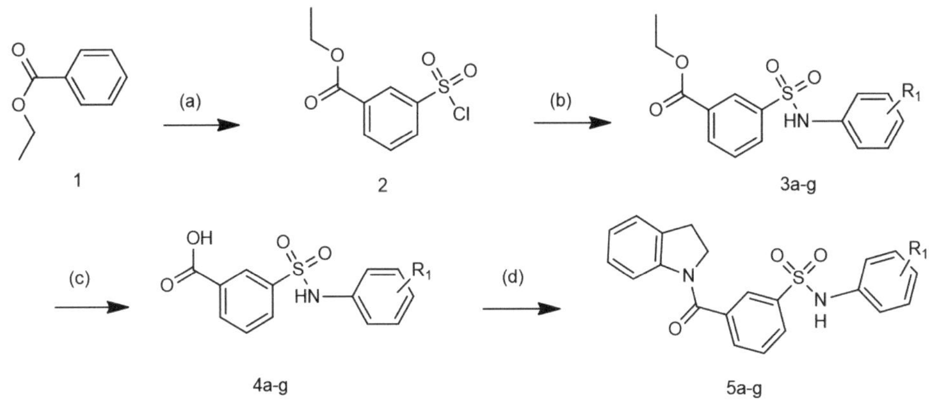

	R_1
3a, 4a, 5a -	2-CH_3
3b,4b, 5b -	2-CH_3-CH_2
3c, 4c, 5c -	2-CF_3
3d, 4d, 5d -	2-$C(CH_3)_3$
3e, 4e, 5e -	Indoline
3f, 4f, 5f -	2-CH_3,4-$C(CH_3)_3$
3g, 4g, 5g -	H

Figure 1.
Synthesis of N-(substituted phenyl)-3-(indoline-1-carbonyl)benzenesulfonamide. Reagents and conditions: (a) sulfonyl chloride, dichloromethane (DCM) 0°C-rt; (b) substituted amine, pyridine, DCM, 0°C-rt; (c) lithium hydroxide (LiOH), tetrahydrofuran (THF), water (H_2O), rt.; (d) indoline, 1-ethyl-3-(3-dimethylaminopropyl)carbodiimide (EDCI), diisopropylethylamine (DIPEA), DCM, rt.

Sr. no.	Coupling reagent	Base	Solvent	Time (h)	Yield (%)
1.	HATU (1.1 equiv.)	TEA (1.2 equiv.)	DMF	14	57
2.	HATU (1.1 equiv.)	DIPEA (1.2 equiv.)	DMF	14	55
3.	PyBOP (1.1 equiv.)	TEA (1.2 equiv.)	THF	14	45
4.	PyBOP (1.1 equiv.)	DIPEA (1.2 equiv.)	THF	14	50
5.	EDCI (1.5 equiv.)	TEA (2.5 equiv.)	DMF	14	62
	HOBt (1.5 equiv.)				
6.	EDCI (1.5 equiv.)	DIPEA (2.5 equiv.)	DMF	14	72
	HOBt (1.5 equiv.)				
7.	EDCI (1.5 equiv.)	TEA (4 equiv.)	DMF	14	78
8.	HOBt (1.5 equiv.)	DIPEA (4 equiv.)	DMF	14	67
9.	T3P (1.2 equiv.)	TEA (2.5 equiv.)	DCM	10	50
10.	T3P (1.2 equiv.)	DIPEA (2.5 equiv.)	DCM	10	60
11.	EDCI (1.5 equiv.)	DIPEA (2.5 equiv.)	DCM	10	95
		Acid (1 equiv.) and indoline (1.2 equiv.)			

Table 1.
Optimization of peptide coupling reaction (5a-g).

For the synthesis of compounds from **3a-g** by sulfonamide coupling, different substituted amines were coupled with **2** in presence of pyridine as base and DCM as solvent at room temperature for 4 h. The reaction mass was treated with cold 2N aqueous HCl and stirred for 30 min., the solid precipitates out in most of cases which was filtered and washed with cold diethyl ether and cold pentane, all the intermediates obtained were white solids. For intermediates **3a-g** the reaction yield was 85–95%.

For synthesis of **4a-4g** requires hydrolysis of **3a-g** using lithium hydroxide, tetrahydrofuran and water at room temperature for 10 h. Work up of reactions were modified, and wash in basic conditions and later acidifying it to get desired product as white solids with required purity. The acids obtained were in pure state so that it can be directly used for next amide coupling with indoline. All reaction intermediates **4a-g** yield up to 80–85%.

For synthesis of **5a-g** we have done series of screenings by varying different coupling reagents, different bases, solvents and time. We have varied the equivalents of reagents and bases used to get better yield and purity by avoiding column purifications. The results of screenings are explained in **Table 1**. In entry 1 and 2 we have used 1-[bis(dimethylamino)methylene]-1H-1,2,3-triazolo[4,5-b]pyridinium 3-oxid hexafluorophosphate (HATU) as coupling reagent and DMF as solvent we have varied bases triethylamine and diisopropylethylamine after 14 h we got product 57 and 55% respectively. In entries 3 and 4 we have used benzotriazol-1-yl-oxytripyrrolidinophosphonium hexafluorophosphate (PyBOP) as coupling reagent and THF as solvent and TEA and DIPEA as base to obtain yields 45 and 50% respectively. In entries 5, 6, 7, and 8 we have used the EDCI and hydroxybenzotriazole (HOBt) as coupling reagents with DMF as solvent along with TEA and DIPEA as base in entries 5 and 6 we have used 2.5 equiv. of base and in entries 7 and 8 we have increased base as 4 equiv. for 14 h. The yields obtained are 62, 72, 78, and 67% respectively. Same results like entry 1 and 2 are obtained in entry 9 and 10 when we used propylphosphonic anhydride (T3P) as coupling reagent and TEA and DIPEA as bases in DCM to get 50 and 60% yield respectively. In entry 11 it was observed that when EDCI (1.5 equiv.) when used along with DIPEA (2.5 equiv.) in DCM the yields was 95%, highlighted bold in **Table 1**. Work up requires extraction, and later on washing with 2N aqueous hydrochloric acid (HCl) to obtain solid compounds. Which was washed with 5% DCM: hexane, cold diethyl ether and cold pentane gives the desired compounds in yield highest yields and with 95% and above purity for **5a-g**. We have not used HOBt in entry 11 and 90% yield obtain after 10 h only. The advantage of peptide coupling screenings are no need of column chromatography, no costly reagents required, no prep purification required. All obtained compounds are with 95% and above purity and are directly used for anticancer testing.

2.2 Biological evaluation: anticancer activity

The synthesized compounds were evaluated for their *in vitro* anticancer activity against human lung cancer cell line (A549), cervical (HeLa) cancer cell line, breast cancer cell line (MCF-7) and prostate cell line (DU-145) using 5-fluorouracil as reference drug [31].

5-Flourouracil is used for anal, breast, colorectal, esophageal, stomach, pancreatic and skin cancers mainly. The response parameter calculated was the IC_{50} value, which corresponds to the concentration required for 50% inhibition of cell viability. The results are presented in **Table 2**, where all compounds exhibit moderate to good activity compared to 5-fluorouracil as positive control. In the case of the human lung cancer cell line (A549) compounds **4a, 4b, 4d, 4f, 5d,** and **5g** were the most potent, with IC_{50} values ranging from 1.98 to 2.82 μM. On the HeLa cell line the compounds which showed potent activity were **4b, 4d, 5d,** and **5g** (IC_{50} = 1.99–2.92 μM). In case of the MCF-7 breast cancer cell line, the potent compounds were **4d, 5d,** and **5g** with IC_{50} activity of 2.12–2.52 μM. Lower activity was observed for the synthesized compounds on the Du-145 prostate cancer cell line, where the most potent candidates were compounds **5g** with IC_{50} activity in the range of 2.12 μM. Generally, the lung

Compound	A549 (lung cancer cell)	HeLa (cervical cancer cell)	MCF-7 (breast cancer cell)	Du-145 (prostate cancer cell)
4a	1.98 ± 0.12	3.83 ± 0.16	3.52 ± 0.06	3.86 ± 0.16
4b	2.81 ± 0.13	2.92 ± 0.08	2.32 ± 0.22	3.82 ± 0.12
4c	4.81 ± 0.12	6.32 ± 0.04	4.32 ± 0.06	3.73 ± 0.12
4d	2.82 ± 0.11	1.99 ± 0.22	2.36 ± 0.12	3.52 ± 0.11
4e	3.86 ± 0.08	4.38 ± 0.06	3.63 ± 0.12	6.52 ± 0.22
4f	2.72 ± 0.11	3.87 ± 0.08	4.12 ± 0.06	3.86 ± 0.22
4g	3.14 ± 0.14	3.98 ± 0.12	4.86 ± 0.11	4.57 ± 0.11
5a	8.48 ± 0.14	9.12 ± 0.08	7.82 ± 0.08	9.12 ± 0.06
5b	3.82 ± 0.08	4.13 ± 0.12	3.13 ± 0.11	3.52 ± 0.08
5c	4.13 ± 0.12	5.16 ± 0.08	6.12 ± 0.12	4.52 ± 0.11
5d	2.06 ± 0.12	2.12 ± 0.08	2.52 ± 0.16	5.12 ± 0.08
5e	2.52 ± 0.11	3.52 ± 0.11	4.48 ± 0.08	4.08 ± 0.11
5f	4.48 ± 0.08	4.98 ± 0.11	5.17 ± 0.22	5.18 ± 0.18
5g	2.73 ± 0.08	2.12 ± 0.12	2.12 ± 0.08	2.12 ± 0.04
5-FU	1.61 ± 0.12	1.72 ± 0.18	1.81 ± 0.10	1.89 ± 0.12

Table 2.
In vitro *anticancer screening of the synthesized compounds against four cell lines, data are expressed as* IC_{50} *(μM) SD (n = 3).*

(A549) and cervical (HeLa) cancer cell lines were the most sensitive to the synthesized compounds. With regard to broad spectrum anticancer activity, close examination of the data presented in **Table 2**, reveals that compounds **4b, 4d,** and **5g** were the most active, showing effectiveness toward the four cell lines. The structure activity relationship (SAR) can be explained on the basis of substitutions on both the aromatic rings less hindered substitution like methyl and ethyl on ortho and para position of rings increases the anticancer activity in all four cell lines, interestingly ortho triflouromethyl and indoline group decreases the anticancer activity and despite steric hindrance **4b**, **4d, 5d,** and **5g** shows promising activity because of electron donating tendency. Most of the compounds show promising anticancer activity with electron donating groups on the ring than electron withdrawing groups.

2.3 General experimental procedure for the synthesis of N-(substituted phenyl)-3-(indoline-1-carbonyl)benzenesulfonamide (5a-g)

2.3.1 Step-1: preparation of ethyl 3-(chlorosulfonyl)benzoate (2)

To a stirred solution of ethyl benzoate (10 g, 67 mmol) in DCM (25 mL). RM was cooled to 0°C and chloro sulfonic acid (9 g, 73 mmol) was added drop wise and stirred for 1 h at same temperature followed by stirring at room temperature for 1 h. After completion of reaction, evaporate reaction mixture under reduced pressure and obtained gummy material is washed with excess of hexane and it is crystalized from 20% ethyl acetate: hexane mixture to obtain white solid as ethyl 3-(chlorosulfonyl)benzoate **(2)** which is used further for sulfonamide coupling reaction. Yield 54 g (81%).

2.3.2 Step-2: preparation of ethyl 3-(N-(o-tolyl)sulfamoyl)benzoate (3a-g)

To a stirred solution of ethyl 3-(chlorosulfonyl)benzoate **(2)** (3 g, 10.1 mmol) in DCM (5 ml) was added pyridine (5 ml) the mixture was stirred at room temperature for 10 min. RM was cooled to 0°C and 2-methyl aniline (1.6 g, 15.16 mmol) was added drop wise followed by stirring at room temperature for 3 h. The reaction was monitored by TLC and LCMS, after completion of reaction poured reaction mass on cold 2N aqueous HCl (10 ml) and stirred RM it for 30 min. Precipitation formed in RM. Filtered the obtained solid and wash it with excess of water and cold diethyl ether (10 ml) and cold pentane (10 ml) to obtain ethyl 3-(N-(o-tolyl)sulfamoyl) benzoate **2** as white solid. Yield 2.8 g (90%).

2.3.3 Step-3: preparation of 3-(N-(o-tolyl)sulfamoyl)benzoic acid (4a-g)

To a stirred solution of ethyl 3-(N-(o-tolyl)sulfamoyl)benzoate **(3a-g)** (2 g, 5.40 mmol) in THF (10 ml) added water (2 ml), and lithium hydroxide (0.377 g, 18.2 mmol) and stirred reaction mixture for 4 h. Progress reaction was monitored by TLC and LCMS. After the completion of reaction evaporate reaction mixture under reduced pressure to obtain gummy material. Added 10 ml of water in it and extracted it with diethyl ether (10 ml). Collected aqueous layer and adjust its pH to 4 by using 6N aqueous HCl. Precipitation occurs stirred it for 30 min. Filtered the obtained solid and wash it with excess of water, cold diethyl ether (10 ml) and cold pentane (10 ml) to obtain desired 3-(N-(o-tolyl)sulfamoyl)benzoic acid **4a** as white solids. Yield 1.6 g (90%).

2.3.4 Step-4: N-(substituted phenyl)-3-(indoline-1-carbonyl)benzenesulfonamide (5a-g)

The compound 3-(N-(o-tolyl)sulfamoyl)benzoic acid **4a-g** (0.2 g, 0.65 mmol) was treated with EDCI (0.188 g, 0.98 mmol), DIPEA (0.34 ml, 1.96 mmol) in DCM (10 ml). Then added 2,4-dimentyl aniline (0.238 g, 1.96 mmol) and stirred RM at room temperature for 4 h. The reaction was monitored by TLC. Added 10 ml of cold water and stirred for 10 min, then extracted it with 10 ml of DCM. Collected organic layer wash it with 1N aqueous HCl and washed with brine (10 ml). To evaporate the organic layer to obtained the compound with 90% purity (**5a-g**). Purification done by washing with 5:95% of DCM: hexane. Obtained solid washed with cold diethyl ether (20 ml) and cold pentane (20 ml) to obtain compounds (**5a-g**). N-(2,4-dimethylphenyl)-3-(N-(o-tolyl)sulfamoyl)benzamide (**5a**): (0.240 g, 90%) as white solid, LC-MS *m/z* (%): 395 (M + H). ^{1}H NMR (400 MHz, DMSO-d6) δ 10.05 (s, 1H), 9.70 (s, 1H), 8.26 (s, 1H), 8.22 (d, *J* = 7.6 Hz, 1H), 7.81 (d, *J* = 8 Hz, 1H), 7.7 (d, *J* = 8 Hz, 1H), 7.18–7.13 (m, 2H), 7.1–7.08 (m, 3H), 7.01 (d, *J* = 8.4 Hz, 1H), 6.95–6.92 (m, 1H), 2.28 (s, 3H), 2.15 (s, 3H), 2.02 (s, 3H). HPLC-98.25% RT-5.68 min. ^{13}C NMR ($CDCl_3$, 100 MH_Z): 17.65, 17.79, 20.54, 126.09, 126.38, 126.40, 126.43, 126.58, 129.23, 129.42, 130.82, 130.89, 131.38, 133.42, 133.62, 134.27, 134.65, 135.40, 135.41, 135.45, 141.09, 163.93.

3. Conclusion

An effective method was developed which provided an easy access to a new series N-(substituted phenyl)-3-(indoline-1-carbonyl)benzenesulfonamide (**5a-g**) analogs. The mild reaction conditions, good to excellent yields, ease of workup and easily available substrates make the reactions attractive for the preparation of compounds. The compounds (**4b, 4d, 5d,** and **5g**) show potent anticancer activity in all the four cell lines tested. The compounds are easy, simple and reproducible to synthesize in normal conditions and no additional conditions or expensive chemicals are required for the reaction. The cell-lines with maximum IC_{50} values are the important in the study.

Acknowledgements

The authors are thankful to the Head, Department of Chemistry, Deogiri college, Aurangabad for the laboratory facility.

Author details

Dattatraya Navnath Pansare* and Rohini Narayan Shelke
Department of Chemistry, Deogiri College, Aurangabad, MS, India

*Address all correspondence to: dattatraya.pansare7@gmail.com

References

[1] Walsh CT, Wencewicz TA. Prospects for new antibiotics: A molecule-centered perspective. The Journal of Antibiotics. 2013;**67**:7. DOI: 10.1038/ja.2013.49

[2] Alvarez M, Salas M, Joule JA. Marine, nitrogen-containing heterocyclic natural products—Structures and syntheses of compounds containing indole units. Heterocycles. 1991;**32**(7):1391. DOI: 10.3987/rev-91-429

[3] Ruiz-Sanchis P, Savina SA, Albericio F, Alvarez M. Structure, bioactivity and synthesis of natural products with hexahydropyrrolo[2,3-b]indole. Chemistry—A European Journal. 2011;**17**(5):1388. DOI: 10.1002/chem.201001451

[4] Mantle PG. The role of tryptophan as a biosynthetic precursor of indole-diterpenoid fungal metabolites: Continuing a debate. Phytochemistry. 2009;**70**(1):7. DOI: 10.1016/j.phytochem.2008.11.004

[5] Jones RSG. Tryptamine: A neuromodulator or neurotransmitter in mammalian brain. Progress in Neurobiology. 1982;**19**(1-2):117. DOI: 10.1016/0301-0082(82)90023-5

[6] Folkes LK, Wardman P. Oxidative activation of indole-3-acetic acids to cytotoxic species—A potential new role for plant auxins in cancer therapy. Biochemical Pharmacology;**61**(2):129. DOI: 10.1016/S0006-2952(00)00498-6

[7] Abdel-Rahman AH, Keshk EM, Hanna MA, El-Bady SM. Bioorganic & Medicinal Chemistry. 2004;**12**(9):2483. DOI: 10.1016/j.bmc.2003.10.063

[8] Scozzafava A, Owa T, Mastrolorenzo A, Supuran CT. Anticancer and antiviral sulfonamides. Current Medicinal Chemistry. 2003;**10**:925. DOI: 10.2174/0929867033457647

[9] Scozzafava A, Briganti F, Ilies MA, Supuran CT. Carbonic anhydrase inhibitors: Synthesis of membrane-impermeant low molecular weight sulfonamides possessing in vivo selectivity for the membrane-bound versus cytosolic isozymes. Journal of Medicinal Chemistry. 2000;**43**:292. DOI: 10.1021/jm990479+

[10] Scozzafava A, Supuran CT. Protease inhibitors. Part 8: Synthesis of potent *Clostridium histolyticum* collagenase inhibitors incorporating sulfonylated L-alanine hydroxamate moieties. Bioorganic & Medicinal Chemistry. 2000;**8**:637. DOI: 10.1016/S0968-0896(99)00316-8

[11] Levitt G. Agricultural sulfonamides. U. S. Patent No. 4, 383, 113; 1983

[12] Drews J. Drug discovery: A historical perspective. Science. 2000;**287**:1960. DOI: 10.1126/science.287.5460.1960

[13] Wells GJ, Tao M, Josef KA, Bihovsky R. 1,2-Benzothiazine 1,1-dioxide P(2)-P(3) peptide mimetic aldehyde calpain I inhibitors. Journal of Medicinal Chemistry. 2001;**44**:3488. DOI: 10.1021/jm010178b

[14] Brooke EW, Davies SG, Mulvaney WA. Synthesis and in vitro evaluation of novel small molecule inhibitors of bacterial arylamine N-acetyltransferases (NATs). Bioorganic & Medicinal Chemistry Letters. 2003;**13**:2527. DOI: 10.1016/S0960-894X(03)00484-0

[15] Bihovsky R, Tao M, Mallamo JP, Wells GJ. 1,2-Benzothiazine 1,1-dioxide alpha-ketoamide analogues as potent calpain I inhibitors. Bioorganic & Medicinal Chemistry Letters. 2004;**14**:1035. DOI: 10.1016/j.bmcl.2003.11.037

[16] Zia-ur-Rehman M, Choudary JA, Ahmad S, Siddiqui HL. Synthesis of potential biologically active 1,2-benzothiazin-3-yl-quinazolin-4(3H)-ones. Chemical & Pharmaceutical Bulletin. 2006;**54**:1175

[17] Hendricks RT, Spencer SR, Blake JF. 3-Hydroxyisoquinolines as inhibitors of HCV NS5b RNA-dependent RNA polymerase. Bioorganic & Medicinal Chemistry Letters. 2009;**19**:410. DOI: 10.1016/j.bmcl.2008.11.060

[18] Kwon Y, Song J, Lee H, Kim EY, Lee K, Lee SK, et al. Design, synthesis, and biological activity of sulfonamide analogues of antofine and cryptopleurine as potent and orally active antitumor agents. Journal of Medicinal Chemistry. 2015;**58**:7749. DOI: 10.1021/acs.jmedchem.5b00764

[19] Sławinski J, Szafranski K, Vullo D, Supuran CT. Carbonic anhydrase inhibitors. Synthesis of heterocyclic 4-substituted pyridine-3-sulfonamide derivatives and their inhibition of the human cytosolic isozymes I and II and transmembrane tumor-associated isozymes IX and XII. European Journal of Medicinal Chemistry. 2013;**69**:701. DOI: 10.1016/j.ejmech.2013.09.027

[20] Mirjafary Z, Ahmadi L, Moradi M, Saeidian H. A copper(II)-thioamide combination as a robust heterogeneous catalytic system for green synthesis of 1,4-disubstituted-1,2,3-triazoles under click conditions. RSC Advances. 2015;**5**:78038. DOI: 10.1039/C5RA16581D

[21] Pingaew R, Prachayasittikul V, Mandi P, Nantasenamat C, Prachayasittikul S, Ruchirawat S, et al. Synthesis and molecular docking of 1,2,3-triazole-based sulfonamides as aromatase inhibitors. Bioorganic & Medicinal Chemistry. 2015;**23**:3472. DOI: 10.1016/j.bmc.2015.04.036

[22] Wang XL, Wan K, Zhou CH. Synthesis of novel sulfanilamide-derived 1,2,3-triazoles and their evaluation for antibacterial and antifungal activities. European Journal of Medicinal Chemistry. 2010;**45**:4631. DOI: 10.1016/j.ejmech.2010.07.031

[23] Zhang Z, Jeyakkumar P, Kumar KV, Zhou CH. Synthesis of novel sulfonamide azoles via C-N cleavage of sulfonamides by azole ring and relational antimicrobial study. New Journal of Chemistry. 2015;**39**:5776. DOI: 10.1039/C4NJ01932F

[24] Pansare DN, Shelke RN, Pawar CD. A facile synthesis of (Z)-2-((5-(4-chlorobenzylidene)-4-oxo-4, 5-dihydrothiazol-2-yl)amino) substituted acid using microwave irradiation and conventional method. Letters in Organic Chemistry. 2017;**14**(7):517. DOI: 10.2174/1570178614666170524142722

[25] Pansare DN, Shelke RN, Shinde DB. A facial synthesis and anticancer activity of (Z)-2-((5-(4-nitrobenzylidene)-4-oxo-4,5-dihydrothiazol-2-yl)amino) substituted acid. Journal of Heterocyclic Chemistry. 2017;**54**(6):3077. DOI: 10.1002/jhet.2919

[26] Pansare DN, Shinde DB. A facile synthesis of novel series (Z)-2-((4-oxo-5-(thiophen-2-yl methylene)-4,5-dihydro thiazol-2-yl)amino) substituted acid. Journal of Saudi Chemical Society. 2017;**21**:434. DOI: 10.1016/j.jscs.2015.10.005

[27] Pawar CD, Chavan SL, Pawar UD, Pansare DN, Deshmukh SV, Shinde DB. Synthesis, anti-proliferative activity, SAR and kinase inhibition studies of thiazol-2-yl-substituted sulfonamide derivatives. Journal of the Chinese Chemical Society. 2018. DOI: 10.1002/jccs.201800312

[28] Pawar CD, Pansare DN, Shinde DB. (Substituted)-benzo[b]

thiophene-4-carboxamide synthesis and anti-proliferative activity study. Letters in Drug Design & Discovery. 2018. DOI: 10.2174/1570180815666181004114125

[29] Pawar CD, Pansare DN, Shinde DB. Synthesis of new 3-(substituted phenyl)-N-(2-hydroxy-2-(substituted phenyl)ethyl)-N-methylthiophene-2-sulfonamide derivatives as antiproliferative agents. European Journal of Chemistry. 2018;**9**(1):13. DOI: 10.5155/eurjchem.9.1.13-21.1669

[30] Pawar CD, Pansare DN, Shinde DB. Synthesis and antiproliferative activity of 3-(substituted)-4,5,6,7-tetrahydro-6-(substituted)-1H-pyrazolo[3,4-c]pyridine derivatives. European Journal of Chemistry. 2017;**8**(4):400. DOI: 10.5155/eurjchem.8.4.400-409.1645

[31] Skehan P, Storeng R, Scudiero D, Monks A, McMohan J, Vistica D, et al. New colorimetric cytotoxicity assay for anticancer-drug screening. Journal of the National Cancer Institute. 1990;**82**:1107

Chapter 9

Synthesis and Biological Evaluation of Novel Phosphonyl Thiazolo Pyrazoles

Avula Srinivas

Abstract

A series of novel dimethyl 7-((2S,3S)-3-((1-(4-chlorophenyl)-1H-1,2,3-triazol-4-yl)methoxy)-3,6-dihydro-2H-pyran-2-yl)-4-(4-fluorophenyl)-9-oxo-8-phenyl-6-thia-1,2,8-triazaspiro[4.4]non-2-en-3-ylphosphonate **11a–g** were synthesized by the reaction of chalcone derivatives of (E)-5-benzylidene-2-((2S,3S)-3-((1-(4-chlorophenyl)-1H-1,2,3-triazol-4-yl)methoxy)-3,6-dihydro-2H-pyran-2-yl)-3-phenylthiazolidin-4-one **10a–g** with Bestmann-Ohira reagent. The chemical structures of newly synthesized compounds were elucidated by IR, NMR, MS, and elemental analysis. The compounds **11a–g** were evaluated for their nematicidal activity against *Dietylenchus myceliophagus* and *Caenorhabditis elegans*, and compounds **11b**, **11c**, **11g**, and **11f** showed appreciable nematicidal activity.

Keywords: phosponylpyrazoles, Bestmann-Ohira reagent, click reaction, Knoevenagel condensation, cyclisation, nematicidal activity

1. Introduction

1,2,3-Triazoles are one of the most important classes of heterocyclic organic compounds, which are reported to present in a plethora of biological activities for diverse therapeutic areas [1–12]. The 1,2,3-triazole motif is associated with diverse pharmacological activities such as antibacterial, antifungal, hypoglycemic, antihypertensive and analgesic properties [13–15]. Polysubstituted five-membered aza heterocyclic's rank the most potent glycosidase inhibitors [16–19]. Further, this nucleus in combination with or in linking with various other classes of compounds such as amino acids, steroids, aromatic compounds, carbohydrates etc. became prominent in having various pharmacological properties [20]. 1,2,3-Triazole modified carbohydrates have became easily available after the discovery of the Cu(I) catalyzed azide-alkynes 1,3-dipolar cycloaddition reaction [21–25] and quickly became a prominent class of non-natural sugars. The chemistry and biology of triazole modified sugars is dominated by triazole glycosides [26]. Therefore, the synthesis and investigation of biological activity of 1,2,3-triazole glycosides is an important objective, which also received the considerable attention by the medicinal chemists.

Thiazoles are familiar group of heterocyclic compounds possessing a wide variety of biological activities and their utility as medicine is very much established [27]. Thiazole nucleus is also an integral part of all the available penicillins

IntechOpen

which have revolutionized the therapy of bacterial diseases [28]. The chemistry of thiazolidinone ring system is one of considerable interest as it is the core structure in various synthetic pharmaceuticals displaying a broad spectrum of biological activities [29]. The thiazolidinone nucleus also appears frequently in the structure of various natural products notably thiamine, compounds possessing cardiac and glycemic benefits such as troglitazone [30] and many metabolic products of fungi and primitive marine animals, including 2-(aminoallyl)-thiazole-4-carboxylic acids [31]. Numerous thiazolidinone derivatives have shown significant bio activities such as antidiarrhoeal [32], anticonvulsant [33], antimicrobial [34], antidiabetic [35], antihistaminic [36], anticancer [37], anti HIV [38], Ca^{2+} channel blocker [39], PAF antagonist [40], cardioprotective [41], antiischemic [42], COX inhibitory [43], antiplatelet activating factor [44], non-peptide thrombin receptor antagonist [45], tumor necrosis factor-α-antagonist [46] and nematicidal activities. Organophosphorus compounds continue to attract much attention because of their various potent biological activities [47, 48] in particular, phosphonates are important synthetic derivatives which can have often act as phosphate and carboxylic acid mimics, and interfere with enzymatic processes. Much of this activity has been attributed to the relatively inert nature of the C—P bond [47, 48], which is not easily hydrolyzed as compared to the P—O bond found in phosphates. The synthesis and biological activities of important natural and nonnatural phosphonate derivatives, including phosphonated aza heterocyclics and nucleotides, have been reviewed [49–51]. In view of the importance of heterocyclics bearing a phosphonate group, new synthetic methods that would allow straightforward access to these versatile building blocks are needed [47, 48, 52]. Among the various bioactive heterocyclics the pyrazole moiety remains of great interest because of its wide applications in the pharmaceutical and agrochemical industry [53, 54]. In addition, pyrazoles also play a central role in coordination chemistry [55].

Nematodes are tiny worms, some of them are plant parasites, and can play an important role in the predisposition of the host plant to the invasion by secondary pathogens [56]. Plants attacked by nematodes show retarded growth and development, as well as loss in the quality and quantity of the harvest. The nematicide use is slated for reduction due to environmental problems, and human and animal health concern. For example, effective nematicides such as dibromochloropropane (DBCD) and ethylene dibromide (EDB) have been withdrawn from the market due to their deleterious effects on human and the environment. Methyl bromide, the most effective and widely used fumigant for soil borne pests including nematodes, has already been banned.

The use of nonfumigant nematicides, based on organophosphates and carbamates, is expected to increase the withdrawal of methyl bromide, which will bring about new environmental concerns. In fact, the highly toxic aldicarb used to control insects and nematodes has been detected in ground water [57]. Therefore alternative nematode control methods or less toxic nematicides need to be developed [58]. One way of searching for such nematicidal compounds is to screen naturally occurring compounds in plants. Several such compounds, e.g., alkaloids, phenols, sesquiterpenes, diterpenes, polyacetylenes, and thienyl derivatives have nematicidal activity [59]. For example, α-terthienyl is a highly effective nematicidal compound [60]. Other compounds with nematicidal activity have been isolated from plants, mainly from the family *Asteraceae* [59]. However, compounds of plant origin and their analogs have not been developed into commercial nematicides; hence there is a need to develop commercial synthesis.

Following the successful introduction of nematicidal agents, inspired by the biological profile of triazoles, thiazoles, Phosponylpyrazoles. In continuation of our work on biological active molecules [61–69] it was thought to interest to accommodate all those moieties in single molecular frame work. In this article we wish to report the synthesis of a new class of hybrid heterocyclic's **11a–g** in good yields and their evaluated nematicidal activity.

2. Result and discussion

The key intermediate, **8** required for the synthesis of title compound was prepared according to the procedure outlined in **Figure 1**. Diacetyl-D-glucal **(2)** prepared from 3,4,6-tri-O-acetyl D-glucal by treating with triethyl silane and boron trifluoride diethyl etherate, de acylation of **2**, with NaOMe in methanol at 0°C for 1 hour gave **3** (77%), which on subsequent treatment with TBDMSCl in dichloromethane in presence of NEt_3 for 12 hours afforded TBS ether **4** (80%), on treatment with propargyl bromide in toluene in presence of tetra butyl ammonium hydrogen sulfate produced di ether **5**. After deprotection of TBS ether the propargyl ether converted into triazole **7** (82%) by using 1,3-dipolar cycloaddition with *p*-chloro phenyl azide was carried out at ambient temperature in the presence of $CuSO_4$ and sodium ascorbate in a mixture of 1:1 CH_2Cl_2-H_2O. Oxidation of **7** with IBX in

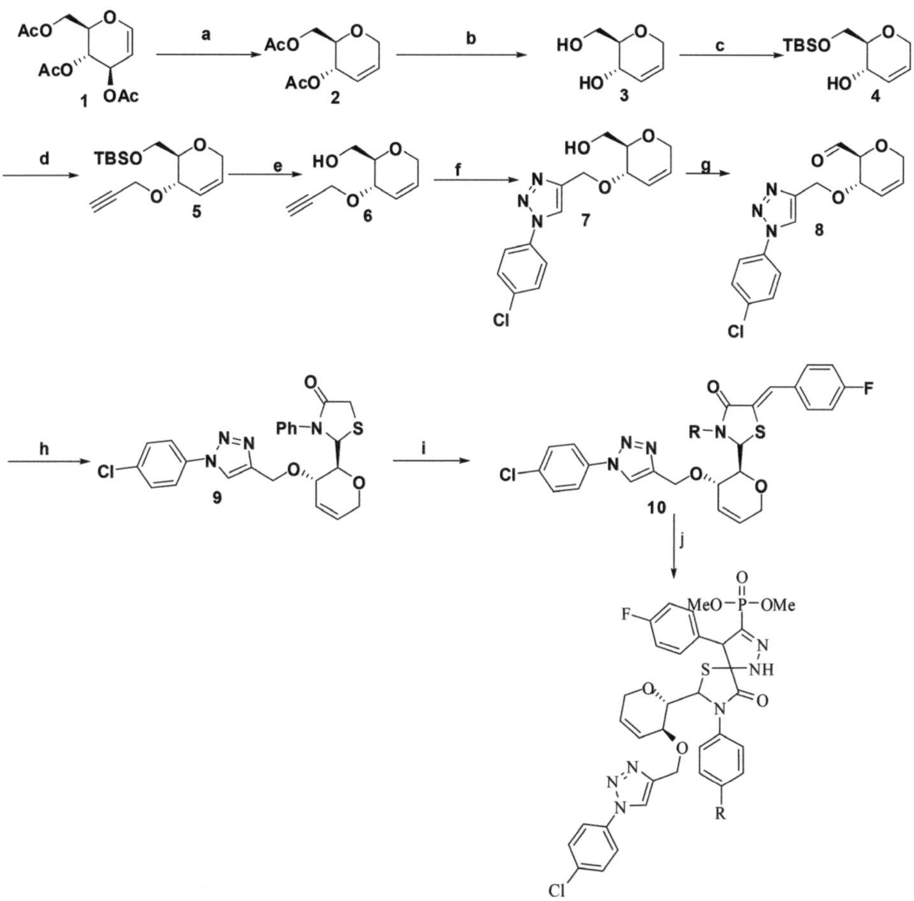

Figure 1.
R= (a) C_6H_5; (b) 4-Cl-C_6H_5; (c) 4-NO_2-C_6H_5; (d) 2-CH_3-C_6H_5; (e) 4-CH_3-C_6H_5; (f) 3-OH-C_6H_5; (g) 2-OH-C_6H_5.

acetonitrile afforded compound **8**. Subsequently one pot synthesis of triazole linked thiazolidinone glycosides was carried out by the condensation reaction between **8**, primary aromatic amine and a thio glycolic acid in presence of $ZnCl_2$ under microwave irradiation (**Figure 1**). The reaction is completed in only 5–10 minutes and the compounds, isolated by conventional work-up, (**9a–g**) are obtained in satisfactory yields, Compound **9a–g** was then reacted with *p*-fluoro benzaldehyde in presence of anhydrous NaOAc in glacial AcOH at reflux temperature gave chalcone derivatives of triazole linked thiazolidinone glycosides **10a–g**, on cyclocondensation under conventional and microwave irradiation with Bestmann-Ohira reagent in presence of anhydrous KOH gave compounds **11(a–g)**. The structures of synthesized compounds were confirmed by IR, NMR, MS and elemental analysis. Further the compounds were subject to nematicidal activity testing.

3. Nematicidal activity

The compounds synthesized **10a-g** in this study were also screened for their nematicidal activity against *Dietylenchus myceliophagus* and *Caenorhabditis elegans* by aqueous *in vitro* screening technique [70] at various concentrations. The nematicidal activity of each test compound was compared with the standard drug *Levamisole*. The results have been expressed in terms of LD_{50} i.e., median lethal dose at which 50% nematodes became immobile (dead). The screened data reveal that, compounds **11b**, **11c**, **11f** and **11g** are the most effective against *Dietylenchus myceliophagus* and *Caenorhabditis elegans* the other test compounds showed moderate activity. The LD_{50} values of the test compounds screened are presented in **Table 1**.

Compound	**LD_{50} value (*ppm*)**	
	D. myceliophagus	***C. elegans***
11a	740	860
11b	220	280
11c	320	270
11d	501	540
11e	960	900
11f	209	210
11g	310	360
Levamisole	160	180

Table 1.
*Nematicidal activity of **11(a–g)**.*

4. Experimental

Commercial grade reagents were used as supplied. Solvents except analytical reagent grade were dried and purified according to literature when necessary. Di-methyl 2-oxopropyl phosphonate was purchased from Aldrich for the synthesis o Bestmann-Ohira reagent. Reaction progress and purity of the compounds were checked by thin-layer chromatography (TLC) on pre-coated silica gel F254 plates from Merck and compounds visualized either by exposure to UV light or dipping in 1% aqueous potassium permanganate solution. Silica gel chromatographic columns

(60–120 mesh) were used for separations. Optical rotations were measured on a Perkin-Elmer 141 polarimeter by using a 2 ml cell with a path length of 1 dm with $CHCl_3$ or $CDCl_3$ as solvent. All melting points are uncorrected and measured using Fisher-Johns apparatus. IR spectra were recorded as KBr disks on a Perkin-Elmer FTIR spectrometer. Micro wave reactions are carried out in mini lab microwave catalytic reactor (ZZKD, WBFY-201). The ^{1}HNMR and ^{13}C NMR spectra were recorded on a Varian Gemini spectrometer (300 MHz for ^{1}H and 75 MHz for ^{13}C). Chemical shifts are reported as δ ppm against TMS as internal reference and coupling constants (*J*) are reported in Hz units. Mass spectra were recorded on a VG micro mass 7070H spectrometer. Elemental analysis (C, H, N) determined by a Perkin-Elmer 240 CHN elemental analyzer, were within ±0.4% of theoretical.

***((2R,3S)-3-acetoxy-3,6-dihydro-2H-pyran-2-yl)methyl acetate* (2)**: Tri-*O*-acetyl-D-glucal (**1**) (3.0 g, 11.02 mmol) was dissolved in anhydrous dichloromethane (5 ml). The solution was cooled to 0°C, triethyl silane (1.53 g, 13.22 mmol) was added and the mixture was stirred for 5 minutes. Next boron tri fluoride diethyl etherate (690 μl of a 40 w% solution in diethyl ether, 11.02 mmol) was added drop wise and the reaction mixture was stirred for 90 minutes. The mixture was poured into a saturated solution of $NaHCO_3$. The organic layer was washed with water, dried over Na_2SO_4 and concentrated under reduced pressure. Column chromatography on silica gel (PE/EtOAc, 3:1) yielded the title compound (2.24 g, 10.42 mmol, 95%) as a colorless syrup. **[α]D20**: +115.5 (*c* = 1.00, $CHCl_3$).^{1}HNMR (300 MHz, $CDCl_3$): δ 5.87-5.84 (m, 2H, =CH),4.95 (t,1H, OCH),4.03-3.99 (m, 1H, CH),4.12-4.09 (m,4H, OCH_2), 2.20 (s, 6H, $COCH_3$);^{13}C NMR (75 MHz, $CDCl_3$): δ170.2, 127.2, 125.8, 73.6, 65.1, 64.0, 62.5, 21.1; MS: m/z (M^++H) 215. Anal. Calcd for $C_{10}H_{14}O_5$: C, 56.07; H, 6.59; Found: C, 55.82; H, 6.35.

(2R,3S)-2-((tert-butyldimethylsilyloxy)methyl)-3,6-dihydro-2H-pyran-3-ol (4): Diacetate **2** (17.22 mmol) was treated by a catalytic amount of sodium methoxide in methanol (100 ml) at room temperature. After evaporation of the solvent, the free hydroxyl unsaturated glycoside was obtained in quantitative yield and used without further purification. This diol was treated with 2.50 equiv. of TBD MSCl (3.14 g, 21.14 mmol), 2.6 equiv. of NEt_3 (3.2 ml, 22.4 mmol), and 0.05 equiv. of imidazole (30 mg, 0.43 mmol) in CH_2Cl_2 (30 ml) at room temperature for *ca*. 24 hours (until TLC analysis showed no more starting material). After addition of 25 ml of water and extraction with 3–30 ml of CH_2Cl_2, the organic layer was dried. After evaporation of the solvent under reduced pressure, the residue was purified by column chromatography using petroleum ether/ethyl acetate as the eluent yielded the title compound (1.94 g, 10.42 mmol, 85%) as a colorless syrup. ^{1}HNMR (300 MHz, $CDCl_3$): δ 6.0–5.82 (m, 2H, =CH), 5.42 (d, *J* = 6.5 Hz, 1H, CH),4.50 (brs, 1H, OH), 4.20–4.12(m,1H, CH), 3.91–3.80(m,4H, CH_2), 0.98 (s, 9H, t-Bu), 0.24 (s, 6H, CH_3); ^{13}C NMR (75 MHz, $CDCl_3$): δ 127.5, 125.6, 84.6, 81.5, 73.6, 62.7, 25.6, 18.1; MS: m/z (M^++Na) 267. Anal. Calcd for $C_{12}H_{24}O_3Si$: C, 58.97; H, 9.90; Found: C, 58.62; H, 9.75.

***tert-butyldimethyl(((2R,3S)-3-(prop-2-ynyloxy)-3,6-dihydro-2H-pyran-2-yl)methoxy)silane* (5)**: To a solution of alcohol **4** (400 mg, 1.63 mmol, 1.0 equiv) in toluene (1.6 ml) was added a 35% aqueous solution of NaOH (1.6 ml), propargyl bromide (80% solution in toluene, 363 μl, 2.4 mmol, 1.5 equiv), and *n*-Bu_4NHSO_4 (280 mg, 0.82 mmol, 0.5 equiv). After 6 hours of vigorous stirring at room temperature, Et_2NH (1.6 ml) was added. The reaction mixture was stirred for 1 hour, poured into ice water, cautiously neutralized by addition of a 3 M solution of hydrochloric acid, and extracted with EtOAc. The combined organic extracts were washed with brine, dried over $MgSO_4$, filtered, and concentrated under reduced pressure. The crude material was purified by flash chromatography on silica gel (hexane/EtOAc 85:15) to afford propargyl ether as colorless oil (0.345 g, 75%). ^{1}HNMR (300 MHz,

$CDCl_3$): δ 6.03–5.80 (m, 2H, =CH), 4.69 (t, *J* = 3.9 Hz 1H, CH), 3.68 (dd, *J* = 8.9 Hz, 4.1 Hz, 1H, OCH), 3.99–3.89(m, 6H, CH_2), 3.20 (s, 1H, CH), 0.96 (s, 9H, t-Bu), 0.23 (s, 6H, CH_3); ^{13}C NMR (75 MHz, $CDCl_3$): δ 127.2, 124.9, 78.0, 76.2, 74.2, 64.2, 63.2, 58.5, 25.3, 18.5; MS: m/z (M^++H) 283. Anal. Calcd for $C_{15}H_{26}O_3Si$: C, 63.78; H, 9.28; Found: C, 63.62; H, 8.95.

***((2R,3S)-3-(prop-2-ynyloxy)-3,6-dihydro-2H-pyran-2-yl)methanol* (6)**: To a stirred solution of 5 (0.325 g) in Tetra hydro furan catalytic amount of TBAF was added and stirred the reaction mixture at room temperature for 15 minutes, extracted the product with Ethyl acetate (20 ml). The combined organic extracts were washed with brine, dried over $MgSO_4$, filtered, and concentrated under reduced pressure. The crude material was purified by flash chromatography on silica gel (60–120 mesh, hexane/EtOAc 70, 0) to afford alcohol as yellow oil (0.285 g, 85%) ^{1}HNMR (300 MHzCDCl$_3$) 5.95–5.75 (m, 2H, =CH), 4.65(d, *J* = 3.9 Hz, 1H, CH), 4.52 (brs, 1H, OH), 4.09–4.11 (m, 4H, OCH_2), 3.64 (dd, *J* = 4.1 Hz, 8.9 Hz, 1H, OCH), 3.76 (d, *J* = 6.8 Hz, 2H, OCH_2), 3.28 (s, 1H, CH): ^{13}C NMR (75 MHz, $CDCl_3$): δ 127.2, 125.6, 78.3, 76.1, 74.1, 64.2, 61.4, 58.0; MS: m/z (M^++H) 169. Anal. Calcd for $C_9H_{12}O_3$: C, 64.27; H, 7.10; Found: C, 64.02; H, 6.95.

***((2R,3S)-3-((1-(4-chlorophenyl)-1H-1,2,3-triazol-4-yl)methoxy)-3,6-dihydro-2H-pyran-2-yl)methanol* (7)**: To a solution containing alkyne **6** (0.250 g, 0.778 mmol), p-chloro phenyl azide (0.130 g, 0.849 mmol) in dichloromethane (10 ml) and water (10 ml) were added $CuSO_4.5H_2O$ (0.110 g) and sodium ascorbate (0.114 g). The resulting suspension was stirred at room temperature for 6 hours. After this time, the mixture was diluted with 5 ml dichloromethane and 5 ml water. The organic phase was separated, dried with sodium sulfate and concentrated at reduced pressure the crude product was purified by column chromatography on silica gel (60–120 mesh, hexane/EtOAc 65:35) to afford 7 (0.290 g, 75%) as a white powder. Mp: 149–1510°C. ^{1}HNMR (300 MHz, $CDCl_3$): δ8.05 (s, 1H, Ar-H), 7.56 (d, *J* = 9.2 Hz, 2H, Ar-H),7.45 (d, *J* = 8.9 Hz, 2H, Ar-H), 5.85–5.79 (m, 2H, =CH), 4.59 (s, 2H, OCH_2), 4.50 (brs, 1H, OH), 3.88–3.99 (m, 4H, OCH_2), 3.8–3.75 (m, 2H, OCH): ^{13}CNMR (75 MHz, $CDCl_3$): δ140.9, 134.5, 134.1, 128.4, 127.5, 125.4, 122.1, 11.5, 78.6, 68.5, 65.7, 64.2, 62.4: MS: m/z (M^++H) 322. Anal. Calcd for $C_{15}H_{16}ClN_3O_3$: C, 55.90; H, 5.01, N, 13.06; Found: C, 55.65, H, 4.95. N, 12.86.

***(R)-2-((2S,3S)-3-((1-(4-chlorophenyl)-1H-1,2,3-triazol-4-yl)methoxy)-3,6-dihydro-2H-pyran-2-yl)-3-phenylthiazolidin-4-one* 9(a-g)**: To a solution of alcohol 7 (0.120 g, 0.465 mmol) in CH_2Cl_2 (5 ml), catalytic amount of IBX was added at 0°C and stirred at room temperature for 30 minutes. The reaction mixture was filtered and washed with CH_2Cl_2 (2 × 10 ml). It was dried (Na_2SO_4) and evaporated to give aldehyde 7 (0.110 g) in quantitative yield as a yellow liquid, which was used as such for the next reaction.

To a stirred mixture of **8** (0.110 g, 0.373 mmol), aromatic amine (0.373 mmol) and anhydrous thioglycolic acid (0.140 g, 0.211 mmol) in dry toluene (5 ml), $ZnCl_2$ (0.100 g, 0.751 mmol) was added after 2 minutes and irradiated in microwave bath reactor at 280 W for 4–7 minutes at 110°C. After cooling, the filtrate was concentrated to dryness under reduced pressure and the residue was taken-up in ethyl acetate. The ethyl acetate layer was washed with 5% sodium bicarbonate solution and finally with brine. The organic layer was dried over Na_2SO_4 and evaporated to dryness at reduced pressure. The crude product thus obtained was purified by column chromatography on silica gel (60–120 mesh) with hexane-ethyl acetate as eluent.

***(R)-2-((2S,3S)-3-((1-(4-chlorophenyl)-1H-1,2,3-triazol-4-yl)methoxy)-3,6-dihydro-2H-pyran-2-yl)-3-phenylthiazolidin-4-one* (9a)**: mp: 157–159°C. Yield—75%. ^{1}HNMR (300 MHz, $CDCl_3$): δ8.04 (s, 1H, Ar-H), 7.50

(d, *J* = 9.2 Hz, 2H, Ar-H), 7.40 (d, *J* = 8.9 Hz, 2H, Ar-H), 7.10–6.20 (m, 5H, Ar-H), 5.80–5.71 (m, 2H, =CH), 4.90 (d, *J* = 5.2 Hz, 1H, CH-S), 4.52 (s, 2H, OCH_2), 4.09–3.94 (m, 2XCH), 3.79 (d, *J* = 6.6 Hz, 2H, OCH_2), 3.72 (s, 2H, CH_2): ^{13}CNMR (75 MHz, $CDCl_3$): δ170.4, 144.1, 141.8, 134.1, 128.2, 125.6, 122.4, 119.4, 85.6, 72.6, 66.4, 64.0, 51.4, 33.9: MS: m/z (M^++H) 469. Anal. Calcd for $C_{23}H_{21}ClN_4O_3S$: C, 58.91; H, 4.51, N, 11.95; Found: C, 58. 68, H, 4.35. N, 11.66.

(R)-3-(4-chlorophenyl)-2-((2S,3S)-3-((1-(4-chlorophenyl)-1H-1,2,3-triazol-4-yl)methoxy)-3,6-dihydro-2H-pyran-2-yl)thiazolidin-4-one **(9b)**: mp: 226–228°C Yield—69%. ^{1}HNMR (300 MHz, $CDCl_3$): 8.05 (s, 1H, Ar-H), 7.54 (d, *J* = 9.4 Hz, 4H, Ar-H), 7.42 (d, *J* = 8.6 Hz, 4H, Ar-H), 5.84–5.75 (m, 2H, =CH), 4.94 (d, *J* = 5.2 Hz, CH-S), 4.50 (s, 2H, OCH_2), 4.06–3.96 (m, 2H, 2XCH), 3.80 (t, 2H, OCH_2), 3.72 (s, 2H, CH_2): ^{13}CNMR (75 MHz, $CDCl_3$): δ 170.5, 144.2, 139.2, 134.2, 129.2, 125.5, 122.2, 119.4, 85.4, 72.8, 65.4, 63.4, 51.2, 34.1: MS: m/z (M^++Na) 525. Anal. Calcd for $C_{23}H_{20}Cl_2N_4O_3S$: C, 54.88; H, 4.00, N, 11.13; Found: C, 54.58, H, 3.75. N, 10.86.

(R)-2-((2S,3S)-3-((1-(4-chlorophenyl)-1H-1,2,3-triazol-4-yl)methoxy)-3,6-dihydro-2H-pyran-2-yl)-3-(4-nitrophenyl)thiazolidin-4-one **(9c)**: mp: 211–213°C, Yield—71%. ^{1}HNMR (300 MHz, $CDCl_3$): δ8.26 (d, *J* = 8.7 Hz, 2H, Ar-H), 8.03 (s, 1H, Ar-H), 7.61 (d, *J* = 9.4 Hz, 2H, Ar-H), 7.46 (d, *J* = 8.5 Hz, 2H, Ar-H), 6.84 (d, *J* = 9.8 Hz, 2H, Ar-H), 5.86–5.79 (m, 2H, =CH), 4.96 (d, *J* = 5.2 Hz, CH-S), 4.55 (s, 2H, OCH_2), 4.05–3.95 (m, 2H, 2XCH), 3.85 (d, *J* = 6.9 Hz, 2H, OCH_2), 3.82 (s, 2H, CH_2): ^{13}CNMR (75 MHz, $CDCl_3$): δ171.5, 144.0, 141.8, 134.2, 128.5, 125.4, 119.5, 85.4, 72.4, 65.9, 63.6, 51.5, 34.6: MS: m/z (M^++H) 514. Anal. Calcd for $C_{23}H_{20}ClN_5O_5S$: C, 53.75; H, 3.92, N, 13.63; Found: C, 53.58, H, 3.75. N, 13.39.

(R)-2-((2S,3S)-3-((1-(4-chlorophenyl)-1H-1,2,3-triazol-4-yl)methoxy)-3,6-dihydro-2H-pyran-2-yl)-3-o-tolylthiazolidin-4-one **(9d)**: mp: 191–193°C, Yield—65%. ^{1}HNMR (300 MHz, $CDCl_3$): δ8.08 (s, 1H, Ar-H), 7.56 (d, *J* = 9.2 Hz, 2H, Ar-H), 7.49 (d, *J* = 8.7 Hz, 2H, Ar-H), 7.45–7.39 (m, 4H, Ar-H), 5.76 (m, 2H, =CH), 4.93 (d, *J* = 5.2 Hz, 1H, CHS), 4.60 (s, 2H, OCH_2), 4.05–3.96 (m, 2H, CH), 3.90 (t, 2H, OCH_2), 3.81 (s, 2H, CH_2), 2.1 (s, 3H, CH_3): ^{13}CNMR (75 MHz, $CDCl_3$): δ170.5, 144.2, 138.2, 134.2, 130.7, 128.6, 125.6, 122.0, 119.5, 116.5, 85.4, 72.6, 65.8, 63.4, 52.0, 32.3, 17.5: MS: m/z (M^++H) 483. Anal. Calcd for $C_{24}H_{23}ClN_4O_3S$: C, 59.68; H, 4.80, N, 11.60; Found: C, 59.48, H, 4.55. N, 11.49.

(R)-2-((2S,3S)-3-((1-(4-chlorophenyl)-1H-1,2,3-triazol-4-yl)methoxy)-3,6-dihydro-2H-pyran-2-yl)-3-p-tolylthiazolidin-4-one **(9e)**: mp: 195–198°C Yield—79%. ^{1}HNMR (300 MHz, $CDCl_3$): δ8.05 (s, 1H, Ar-H), 7.51 (d, *J* = 9.2 Hz, 2H, Ar-H), 7.45 (d, *J* = 8.7 Hz, 2H, Ar-H), 7.25 (d, *J* = 8.2 Hz, 2H, Ar-H), 6.84 (d, *J* = 9.4 Hz, 2H, Ar-H), 5.72–5.68 (m, 2H, =CH), 4.95 (s, 1H, CHS), 4.59 (s, 2H, OCH_2), 4.04–3.99 (m, 2H, CH), 3.98 (t, 2H, OCH_2), 3.90 (s, 2H, CH_2), 2.32 (s, 3H, CH_3): ^{13}CNMR (75 MHz, $CDCl_3$): δ170.5, 144.2, 138.6, 136.2, 14.1, 133.2, 129.4, 127.5, 122.5, 119.5, 85.4, 72.0, 66.4, 63.5, 51.5, 34.0, 21.4: MS: m/z (M^++H) 483. Anal. Calcd for $C_{24}H_{23}ClN_4O_3S$: C, 59.68; H, 4.80, N, 11.60; Found: C, 59.58, H, 4.65. N, 11.43.

(R)-2-((2S,3S)-3-((1-(4-chlorophenyl)-1H-1,2,3-triazol-4-yl)methoxy)-3,6-dihydro-2H-pyran-2-yl)-3-(3-hydroxyphenyl)thiazolidin-4-one **(9f)**: mp: 218–219°C, Yield—85%. ^{1}H-NMR (300 MHz, $CDCl_3$): δ9.40 (brs, 1H, Ph-OH), 8.08 (s, 1H, Ar-H), 7.58 (d, *J* = 9.3 Hz, 2H, Ar-H), 7.49 (d, *J* = 8.6 Hz, 2H, Ar-H), 6.83–6.76 (m, 4H, Ar-H), 5.72–5.68 (m, 2H, =CH), 4.94 (d, J = 5.2 Hz, 1H, CHS), 4.64 (s, 2H, OCH_2), 4.12 (t, 2H, OCH_2), 4.01–3.94 (m, 2H, CH), 3.92 (s, 2H, CH_2): ^{13}CNMR (75 MHz, $CDCl_3$): δ 170.5, 158.2, 143.8, 134.5, 130.4, 128.6, 125.6, 122.4, 119.5, 114.8, 106.5, 85.4, 72.5, 66.4, 63.4, 51.5, 34.1: MS: m/z (M^++Na) 507. Anal. Calcd for $C_{23}H_{21}ClN_4O_4S$: C, 59.96; H, 4.36, N, 11.55; Found: C, 59.28, H, 4.65. N, 11.43.

***(R)-2-((2S,3S)-3-((1-(4-chlorophenyl)-1H-1,2,3-triazol-4-yl)methoxy)-3,6-dihydro-2H-pyran-2-yl)-3-(4-hydroxyphenyl)thiazolidin-4-one* (9g)**: mp: 273–275°C, Yield—82%. ^{1}H-NMR (300 MHz, $CDCl_3$): δ9.42 (brs, 1H, Ph-OH), 8.05 (s, 1H, Ar-H), 7.56 (d, *J* = 9.2 Hz, 2H, Ar-H), 7.46 (d, *J* = 8.4 Hz, 2H, Ar-H), 7.32 (d, *J* = 8.6 Hz, 2H, Ar-H), 7.02 (d, *J* = 8.8 Hz, 2H, Ar-H), 5.89–5.80 (m, 2H, =CH), 4.96 (d, *J* = 5.4 Hz, 1H, CHS), 4.66 (s, 2H, OCH_2), 4.09 (d, *J* = 2H, OCH_2), 4.04–3.98 (m, 2H, CH), 3.94 (s, 2H, CH_2): ^{13}CNMR (75 MHz, $CDCl_3$): δ170.9, 154.1, 144.4, 134.9, 134.8, 128.8, 127.2, 125.6, 123.2, 119.4, 116.4, 85.4, 72.6, 66.5, 64.0, 51.6, 34.5: MS: m/z (M^++H) 485. Anal. Calcd for $C_{23}H_{21}ClN_4O_4S$: C, 59.96; H, 4.36, N, 11.55; Found: C, 59.38, H, 4.75. N, 11.33.

General procedure for the synthesis of (10a-g): A mixture of compound **9a** (0.01 mol), *p*-fluoro benzaldehyde (0.02 mol) and sodium acetate (0.01 mol) in anhydrous glacial acetic acid (20 ml), was refluxed for 3 hours. The reaction mixture was concentrated and then poured into ice cold water, the solid thus separated, was filtered, washed with water and crystallized from glacial acetic acid. To afford pure **10a** as yellow solid.

***(R,Z)-2-((2S,3S)-3-((1-(4-chlorophenyl)-1H-1,2,3-triazol-4-yl)methoxy)-3,6-dihydro-2H-pyran-2-yl)-5-(4-fluorobenzylidene)-3-phenylthiazolidin-4-one* (10a)**: mp: 235–237°C, Yield—85%. ^{1}HNMR (300 MHz, $CDCl_3$): δ8.07 (s, 1H, Ar-H), 7.80 (s, 1H, CH=C), 7.72 (d, *J* = 9.6 Hz, 2H, Ar-H), 7.40 (d, *J* = 9.2 Hz, 2H, Ar-H), 7.45 (d, *J* = 8.9 Hz, 2H, Ar-H), 7.19 (d, *J* = 8.2 Hz, 2H, Ar-H), 7.02–6.80 (m, 5H, Ar-H), 5.80–5.74 (m, 2H, =CH), 4.90 (d, *J* = 5.2 Hz, 1H, CH—S), 4.52 (s, 2H, OCH_2), 4.09–3.94 (m, 2H, 2XCH), 3.79 (d, *J* = 6.6 Hz, 2H, OCH_2): ^{13}CNMR (75 MHz, $CDCl_3$): δ170.4, 162.1, 144.1, 141.8, 139.8, 134.1, 130.4, 128.2, 125.6, 124.6, 122.4, 119.4, 115.5, 85.6, 72.6, 66.4, 64.0, 51: MS: m/z (M^++H) 575. Anal. Calcd for $C_{30}H_{24}ClFN_4O_3S$: C, 62.66; H, 4.21, N, 9.74; Found: C, 62. 48, H, 4.15. N, 9.56.

***(R,Z)-3-(4-chlorophenyl)-2-((2S,3S)-3-((1-(4-chlorophenyl)-1H-1,2,3-triazol-4-yl)methoxy)-3,6-dihydro-2H-pyran-2-yl)-5-(4-fluorobenzylidene) thiazolidin-4-one* (10b)**: mp: 216–218°C. Yield—72%. ^{1}HNMR (300 MHz, $CDCl_3$): 8.09 (s, 1H, Ar-H), 7.75 (s, 1H, CH=C), 7.62 (d, *J* = 9.5 Hz, 2H, Ar-H), 7.52 (d, *J* = 9.4 Hz, 4H, Ar-H), 7.40 (d, *J* = 8.6 Hz, 4H, Ar-H), 7.19 (d, *J* = 8.1 Hz, 2H, Ar-H), 5.84–5.75 (m, 2H, =CH), 4.94 (d, *J* = 5.2 Hz, 1H, CH-S), 4.52 (s, 2H, OCH_2), 4.06–3.94 (m, 2H, 2XCH), 3.80 (t, 2H, OCH_2): ^{13}CNMR (75 MHz, $CDCl_3$): δ170.5, 162.1, 144.2, 139.2, 134.2, 130.4, 129.2, 125.5, 124.1, 122.2, 119.4, 85.4, 72.8, 65.4, 63.4, 51.2: MS:m/z(M^++Na)632. Anal. Calcd for $C_{30}H_{23}Cl_2FN_4O_3S$: C, 59.12; H, 3.80, N, 9.19; Found: C, 59.01, H, 3.45. N, 8.96.

***(R,Z)-2-((2S,3S)-3-((1-(4-chlorophenyl)-1H-1,2,3-triazol-4-yl)methoxy)-3,6-dihydro-2H-pyran-2-yl)-5-(4-fluorobenzylidene)-3-(4-nitrophenyl) thiazolidin-4-one* (10c)**: mp: 221–223°C Yield—75%. ^{1}HNMR (300 MHz, $CDCl_3$): δ8.29 (d, *J* = 8.7 Hz, 2H, Ar-H), 8.09 (s, 1H, Ar-H), 7.69 (d, *J* = 9.1 Hz, 2H, Ar-H), 7.65 (s, 1H, CH=C), 7.61 (d, *J* = 9.4 Hz, 2H, Ar-H), 7.46 (d, *J* = 8.5 Hz, 2H, Ar-H), 7.18 (d, *J* = 8.3 Hz, 2H, Ar-H), 6.84 (d, *J* = 9.8 Hz, 2H, Ar-H), 5.86–5.79 (m, 2H, =CH), 4.96 (d, *J* = 5.2 Hz, CH-S), 4.55 (s, 2H, OCH_2), 4.05–3.95 (m, 2H, 2XCH), 3.85 (d, *J* = 6.9 Hz, 2H, OCH_2): ^{13}CNMR (75 MHz, $CDCl_3$): δ171.5, 162.1, 144.0, 141.8, 134.2, 130.4, 128.5, 125.4, 119.5, 115.4, 85.4, 72.4, 65.9, 63.6, 51.5: MS: m/z (M^++H) 620. Calcd for $C_{30}H_{23}ClFN_5O_5S$: C, 58.11; H, 3.74, N, 11.29; Found: C, 57.98, H, 3.55. N, 11.09.

***(R,Z)-2-((2S,3S)-3-((1-(4-chlorophenyl)-1H-1,2,3-triazol-4-yl)methoxy)-3,6-dihydro-2H-pyran-2-yl)-5-(4-fluorobenzylidene)-3-o-tolylthiazolidin-4-one* (10d)**: mp: 201–203°C, Yield—85%. ^{1}HNMR (300 MHz, $CDCl_3$): δ8.08 (s, 1H, Ar-H), 7.69 (d, *J* = 8.5 Hz, 2H, Ar-H), 7.62 (s, 1H, CH=C), 7.56 (d, *J* = 9.2 Hz, 2H, Ar-H), 7.49 (d, *J* = 8.7 Hz, 2H, Ar-H), 7.45–7.39 (m, 4H, Ar-H), 7.10 (d, *J* = 9.1 Hz, 2H, Ar-H), 5.76 (m, 2H, =CH), 4.93 (d, *J* = 5.2 Hz, 1H, CHS), 4.60 (s, 2H, OCH_2), 4.05–3.96 (m, 2H, CH), 3.90 (t, 2H, OCH_2), 2.1 (s, 3H, CH_3): ^{13}CNMR (75 MHz, $CDCl_3$): δ170.8, 162.9, 144.6, 137.2, 133.2, 130.6, 130.4, 128.2,

125.9, 122.7, 119.2, 116.2, 115.4, 84.4, 72.1, 65.3, 63.1, 52.5, 32.0, 17.5: MS: m/z (M^++H) 589. Anal. Calcd for $C_{31}H_{26}ClFN_4O_3S$: C, 63.21; H, 4.45, N, 9.51; Found: C, 62.75, H, 4.25. N, 9.29.

***(R,Z)-2-((2S,3S)-3-((1-(4-chlorophenyl)-1H-1,2,3-triazol-4-yl)methoxy)-3,6-dihydro-2H-pyran-2-yl)-5-(4-fluorobenzylidene)-3-p-tolylthiazolidin-4-one* (10e)**: mp: 205–215°C, Yield—66%. ^{1}HNMR (300 MHz, $CDCl_3$): δ8.02 (s, 1H, Ar-H), 7.69 (s, 1H, CH=C), 7.65 (d, *J* = 9.1 Hz, 2H, Ar-H), 7.54 (d, *J* = 9.2 Hz, 2H, Ar-H), 7.42 (d, *J* = 8.7 Hz, 2H, Ar-H), 7.35 (d, *J* = 8.2 Hz, 2H, Ar-H), 7.18 (d, *J* = 8.8 Hz, 2H, Ar-H), 6.80 (d, *J* = 9.4 Hz, 2H, Ar-H), 5.70–5.69 (m, 2H, =CH), 4.94 (s, 1H, CHS), 4.55 (s, 2H, OCH_2), 4.04–3.98 (m, 2H, CH), 3.96 (t, 2H, OCH_2), 2.32 (s, 3H, CH_3): ^{13}CNMR (75 MHz, $CDCl_3$): δ170.1, 162.5, 144.1, 139.5, 137.6, 135.2, 133.2, 130.4, 129.1, 127.5, 124.1, 122.5, 119.5, 115.3, 85.1, 72.5, 66.1, 63.2, 51.2, 21.6: MS: m/z (M^++H) 589. Anal. Calcd for $C_{31}H_{26}ClFN_4O_3S$: C, 63.21; H, 4.45, N, 9.51; Found: C, 62.98, H, 4.25. N, 9.33.

***(R,Z)-2-((2S,3S)-3-((1-(4-chlorophenyl)-1H-1,2,3-triazol-4-yl)methoxy)-3,6-dihydro-2H-pyran-2-yl)-5-(4-fluorobenzylidene)-3-(3-hydroxyphenyl)thiazolidin-4-one* (10f)**: mp: 218–219°C, Yield—82%. ^{1}H-NMR (300 MHz, $CDCl_3$): δ9.42 (brs, 1H, Ph-OH), 8.08 (s, 1H, Ar-H), 7.71 (d, *J* = 9.7 Hz, 2H, Ar-H), 7.65 (s, 1H, CH=C), 7.59 (d, *J* = 9.3 Hz, 2H, Ar-H), 7.44 (d, *J* = 8.6 Hz, 2H, Ar-H), 7.15 (d, *J* = 8.4 Hz, 2H, Ar -H), 6.80–6.78 (m, 4H, Ar-H), 5.70–5.68 (m, 2H, =CH), 4.92 (d, *J* = 5.2 Hz, 1H, CHS), 4.64 (s, 2H, OCH_2), 4.10 (t, 2H, OCH_2), 4.01–3.98 (m, 2H, CH): ^{13}CNMR (75 MHz, $CDCl_3$): δ170.5, 162.1, 158.2, 143.8, 139.8, 134.5, 130.8, 128.6, 125.6, 124.1, 122.4, 119.5, 115.7, 114.8, 106.5, 85.4, 72.5, 66.4, 63.4, 51.5: MS: m/z (M^++H) 591. Anal. Calcd for $C_{30}H_{24}ClFN_4O_4S$: C, 60.96; H, 4.09, N, 9.48; Found: C, 60.58, H, 3.85. N, 9.13.

***(R,Z)-2-((2S,3S)-3-((1-(4-chlorophenyl)-1H-1,2,3-triazol-4-yl)methoxy)-3,6-dihydro-2H-pyran-2-yl)-5-(4-fluorobenzylidene)-3-(4-hydroxyphenyl) thiazolidin-4-one* (10g)**: mp: 283–285°C, Yield—62%. ^{1}H-NMR (300 MHz,$CDCl_3$): δ9.42 (brs, 1H, Ph-OH), 8.05 (s, 1H, Ar-H), 7.85 (d, *J* = 9.3 Hz, 2H, Ar-H), 7.65 (s, 1H, CH=C), 7.56 (d, *J* = 9.2 Hz, 2H, Ar-H), 7.46 (d, *J* = 8.4 Hz, 2H, Ar-H), 7.32 (d, *J* = 8.6 Hz, 2H, Ar-H), 7.19 (d, *J* = 8.3 Hz, 2H, ArH), 7.02 (d, *J* = 8.8 Hz, 2H, Ar-H), 5.89–5.80 (m, 2H, =CH), 4.96 (d, *J* = 5.4 Hz, 1H, CHS), 4.66 (s, 2H, OCH_2), 4.09 (d, *J* = 2H, OCH_2), 4.04–3.98 (m, 2H, CH), ^{13}CNMR (75 MHz, $CDCl_3$): δ170.9, 162.5, 154.1, 144.4, 139.8, 134.9, 134.8, 130.4, 128.8, 127.2, 125.6, 123.2, 119.4, 116.4, 115.9, 85.4, 72.6, 66.5, 64.0, 51.6: MS: m/z (M^++H) 591. Anal. Calcd for $C_{30}H_{24}ClFN_4O_4S$: C, 60.96; H, 4.09, N, 9.48; Found: C, 60.58, H, 3.95. N, 9.23.

General procedure for the synthesis of Pyrazole phosphonates (11a-g): To a stirred mixture of **10a** (1 mmol), and Bestmann-Ohira Reagent (2.5 mmol) in dry EtOH (10 ml) was added KOH (2.5 mmol) at room temperature, after 2 minutes and irradiated in microwave bath reactor at 500 W for 4–7 minutes at 50°C. The crude product thus obtained was purified by column chromatography on silica gel (60–120 mesh) with hexane-ethyl acetate as eluent. Under conventional method the reaction mixture in EtOH (10 ml) was stirred at room temperature for the appropriate time (**Table 2**).

***Dimethyl 7-((2S,3S)-3-((1-(4-chlorophenyl)-1H-1,2,3-triazol-4-yl)methoxy)-3,6-dihydro-2H-pyran-2-yl)-4-(4-fluorophenyl)-9-oxo-8-phenyl-6-thia-1,2,8-triazaspiro[4.4]non-2-en-3-ylphosphonate* (11a)**: 245–247°C, Yield—75%. ^{1}HNMR (300 MHz, $CDCl_3$): δ13.06 (brs, 1H, =NH), 8.03 (s, 1H, Ar-H), 7.70 (d, *J* = 9.6 Hz, 2H, Ar-H), 7.30 (d, *J* = 9.2 Hz, 2H, Ar-H), 7.45 (d, *J* = 8.9 Hz, 2H, Ar-H), 7.19 (d, *J* = 8.2 Hz, 2H, Ar-H), 6.95–6.70 (m, 5H, Ar-H), 5.80–5.74 (m, 2H, =CH), 4.80 (d, *J* = 5.2 Hz, 1H, CH-S), 4.42 (s, 2H, OCH_2), 4.09–3.94 (m, 2H, 2XCH), 3.78 (s, 6H, OCH_3), 3.69 (d, *J* = 6.6 Hz, 2H, OCH_2), 3.52 (s, 1H, CH): ^{13}CNMR (75 MHz, $CDCl_3$): δ170.1, 160.1, 155.2, 144.1, 141.6, 136.2, 134.1, 129.2, 127.5, 125.6, 122.1, 119.1, 115.8, 86.6, 72.9, 63.8, 53.8, 44.5, 34.9: MS: m/z (M^++H) 725. Anal. Calcd for $C_{33}H_{31}ClFN_6O_6PS$: C, 54.66; H, 4.31, N, 11.59; Found: C, 54.48, H, 4.05. N, 11.36.

Compound	R	Mol. formula	Reaction time		Yield %	
			A (hours)	B (minutes)	A	B
11a	C_6H_5	$C_{33}H_{31}ClFN_6O_6PS$	3.5	6	62	89
11b	4-Cl-C_6H_4	$C_{33}H_{30}Cl_2FN_6O_6PS$	2.5	4	60	85
11c	4-NO_2-C_6H_4	$C_{33}H_{30}ClFN_7O_8PS$	2.0	5	61	84
11d	2-CH_3-C_6H_4	$C_{34}H_{33}ClFN_6O_6PS$	3.0	6	65	86
11c	4-CH_3-C_6H_4	$C_{34}H_{33}ClFN_6O_6PS$	3.2	4	69	85
11f	3-OH-C_6H_4	$C_{35}H_{31}ClFN_6O_7PS$	2.0	5	72	89
11g	4-OH-C_6H_4	$C_{35}H_{35}ClFN_6O_7PS$	3.0	4	71	82

A: conventional method; B: microwave irradiation method.

Table 2.
Synthesis of phosphonyl pyrazoles ***11(a–g)***.

Dimethyl8-(4-chlorophenyl)-7-((2S,3S)-3-((1-(4-chlorophenyl)-1H-1,2,3-triazol-4-yl)methoxy)-3,6-dihydro-2H-pyran-2-yl)-4-(4-fluorophenyl)-9-oxo-6-thia-1,2,8-triazaspiro[4.4]non-2-en-3-ylphosphonate (**11b**): mp: 206–208°C, Yield—82%. ^{1}HNMR (300 MHz, $CDCl_3$): δ13.11 (brs, 1H, —NH), 8.19 (s, 1H, Ar-H), 7.60 (d, *J* = 9.5 Hz, 2H, Ar-H), 7.54 (d, *J* = 9.4 Hz, 4H, Ar-H), 7.30 (d, *J* = 8.6 Hz, 4H, Ar-H), 7.22 (d, *J* = 8.1 Hz, 2H, Ar-H), 5.80–5.78 (m, 2H, =CH), 4.92 (d, *J* = 5.2 Hz, 1H, CH-S), 4.52 (s, 2H, OCH_2), 4.06–3.94 (m, 2H, 2XCH), 3.80 (t, 2H, OCH_2), 3.68 (s, 6H, OCH_3), 3.54 (s, 1H, CH): ^{13}CNMR (75 MHz, $CDCl_3$): δ170.9, 162.1, 155.4, 144.2, 139.8, 134.6, 129.5, 125.8, 124.1, 122.0, 119.2, 115.4, 86.1, 72.5, 64.4, 53.5, 44.8, 34.9: MS: m/z (M^++Na) 781. Anal. Calcd for $C_{33}H_{30}Cl_2$ FN6O6PS: C, 52.18; H, 3.98, N, 11.06; Found: C, 51.91, H, 3.65. N, 10.86.

Dimethyl 7-((2S,3S)-3-((1-(4-chlorophenyl)-1H-1,2,3-triazol-4-yl)methoxy)-3,6-dihydro-2H-pyran-2-yl)-4-(4-fluorophenyl)-8-(4-nitrophenyl)-9-oxo-6-thia-1,2,8-triazaspiro[4.4]non-2-en-3-ylphosphonate (**11c**): mp: 231–233°C, Yield—82%. ^{1}HNMR (300 MHz, $CDCl_3$): δ13.06 (brs, 1H, —NH), 8.23 (d, *J* = 8.7 Hz, 2H, Ar-H), 8.06 (s, 1H, Ar-H), 7.65 (d, *J* = 9.1 Hz, 2H, Ar-H), 7.51 (d, *J* = 9.4 Hz, 2H, Ar-H), 7.41 (d, *J* = 8.5 Hz, 2H, Ar-H), 7.10 (d, *J* = 8.3 Hz, 2H, Ar-H), 6.64 (d, *J* = 9.8 Hz, 2H, Ar-H), 5.76–5.59 (m, 2H, =CH), 4.86 (d, *J* = 5.2 Hz, 1H, CH-S), 4.35 (s, 2H, OCH_2), 4.01–3.93 (m, 2H, 2XCH), 3.72 (s, 6H, OCH_3), 3.65 (d, *J* = 6.9 Hz, 2H, OCH_2), 3.45 (s, 1H, CH), ^{13}CNMR (75 MHz, $CDCl_3$): δ171.1, 162.1, 150.0, 147.8, 144.0, 136.8, 131.4, 128.8, 127.2, 122.0, 119.5, 115.4, 86.4, 72.4, 65.9, 63.9, 53.5, 44.5, 34.8: MS: m/z (M^++H) 780. Calcd for $C_{33}H_{30}ClFN_7O_8PS$: C, 51.47; H, 3.93, N, 12.73; Found: C, 51.18, H, 3.55. N, 12.49.

Dimethyl 7-((2S,3S)-3-((1-(4-chlorophenyl)-1H-1,2,3-triazol-4-yl)methoxy)-3,6-dihydro-2H-pyran-2-yl)-4-(4-fluorophenyl)-9-oxo-8-o-tolyl-6-thia-1,2,8-triazaspiro[4.4]non-2-en-3-ylphosphonate (**11d**): mp: 221–223°C, Yield—75%. ^{1}HNMR (300 MHz, $CDCl_3$): δ13.10 (brs, 1H, —NH), 8.02 (s, 1H, Ar-H), 7.59 (d, *J* = 8.5 Hz, 2H, Ar-H), 7.59 (d, *J* = 9.2 Hz, 2H, Ar-H), 7.44 (d, *J* = 8.7 Hz, 2H, Ar-H), 7.42–7.40 (m, 4H, Ar-H), 7.12 (d, *J* = 9.1 Hz, 2H, Ar-H), 5.76 (m, 2H, =CH), 4.92 (d, *J* = 5.2 Hz, 1H, CHS), 4.62 (s, 2H, OCH_2), 4.09–3.99 (m, 2H, CH), 3.74 (s, 6H, OCH_3), 3.62 (s, 1H, CH), 3.80 (t, 2H, OCH_2), 2.12 (s, 3H, CH_3): ^{13}CNMR (75 MHz, $CDCl_3$): δ170.4, 160.1, 155.1, 144.4, 138.6, 136.2, 134.3, 130.7, 128.6, 127.2, 122.0, 119.2, 116.9, 115.4, 86.1, 72.8, 63.8, 53.5, 44.9, 34.8, 17.9: MS: m/z (M^++H) 739. Anal. Calcd for $C_{34}H_{33}ClFN_6O_6S$: C, 55.25; H, 4.50, N, 11.37; Found: C, 55.01, H, 4.25. N, 11.09.

Dimethyl 7-((2S,3S)-3-((1-(4-chlorophenyl)-1H-1,2,3-triazol-4-yl)methoxy)-3,6-dihydro-2H-pyran-2-yl)-4-(4-fluorophenyl)-9-oxo-8-p-tolyl-6-thia-1,2,8-triazaspiro[4.4]non-2-en-3-ylphosphonate **(11e)**: mp: 209–211°C, Yield—76%. ^{1}HNMR (300 MHz, $CDCl_3$): δ13.01 (brs, 1H, —NH), 8.07 (s, 1H, Ar-H), 7.62 (d, J = 9.1 Hz, 2H, Ar-H), 7.50 (d, J = 9.2 Hz, 2H, Ar-H), 7.40 (d, J = 8.7 Hz, 2H, Ar-H), 7.32 (d, J = 8.2 Hz, 2H, Ar-H), 7.18 (d, J = 8.8 Hz, 2H, Ar-H), 6.70 (d, J = 9.4 Hz, 2H, Ar-H), 5.60–5.59 (m, 2H, =CH), 4.90 (s, 1H, CHS), 4.45 (s, 2H, OCH_2), 4.01–3.99 (m, 2H, CH), 3.94 (t, 2H, OCH_2), 3.75 (s, 6H, OCH_3), 3.62 (s, 1H, CH), 2.30 (s, 3H, CH_3): ^{13}CNMR (75 MHz, $CDCl_3$): δ170.9, 160.1, 155.0, 144.1, 138.7, 136.8, 133.4, 130.4, 129.1, 127.2, 122.0, 119.1, 115.3, 86.1, 72.9, 68.1, 63.9, 53.5, 44.5, 34.8, 21.6: MS: m/z (M^++H) 739. Anal. Calcd for $C_{31}H_{26}ClFN_4O_3S$: C, 55.25; H, 4.50, N, 11.37; Found: C, 54.98, H, 4.25. N, 11.03.

Dimethyl 7-((2S,3S)-3-((1-(4-chlorophenyl)-1H-1,2,3-triazol-4-yl)methoxy)-3,6-dihydro-2H-pyran-2-yl)-4-(4-fluorophenyl)-8-(3-hydroxyphenyl)-9-oxo-6-thia-1,2,8-triazaspiro[4.4]non-2-en-3-ylphosphonate **(11f)**: mp: 228–229°C, Yield—88%. ^{1}H-NMR (300 MHz, $CDCl_3$): δ13.09 (brs, 1H, —NH), 9.40 (brs, 1H, Ph-OH), 8.04 (s, 1H, Ar-H), 7.61 (d, J = 9.7 Hz, 2H, Ar-H), 7.52 (d, J = 9.3 Hz, 2H, Ar-H), 7.42 (d, J = 8.6 Hz, 2H, Ar-H), 7.13 (d, J = 8.4 Hz, 2H, Ar-H), 6.70–6.68 (m, 4H, Ar-H), 5.73–5.70 (m, 2H, =CH), 4.82 (d, J = 5.2 Hz, 1H, CHS), 4.54 (s, 2H, OCH_2), 4.14 (t, 2H, OCH_2), 4.0–3.97 (m, 2H, CH), 3.70 (s, 6H, OCH_3), 3.57 (s, 1H, CH): ^{13}CNMR (75 MHz, $CDCl_3$): δ170.2, 156.1, 155.2, 144.8, 136.8, 129.6, 128.2, 127.5, 122.4, 119.4, 115.4, 106.5, 86.4, 72.5, 66.4, 63.4, 53.5, 44.9, 34.3: MS: m/z (M^++H) 741. Anal. Calcd for $C_{33}H_{31}ClFN_6O_7PS$: C, 53.48; H, 4.22, N, 11.34; Found: C, 53.18, H, 4.01. N, 11.13.

Dimethyl 7-((2S,3S)-3-((1-(4-chlorophenyl)-1H-1,2,3-triazol-4-yl)methoxy)-3,6-dihydro-2H-pyran-2-yl)-4-(4-fluorophenyl)-8-(4-hydroxyphenyl)-9-oxo-6-thia-1,2,8-triazaspiro[4.4]non-2-en-3-ylphosphonate **(11g)**: mp: 293–295°C, Yield—69%. ^{1}H-NMR (300 MHz, $CDCl_3$): δ12.85 (brs, 1H, —NH), 9.32 (brs, 1H, Ph-OH), 8.02 (s, 1H, Ar-H), 7.65 (d, J = 9.3 Hz, 2H, Ar-H), 7.59 (d, J = 9.2 Hz, 2H, Ar-H), 7.49 (d, J = 8.4 Hz, 2H, Ar-H), 7.30 (d, J = 8.6 Hz, 2H, Ar-H), 7.16 (d, J = 8.3 Hz, 2H, ArH), 7.0 (d, J = 8.8 Hz, 2H, Ar-H), 5.89–5.82 (m, 2H, =CH), 4.96 (d, J = 5.4 Hz, 1H, CHS), 4.56 (s, 2H, OCH_2), 4.07 (d, J = 2H, OCH_2), 4.02–3.99 (m, 2H, CH), 3.82 (s, 6H, OCH_3), 3.62 (s, 1H, CH), ^{13}CNMR (75 MHz, $CDCl_3$): δ172.9, 160.5, 154.3, 144.6, 136.2, 134.9, 134.3, 130.4, 129.8, 127.2, 125.6, 123.2, 119.8, 116.1, 86.4, 73.6, 66.5, 64.0, 53.6, 44.8, 34.9: MS: m/z (M^++Na) 763. Anal. Calcd for $C_{33}H_{31}ClFN_6O_7PS$: C, 53.48; H, 4.22, N, 11.34; Found: C, 53.18, H, 3.99. N, 11.13.

5. Conclusion

In conclusion, a series of a new class of hybrid heterocyclic's **11a–g** has been synthesized. The nematicidal activity of these compounds was evaluated against *Dietylenchus myceliophagus* and *Caenorhabditis elegans*. Among synthesized compounds **11b, 11c, 11f** and **11g** are the most effective against *Dietylenchus myceliophagus* and *Caenorhabditis elegans* the other test compounds showed moderate activity.

Acknowledgements

The authors are thankful to CSIR-New Delhi for the financial support (Project funding no.: 02 (247)15/EMR-II), Director, CSIR-IICT, Hyderabad, India, for NMR and MS spectral analysis and Principal, Vaagdevi Degree and PG College, Hanamkonda, for his consistent encouragement.

Author details

Avula Srinivas
Department of Chemistry, Vaagdevi Degree and PG College Kishanpura, Warangal, Telangana, India

*Address all correspondence to: avula.sathwikreddy@gmail.com

References

[1] Hani KD, Leigh DA. The application of CuAAC 'click' chemistry to catenane and rotaxane. Chemical Society Reviews. 2010;**39**:1240

[2] Kappa CO, Van der Eycken E. Click chemistry under non-classical reaction conditions. Chemical Society Reviews. 2010,**39**.1280

[3] El-Sagheer AH, Brown T. Current protocols in nucleic acid chemistry. Chemical Society Reviews. 2010;**39**:1388

[4] Qin A, Lam JWY, Tang BZ. Click polymerization. Chemical Society Reviews. 2010;**39**:2522

[5] Meldal M, Tornoe CW. Cu-catalyzed azide-alkyne cycloaddition. Chemical Reviews. 2008;**108**:2952

[6] Nandivada H, Jiang X, Lahann J. Click chemistry: versatility and controlin the hands of materials scientists. Advanced Materials. 2007;**19**:2197

[7] Angell YL, Burgess K. Peptidomimetics via copper-catalyzed azide–alkyne cycloadditions. Chemical Society Reviews. 2007;**36**:1674

[8] Fournier D, Hoogenboom R, chubert USS. Clicking polymers: A straightforward approach to novel macromolecular architectures. Chemical Society Reviews. 2007;**36**:1369

[9] Moses JE, Moorhouse AD. The growing applications of click chemistry. Chemical Society Reviews. 2007;**36**:1249

[10] Lutz JF. 1,3-dipolar cycloadditions of azides and alkynes: A universal ligation tool in polymer and materials science. Angewandte Chemie, International Edition. 2007;**46**:1018

[11] Dondoni A. Click chemistry in the glycosylation reactions. Chemistry, an Asian Journal. 2007;**2**:700

[12] Kolb HC, Sharpless KB. The Growing Impact of Click Chemistry on Drug Discovery. Drug Discovery Today. 2003;**8**:1128

[13] Ch Zhou YW. Recent researches in triazole compounds as medicinal drugs. Current Medicinal Chemistry. 2012;**19**:239

[14] Brick A, Muldoon J, YC Lin JH, Elder DS, Goodsell AJ, Olson VV, et al. Rapid diversity-oriented synthesis in microtiter plates for in situ screening of HIV protease inhibitors. Chembiochem. 2003;**4**:1246

[15] Soltis MJ, Yeh HJ, Cole KA, Whittaker N, Wersto RP, Kohn EC. Drug Metabolism and Disposition. 1996;**24**:799

[16] Fan W-Q, Katritzky AR. 1,2,3-Triazoles. In: Katritzky AR, Rees CW, Scriven V, editors. Comprehensive Heterocyclic Chemistry II. Vol. 4. Oxford: Elsevier; 1996. p. 905

[17] Whiting M, Muldoon J, Lin YC, Silverman SM, Lindstrom W, Olson AJ, et al. Inhibitors of HIV-1 protease by using in situ click chemistry. Angewandte Chemie, International Edition. 2006;**45**:1435

[18] Bourne Y, Kolb HC, Radić Z, Sharpless KB, Taylor P, Marchot P. A One-Pot Procedure for the Synthesis of "Click-Ready" Triazoles from Ketones. Proceedings of the National Academy of Sciences of the United States of America. 2004;**101**:1449

[19] Lewis WG, Green G, Grynszpan FZ, Carlier PR, Taylor P, Finn MG, et al. Angewandte Chemie, International Edition. 2002;**41**:1053

[20] Huisgen R, Padwa A. 1,3-Dipolar Cycloaddition Chemistry. Vol. 1. New York: Wiley; 1984. p. 1

[21] Al Maoudi NA, Al-Soud AY. Tetrahedron Letters. 2002;**43**:4021

[22] Kuijpers MHB, Groothuys S, Keereweer RAB, Quaedflieg PJLM, Blaauw RH, van Delft FL, et al. Expedient synthesis of triazole-linked glycosyl amino acids and peptides. Organic Letters. 2004;**6**:3123

[23] Ch Srinivas X, Fang QW. One-pot synthesis of triazole-linked glycoconjugates. Tetrahedron Letters. 2005;**46**:2331

[24] Hotha S, Anegundi RI, Natu AA. Expedient synthesis of 1,2,3-triazole-fused tetracyclic compounds by intramolecular Huisgen ('click') reactions on carbohydrate-derived azido-alkynes. Tetrahedron Letters. 2005;**46**:4585

[25] Hotha S, Kashyap S. "Click Chemistry" Inspired Synthesis of pseudo-Oligosaccharides and Amino Acid Glycoconjugates. The Journal of Organic Chemistry. 2006;**71**:364

[26] Andrew S, Susan M, Michael F, Mathew W, Penny L, Paul L, et al. Synthesis and biological activity of anticoccidial agents: 5,6-Diarylimidazo[2,1-b][1,3]thiazoles. Bioorganic & Medicinal Chemistry Letters. 2008;**18**:5263

[27] Onca S, Punar M, Eracosy H. Comparative activities of β-lactam antibiotics and quinolones for invasive streptococcus pneumoniae isolates. Chemotherapy. 2004;**50**:98

[28] Dave CV, Shukla MC. Pyridopyrimidines : Part IX - Synthesis and antibacterial activity of 2-methylthio-6-phenylazo-5,7-dimethylpyrido[2,3-d]pyrimidin-4(3H)-ones. Indian Journal of Chemistry. 2000;**39B**:210

[29] Ghazzi MN, Perez E, TK Antonucci H, Driscoll SM, Hunang BWF. Cardiac and glycemic benefits of troglitazone treatment in NIDDM. Diabetes. 1997;**46**:433

[30] Scmidt U, Utz R, Liberknecht A, Griesser H, Potzolli B, Bahr J, et al. Synthesis of troglitazone. Synthesis. 1978:233

[31] Diurno MV, Mazzoni O, Lzzo AA, Bolognese A. Synthesis of 2-(aminoallyl)-thiazole-4-carboxylic acids. II Farmaco. 1992;**52**:237

[32] Ergene N, Gapan G. Anti convulsant activity of Thiazolidinones. II Farmaco. 1994;**49**:237

[33] Viswajanani JS, Ajay S, Smita S, Seema K, Manisha P, Pragya B, et al. Synthesis and antimicrobial activity of novel thiazolidinones. ARKIVOC. 2005:46

[34] Ueno H, Oe T, Snehiro I, Nakamura S. US Patent. 1997. 5594116

[35] Previtera T, Vigortia MG, Bisila M, Orsini F, Benetolla F, Bombieri G. 3,3′-Di [1,3-thiazolidine-4-one] system. VI. Structural and conformational studies on configurational isomers with antihistaminic activity. European Journal of Medicinal Chemistry. 1994;**29**:317

[36] Ebied MY, Fathallah OA, El-Zaheer MI, Kamel MM, Abdon WA, Anwer MM. Synthesis and anti histaminic activity of Thiazolidinones. Bulletin of Faculty of Pharmacy. 1996;**34**:125

[37] Rawal RK, Prabhakar YS, Katti SB, Declercq E. 2-(Aryl)-3-furan-2-ylmethyl-thiazolidin-4-ones as selective HIV-RT Inhibitors. Bioorganic & Medicinal Chemistry. 2005;**13**:6771

[38] Kato T, Ozaki T, Tamura K. Novel calcium antagonists with both calcium overload inhibition and antioxidant Activity. 2. Structure-Activity relationships of thiazolidinone

derivatives. Journal of Medicinal Chemistry. 1999;**42**:3134

[39] Tanabe Y, Suzukamo G, Komuro Y, Imanishi N, Morooka S, Enomoto M, et al. Structure-activity relationship of optically active 2-(3-pyridyl) thiazolidin-4-ones as a PAF antagonists. Tetrahedron Letters. 1991;**32**:379

[40] Kato T, Ozaki T, Ohi N. Improved synthetic methods of CP-060S, a novel cardioprotective drug. Teraherdron Assymetry. 1999;**10**:3963

[41] Adachi Y, Suzuki Y, Homma N, Fukazawa M, Tamura K, Nishie I, et al. The anti-ischemic effects of CP-060S during pacing-induced ischemia in anesthetized dogs. European Journal of Pharmacology. 1999;**367**:267

[42] Ottana R, Mazzon E, Dugo L, Monforte F, Macari F, Sautebin L, et al. Modeling and biological evaluation of 3,3′-(1,2-ethanediyl)bis[2-(4-methoxyphenyl)-thiazolidin-4-one], a new synthetic cyclooxygenase-2 inhibitor. European Journal of Pharmacology. 2002;**448**:71

[43] Tanabe Y, Yamamoto H, Murakami M, Yanagi K, Kubota Y, Okumara H, et al. Synthetic study of the highly potent and selective anti-platelet activating factor thiazolidin-4-one agents and related compounds. Journal of the Chemical Society, Perkin Transactions. 1995;**17**:935

[44] Kato Y, Kita Y, Nisho M, Hirasawa Y, Ito K, Yamanaka T, et al. In vitro antiplatelet profile of FR171113, a novel non-peptide thrombin receptor antagonist. European Journal of Pharmacology. 1999;**384**:197

[45] Voss ME, Carter PH, Tebben AJ, Scherie PA, Brown GD, Thampson LA, et al. Synthesis of thiazolidinones as thrombin receptor antagonist. Bioorganic & Medicinal Chemistry Letters. 2003;**13**:6771

[46] Field SC. Synthesis of natural products containing a C-P bond. Tetrahedron. 1999;**55**:12237

[47] Moonen K, Laureyn CV Stevens I. Synthetic methods for azaheterocyclic phosphonates and their biological activity. Chemical Reviews. 2004;**104**:6177

[48] Piperno A, Chiacchio D, Lannazo D, Romeo R. Synthesis and biological activity of phosphonated nucleosides: Part 1 furanose, carbocyclic and heterocyclic analogues. Current Medicinal Chemistry. 2006;**13**:3675

[49] Qu GR, Xia R, Yang XN, Li JG, Wang DC, Guo HM. Synthesis of novel C6-phosphonated purine nucleosides under microwave irradiation by SNAr–arbuzov reaction. The Journal of Organic Chemistry. 2008;**73**:2416

[50] Alen J, Dobrazanaska L, Dc Borggraeve WM, Comper nolle FJ. Synthesis of C-P bond bearing heterocyclics. The Journal of Organic Chemistry. 2007;**72**:1055

[51] Moriguchi T, Yanagi Y, Kunimori M, Weda T, Sekni MJ. Synthesis and Properties of Aminoacylamido-AMP: Chemical Optimization for the Construction of an N-Acyl Phosphoramidate Linkage. The Journal of Organic Chemistry. 2000;**65**:8229

[52] Lamberth C. Pyrazole Chemistry in Crop Protection. Heterocyclics. 2007;**71**:1467

[53] McDonald E, Jones K, Brough PA, Drysdale MJ, Workman P. Discovery and development of pyrazole-scaffold Hsp90 inhibitors. Current Topics in Medicinal Chemistry. 2006;**6**:1193

[54] Halcrow MA. Pyrazoles and pyrazolides—flexible synthons in self-assembly. Dalton Transactions. 2009:2059

[55] Jayasinghe ULB, Kumarihamy BMM, Bandara AGD, Vasquez EA, Karus W. Nematicidal activity of some sri lankan plants. Natural Product Research. 2003;**17**:259

[56] Xaki MH, Moran D, Harries D. Pesticides in groundwater: The aldicarb story in Suffolk County, NY. American Journal of Public Health. 1982;**72**:1391

[57] Noling JW, Becker JO. The challenge of research and extension to define and implement alternatives to methyl bromide. Journal of Nematology. 1994;**26**:573

[58] Kagan J, Kagan PA, Bushe HE. Light-dependent toxicity of α-terthienyl and anthracene toward late embryonic stages of Rana pipiens. Journal of Chemical Ecology. 1984;**10**:1115

[59] Yuji O, Sengual N, Eli P, Uzi R, Zohara Y, Yitzhak S. Nematicidal activity of essential oils and their components against the root-knot nematode. Phytopathology. 2000;**90**:710

[60] Srinivas A, Sunitha M, Karthik P, Nikitha G, Raju K, Ravinder B, et al. Synthesis, nematicidal and antifungal properties of hybrid heterocyclics. Journal of Heterocyclic Chemistry. 2017;**54**:3250

[61] Srinivas A, Santhosh M, Sunitha M, Karthik P, Srinivas K, Vasumathi Reddy K. Synthesis and biological evaluation of triazole linked thiazolidenone glycosides. Acta Chimica Slovenica. 2016;**63**:827

[62] Srinivas A. Synthesis and Antimicrobial Activity of Bis[4-methoxy-3-(6-aryl-7H-[1,2,4] triazolo[3,4-b][1,3,4]thiadiazin-3-yl)phenyl]methanesand Bis[(triazolo[3,4-b]thiadiazepin-3-yl) phenyl]methanes. Acta Chimica Slovenica. 2016;**63**:344

[63] Srinivas A, Sunitha M. Synthesis of piparonyl triazoles as anti microbial agents. Indian Journal of Chemistry, Section A. 2016;**55B**:102

[64] Srinivas A, Sunitha M. Synthesis of 1,2,3-triazole glycosides as anticancer agents. Indian Journal of Chemistry Section B. 2016;**55B**:231

[65] Reddy CS, Srinivas A, Sunitha M, Nagaraj A. Design and synthesis of novel methylene-bis-fused pyrazoles as biologically active molecules. Journal of Heterocyclic Chemistry. 2010;**47**:1303

[66] Srinivas A, Reddy CS, Nagaraj A. Synthesis, Nematicidal and Antimicrobial Properties of Bis-[4-methoxy-3-[3-(4-fluorophenyl)-6-(4-methylphenyl)-2(aryl)-tetrahydro-2Hpyrazolo[3,4-d] thiazol-5-yl]phenyl]methanes. Chemical & Pharmaceutical Bulletin. 2009;**57**:685

[67] Reddy CS, Srinivas A, Nagaraj A. Synthesis and in vitro study of a new class of methylenebis-4,6-diarylbenzo[d]isoxazoles as potential antifungal agents. Journal of Heterocyclic Chemistry. 2009;**46**:497

[68] Reddy CS, Srinivas A, Nagaraj A. Synthesis of some novel methylene-bis-pyrimidinyl-spiro-4-thiazolidinones as biologically potent agents. Journal of Heterocyclic Chemistry. 2008;**45**:1121

[69] Srinivas A, Nagaraj A, Sanjeeva Reddy CH. Synthesis and biological evaluation of novel methylene-bisthiazolidinone derivatives as potential nematicidal agents. Journal of Heterocyclic Chemistry. 2008;**45**:999

[70] McBeth CW, Bergerson GB. Nematicidal activity of heterocyclics. Phytopathology. 1953;**43**:264

Chapter 10

Quinoline Heterocycles: Synthesis and Bioactivity

Yamajala B.R.D. Rajesh

Abstract

Among heterocyclic compounds, quinoline is a privileged scaffold that appears as an important construction motif for the development of new drugs. Quinoline nucleus is endowed with a variety of therapeutic activities, and new quinolone derivatives are known to be biologically active compounds possessing several pharmacological activities. Many new therapeutic agents have been developed by using quinoline nucleus. Hence, quinoline and its derivatives form an important class of heterocyclic compounds for the new drug development. Numerous synthetic routes have been developed for the synthesis of quinoline and its derivatives due to its wide range of biological and pharmacological activities. The article covers the synthesis as well as biological activities of quinoline derivatives such as antimalarial, anticancer, antibacterial, anthelmintic, antiviral, antifungal, anti-inflammatory, analgesic, cardiovascular, central nervous system, hypoglycemic, and miscellaneous activities.

Keywords: quinoline, heterocyclic compound, quinoline derivatives, synthesis, biological activity

1. Introduction

Quinoline **1** or 1-azanaphthalene or benzo[*b*]pyridine is an aromatic nitrogen-containing heterocyclic compound having a molecular formula of C_9H_7N, and the molecular weight is 129.16. Being a weak tertiary base, it forms salts with acids and exhibits reactions similar to benzene and pyridine. It participates in both electrophilic and nucleophilic substitution reactions.

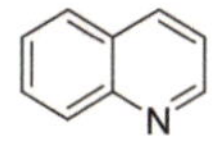

1

Quinoline moiety commonly exists in various natural compounds (*Cinchona* alkaloids), and pharmacological studies have shown that the quinolone ring system is present in many compounds exhibiting a broad range of biological activities. Quinoline has been found to have antibacterial, antifungal, antimalarial, anthelmintic, anticonvulsant, cardiotonic, anti-inflammatory, and analgesic activities.

2. Synthesis

In the literature, a number of established protocols have been reported for the synthesis of quinoline ring, which can be altered to produce a number of differently substituted quinolines. The quinoline ring has been generally synthesized by various conventional named reactions such as Skraup, Doebner-Von Miller, Pfitzinger, Friedlander, Conrad-Limpach, and Combes synthesis (**Figure 1**) [1].

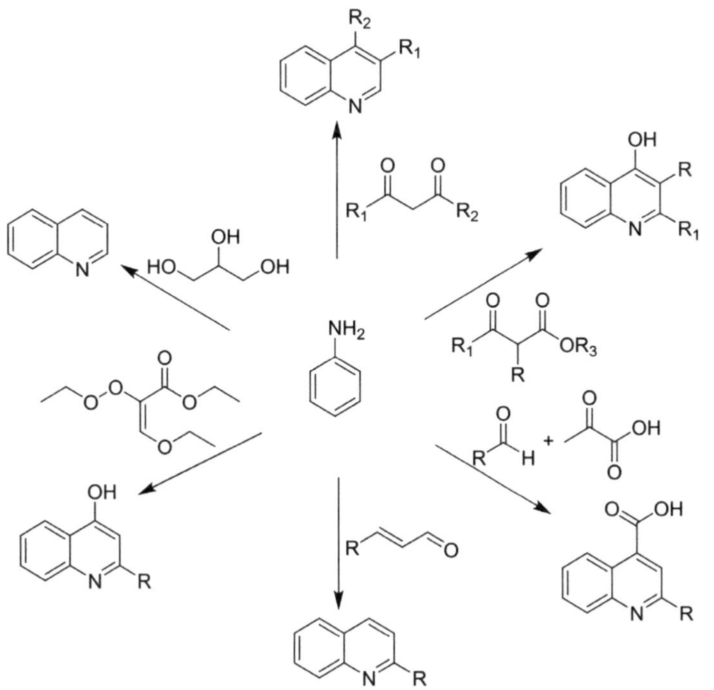

Figure 1.
Conventional methods of synthesis of various substituted quinolines.

Apart from the conventional methods, a vast number of synthetic routes have been developed for the synthesis of quinoline and quinoline derivatives. Chen et al. reported the synthesis of 2,4-disubstituted quinolines, **2** by the condensation of 2-iodoanilines with alkynyl aryl ketones using nickel catalyst [2].

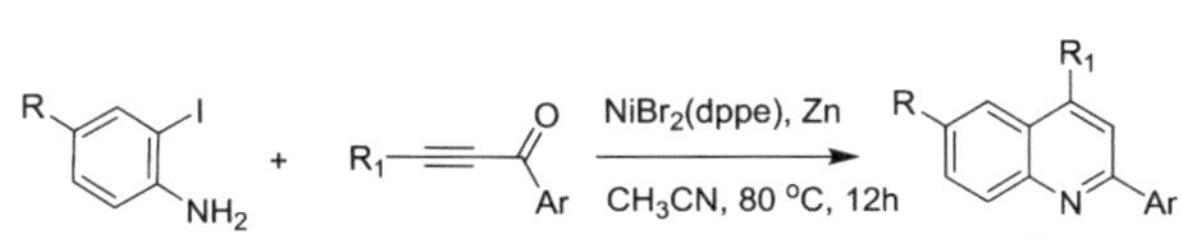

2

2,4-Disubstituted quinolones, **3** have been obtained by the cyclization of 2-aminoaryl ketones with phenylacetylenes. This reaction takes place in ionic liquid medium ([hmim]PF_6) in the presence of zinc trifluoromethanesulfonate catalyst [3]. Lekhok et al. synthesized the same product in the presence of catalytic amount of indium(III) trifluoromethanesulfonate ($In(CF_3SO_3)_3$) under microwave and solvent-free conditions [4].

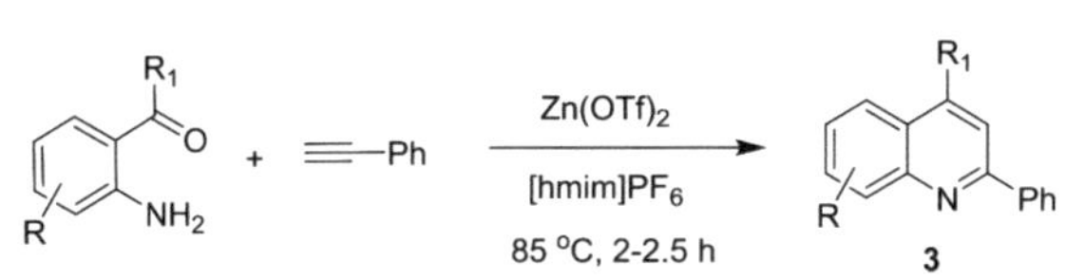

3

2,4-Diphenyl-2-methyl-1,2-dihydroquinoline, **4** has been prepared by the condensation followed by cyclization of aniline and acetophenone. The reaction proceeds with the help of a zeolite catalyst, E_{4a} [5].

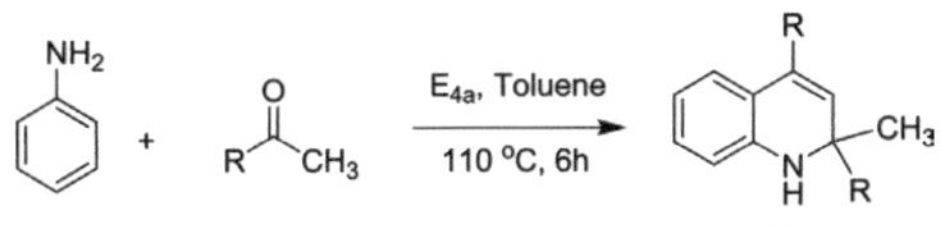

4

2,3,4-Trisubstituted quinolones, **5** have been synthesized by Friedlander annulation of 2-amino substituted aromatic ketones and reactive methylene group containing carbonyl compounds in the presence of ethyl ammonium nitrate (EAN) [6].

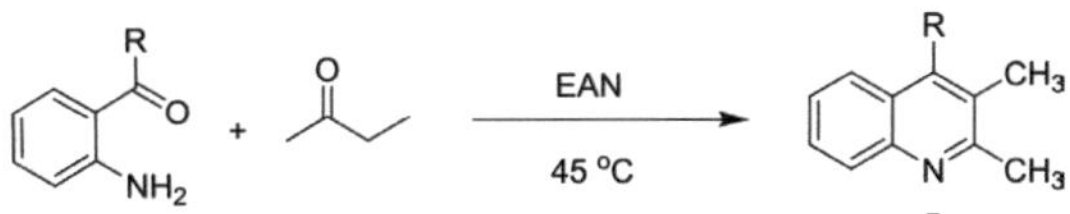

5

By stirring 2-aminoaryl ketones and various α-methylene ketones in the presence of dodecylphosphonic acid (DPA) catalyst in water or solvent-free conditions, poly-substituted quinolones, **6** have been synthesized [7].

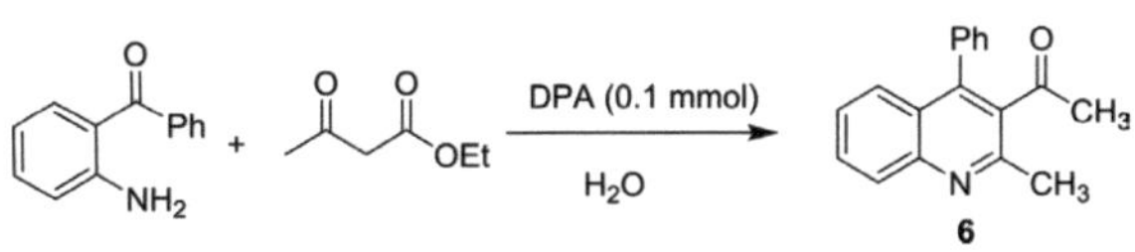

6

2-Aminobenzyl alcohol reacts with ketones or alcohols in the presence of a base, and benzophenone resulted in the formation of poly-substituted quinolones, 7 [8]. Here, benzophenone acts as a hydride scavenger.

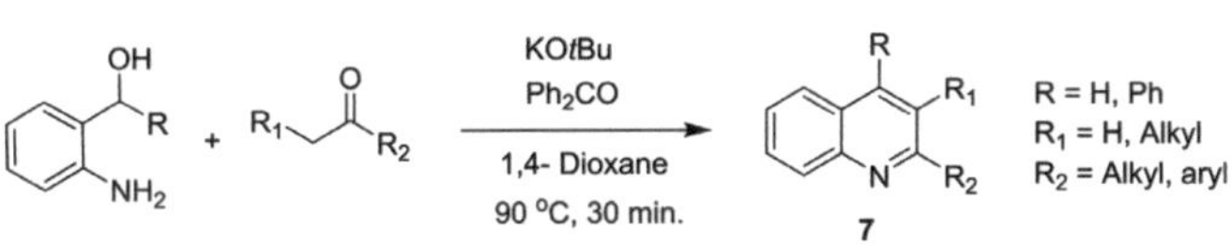

7

Horn et al. reported the synthesis of quinolines, **8** from α, β-unsaturated ketones and *o*-aminophenylboronic acid derivatives [9]. This method is the modification of the conventional Skraup-Doebner-Von Miller synthesis and that the reaction proceeded under basic conditions.

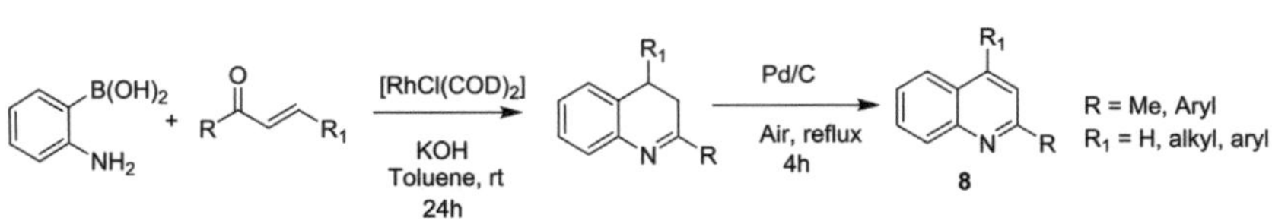

8

3,4-Dihydroquinolin-2-ones, **9** have been synthesized by treating 2-iodoanilines and various acrylates using azobisisobutyronitrile (AIBN) in the presence of tributyltin hydride [10].

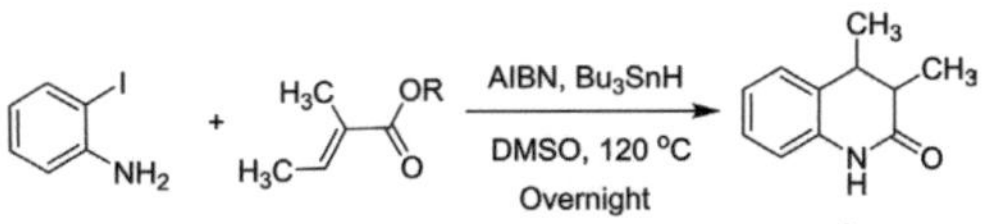

9

Wang et al. developed a method for the synthesis of 2-phenylquinoline-4-carboxylic acids, **10** by the treatment of pyruvic acid with substituted aniline and benzaldehyde in the presence of rare-earth metal catalysts in water under reflux condition [11].

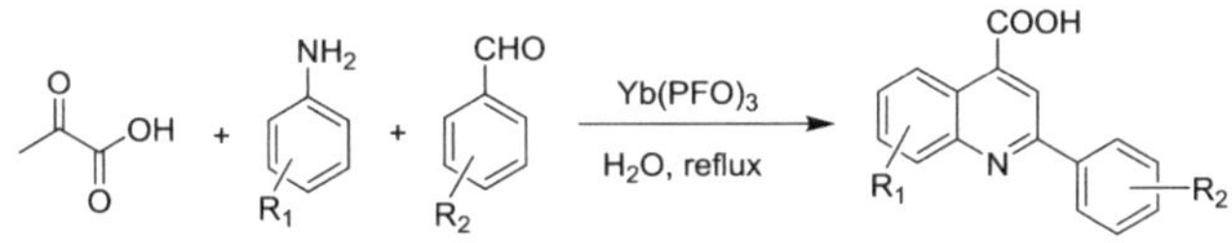

10

Kouznetsov et al. synthesized phenyl-substituted quinolones, **11** by reacting ethyl vinyl ether or ethyl vinyl sulfide with N-arylaldimine in the presence of Lewis acidic catalysts such as boron trifluoride etherate ($BF_3.OEt_2$) to obtain 2,4-substituted tetrahydroquinolines. The tetrahydroquinolines were aromatized to 2-phenyl-substituted quinolines under vacuum distillation in the presence of *p*-TSOH [12].

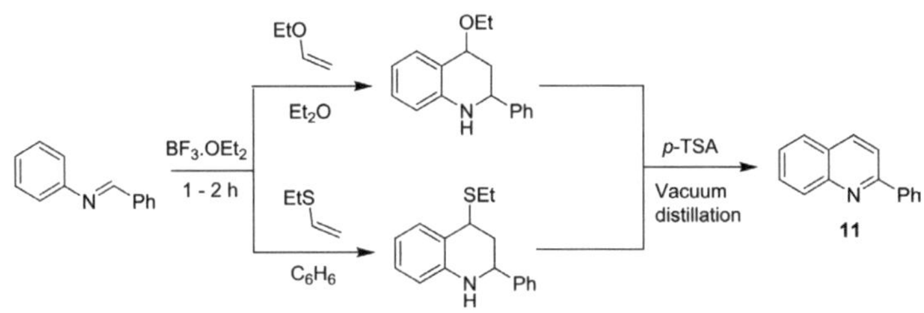

11

Wang et al. reported the synthesis of 2-phenyl-4-alkoxy quinolines, **12** by cyclocondensation of 2-(2-trimethylsilyl)ethynyl) aniline with aromatic aldehydes in the presence of sulfuric acid as catalyst in methanol solvent [13].

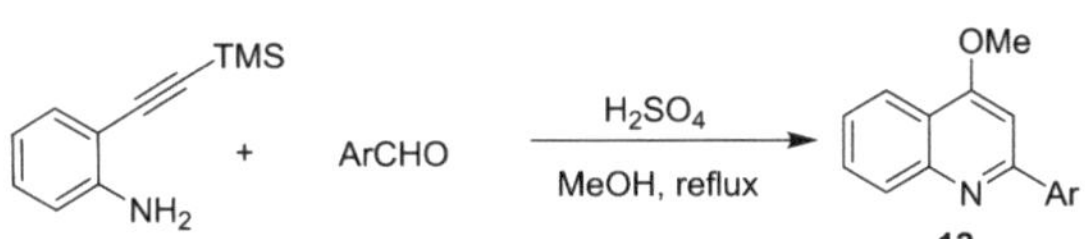

12

Two molecules of *o*-haloacetophenones condensed with urea or primary amines yielded certain halogen-substituted quinolones, **13**. The halogen-substituted quinolines were formed through the cleavage of C(sp^2)–halogen and α-C(sp^3)–H bonds and the formation of new bonds in a selective manner [14].

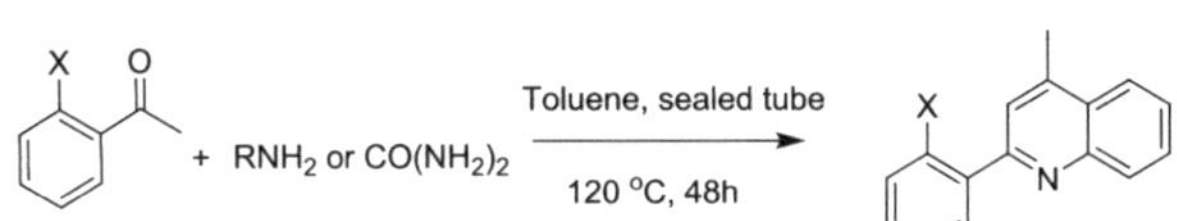

13

A one-pot reaction of 2-aminoaryl ketones with certain arylacetylenes results in the formation of 2,4-disubstituted quinolones, **14**. The reaction was performed in a green synthetic route using potassium dodecatugstocobaltate trihydrate ($K_5CoW_{12}O_{40}.3H_2O$) as a recyclable and eco-friendly catalyst under microwave and solvent-free conditions [15].

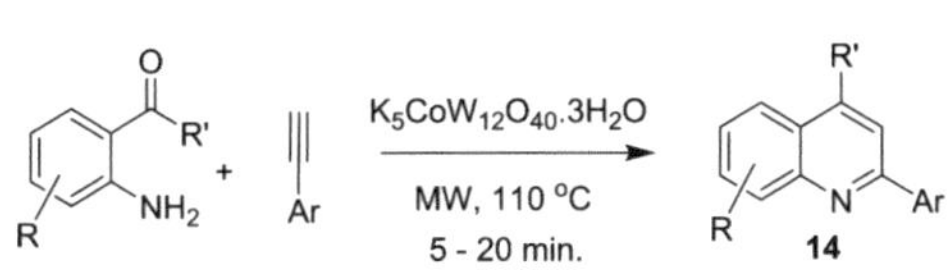

14

Kowsari et al. synthesized certain quinolones, **15** by reacting isatin with aryl methyl ketones in the presence of basic ionic liquids in water [16]. The reaction was conducted under ultrasound green synthetic conditions. The main advantages of this procedure are (i) a green method, (ii) milder and shorter reaction time, and (iii) higher yields and selectivity without a transition metal catalyst.

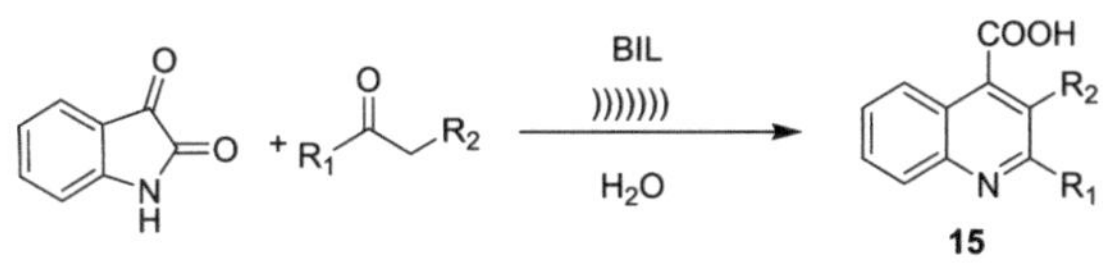

1,4-Diazabicyclo[2.2.2]octane (DABCO) promoted structurally diverse 2-alkoxy- and 2-aryloxy-3-substituted quinolones, **16** that have been synthesized by treating *o*-alkynylaryl isocyanides with alcohols and phenols [17]. DABCO initiates the reaction as a nucleophile and facilitates the formation of the product as a leaving group being replaced by oxygen nucleophiles.

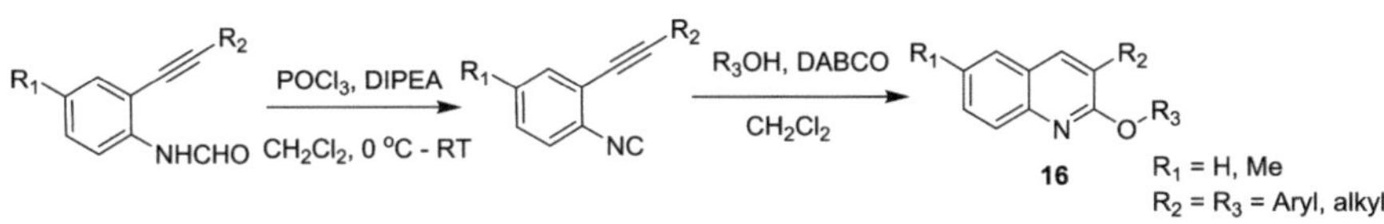

Benzimidoyl chlorides when treated with 1-(1-(allyloxy)prop-2-ynyl)benzene (1,6-enynes) yielded diverse quinoline derivatives, **17** via a domino palladium-catalyzed Sonogashira coupling and followed by cyclization [18].

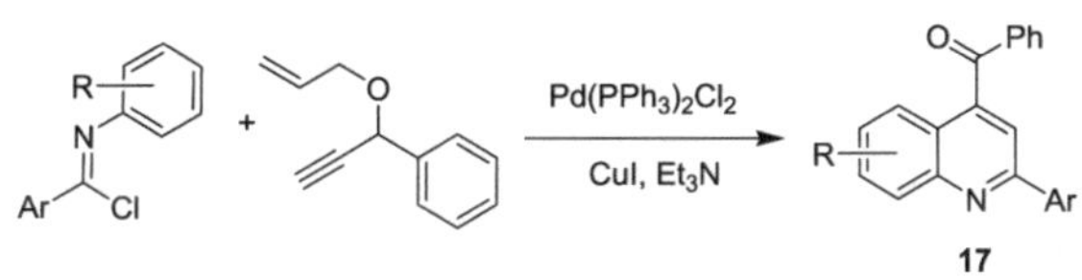

Diversified quinolones, **18** have been synthesized by the intramolecular cyclocondensation of 1-azido-2-(2-propynyl)benzenes using electrophilic reagents (I_2, Br_2, ICl, NBS, NIS, and $HNTf_2$) in nitromethane at 0°C to room temperature. The reaction also proceeds in the presence of $AuCl_3$/$AgNTf_2$ catalysts in THF at 100°C [19].

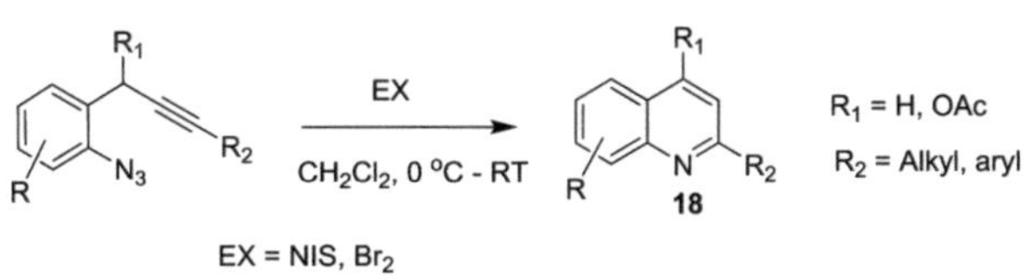

3. Biological activity

3.1 Antimalarial

Quinolines are known for their excellent antimalarial properties. Raynes et al. developed bisquinolines, **19, 20** that exhibit antimalarial activity against chloroquine-resistant and chloroquine-sensitive parasites [20]. Derivatives of ferrochloroquine, **21** were also found to possess antimalarial activity [21]. In these derivatives, the carbon skeleton of chloroquine is replaced by ferrocene group. Modapa et al.

reported that the synthesis of ureido-4-quinolinamides, **22** showed antimalarial activity at MIC 0.25 mg/mL against chloroquine-sensitive *Plasmodium falciparum* strain [22]. Several 7-chloroquinolinyl thioureas, **23, 24** have been synthesized by Mahajan et al. that possess excellent antimalarial properties [23]. Kovi et al. synthesized a chloroquinolyl derivative, **25** that has an excellent antimalarial activity even at very low concentrations [24]. Acharya et al. reported the synthesis and potent antimalarial activity of certain pyridine-quinoline hybrid conjugates, **26, 27** against chloroquine susceptible *P. falciparum* strain [25]. Shiraki et al. produced some 5-aryl-8-aminoquinolines, **28** with good antimalarial activity and had mild hemolytic activity than tafenoquine [26]. Singh et al. developed several antimalarial 4-anilinoquinolines, **29** which showed good antimalarial activity against chloroquine-sensitive *P. falciparum* strains [27]. Novel hybrid conjugates of N-(7-chloroquinolin-4-yl) piperazine-1-carbothioamide and 1,3,5-triazine derivatives, **30** have been synthesized by Bhat et al. These hybrid conjugates possess considerable antimalarial activity against both wild and mutant parasites on changing the pattern of substitution [28]. McNulty et al. developed 4-arylquinoline-2-carboxylate derivatives, **31** which show antiprotozoal activity against the pathogenic parasite *Toxoplasma gondii* [29].

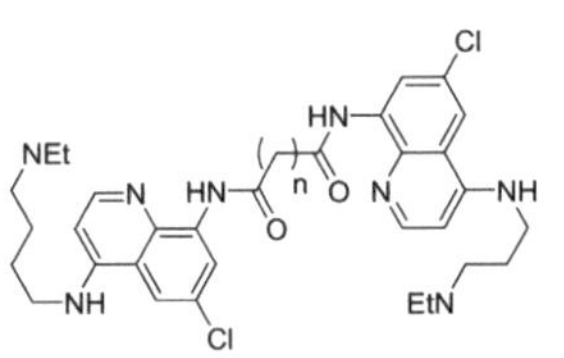

n = 4,6,8

19

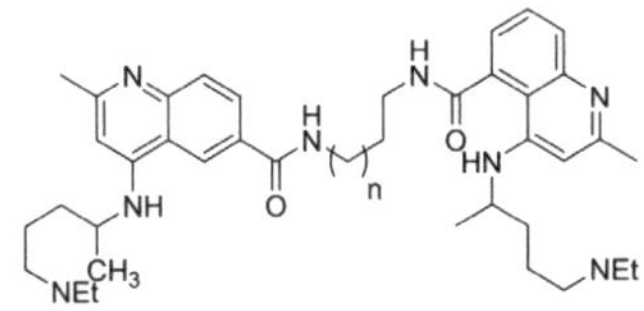

n = 2,4,6,8

20

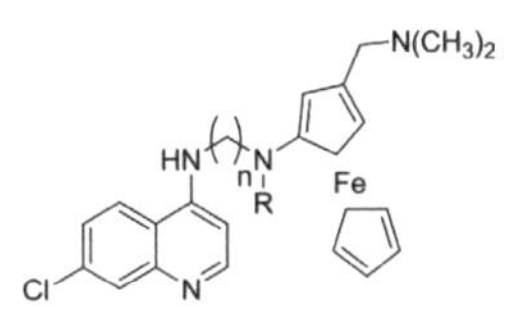

n = 2-6
R = H,CONHBn

21

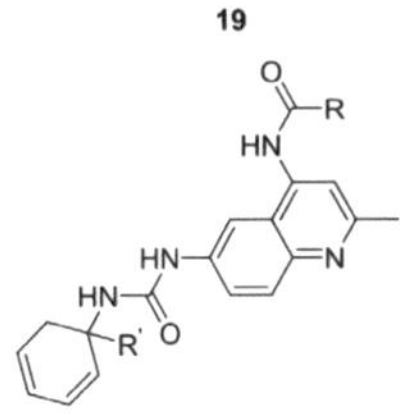

R=Me, Ph, CH_2Cl, 2-ClC_6H_5, 3-ClC_6H_5, 2-Furyl
R'=F, Cl, Br, CF_3

22

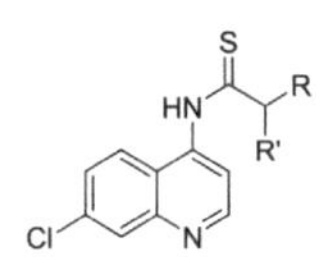

R = $(CH_2)_2OH$, $(CH_2)_3N(Et)_2$, $(CH_2)_3N(Me)_2$, $(CH_2)_2NH_2$
R' = H, C_6H_5, $CH_2C_6H_5$, $COOC_2H_5$

23

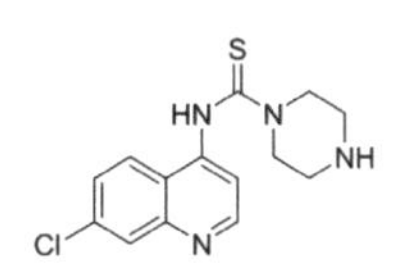

24

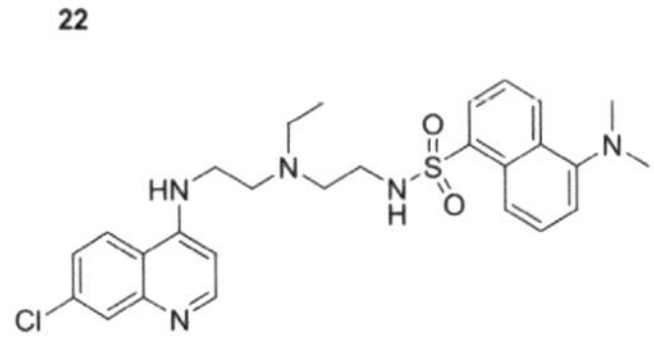

25

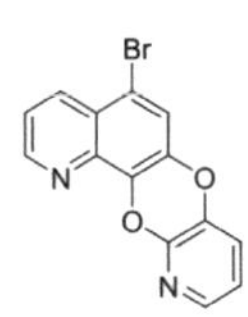

26

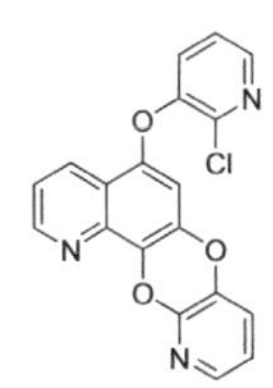

27

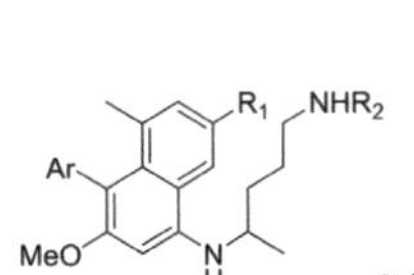

R_1 = OCH_3, CF_3
R_2 = H or $C(O)OC(CH_3)_3$

28

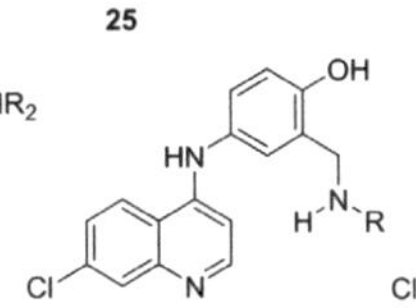

R = H, Phenyl,
Butyl, Isopropyl, n-Butyl

29

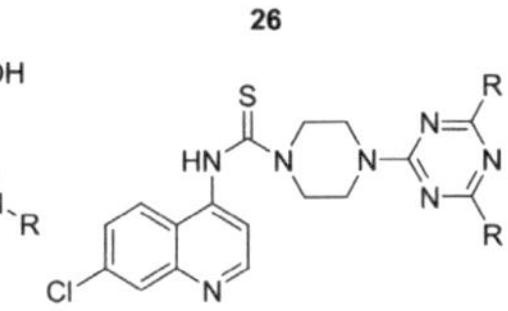

30

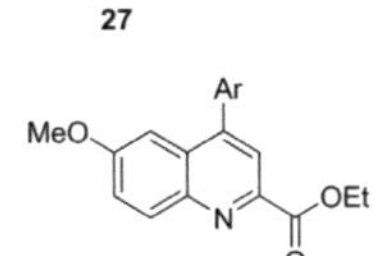

31

3.2 Anti-inflammatory activity

A quinoline derivative, **32** with strong anti-inflammatory activity was synthesized by Baba et al. in adjuvant arthritis rat model [30]. Chen et al. developed

2-(furan-2-yl)-4-phenoxy-quinoline derivatives, **33, 34** that inhibit the lysozyme and β-glucuronidase release [2]. Few quinoline derivatives, **35, 36** have been synthesized and evaluated by Gilbert et al. for treating osteoarthritis and that are amino-acetamide inhibitors of aggrecanase-2 [31].

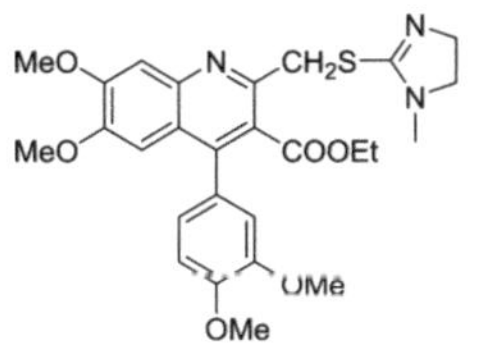

32

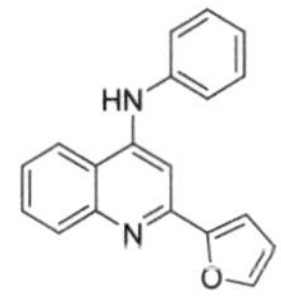

33

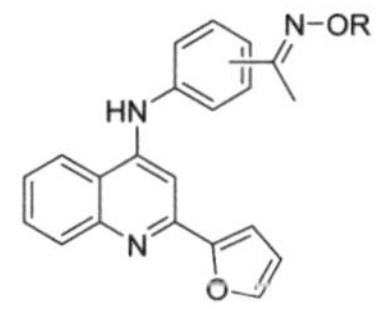

R = H, CH_3

34

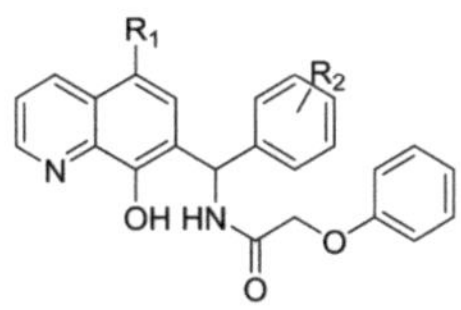

R_1 = H, F, NO_2, Cl
R_2 = H, 4-Cl, 4-CH_3, 4-OCH_3

35

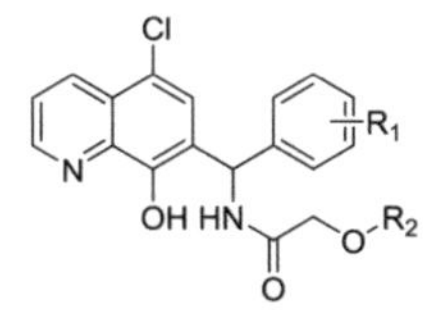

R_1 = H, 3-NO_2, 4-OCH_3
R_2 = Nitrogen Heterocyclic

36

3.3 Analgesic activity

4-Substituted-7-trifluoromethylquinolines **37, 38** have been developed by Abadi et al., and these derivatives were found to possess excellent analgesic activity with nitric oxide releasing characteristics [32]. Gomtsyan et al. synthesized an analgesic active derivative, **39**. The activity is due to its antagonism at vanilloid receptors [33]. Some quinoline derivatives, **40** were synthesized by Manera et al. that show analgesic activity and are selective agonists at cannabinoid CB_2 receptors [34].

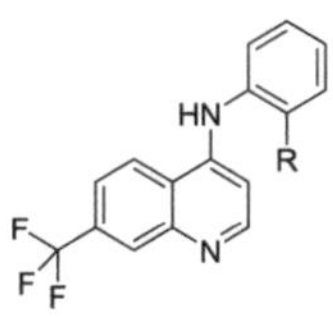

37

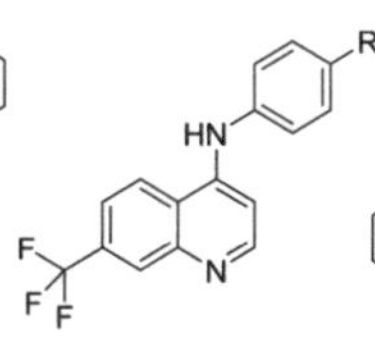

38

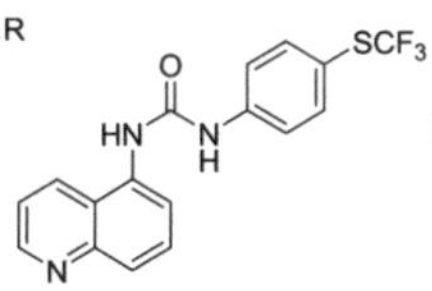

39

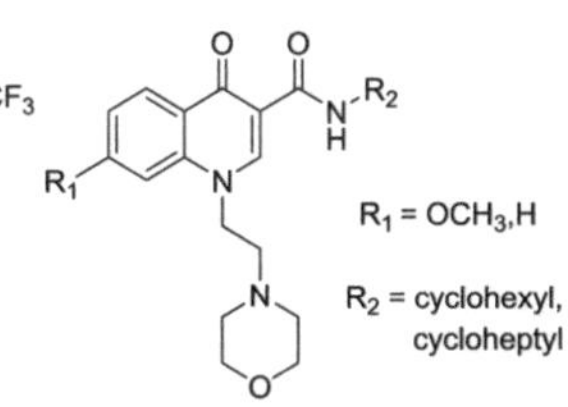

40

3.4 Antibacterial

Ma et al. reported the synthesis and antibacterial evaluation of phenoxy-, phenylthio-, and benzyloxy-substituted quinolones, **41** [35]. A few 8-substituted quinoline carboxylic acids, **42** were synthesized by Sanchez et al. that showed antibacterial activity [36]. Upadhayaya et al. developed 3-benzyl-6-bromo-2-methoxy quinoline derivatives, **43,** and these derivatives are active against *Mycobacterium tuberculosis* H37Rv strain [37]. A few analogues of 7-chloro quinolones, **44** were synthesized by De Souza et al., and these derivatives were found to be effective against multidrug-resistant tuberculosis [38]. Lilienkampf et al. synthesized quinoline-based compounds containing an isoxazole unit and side chain, **45** that was active against *Mycobacterium tuberculosis* [39]. The novel hybrid conjugates of N-(7-chloroquinolin-4-yl) piperazine-1-carbothioamide and 1,3,5-triazine derivatives, **30** synthesized by Bhat et al. also showed excellent antibacterial activity against several Gram-positive and Gram-negative microorganisms [40].

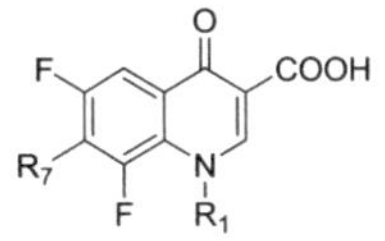

R_1 = Ethyl, Cyclopropyl, FCH_2CH_2
R_2 = Substituted phenyl

41

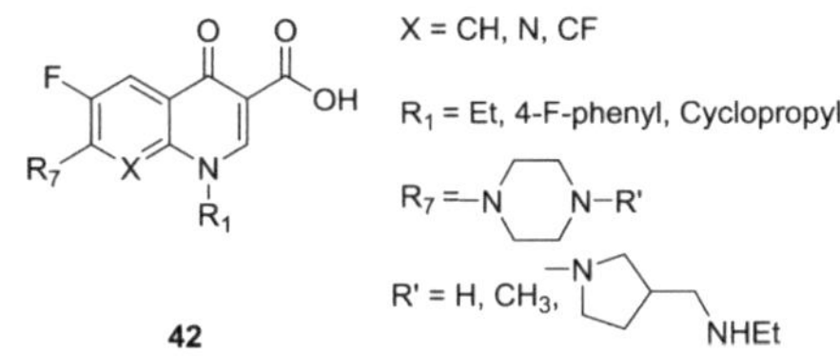

42

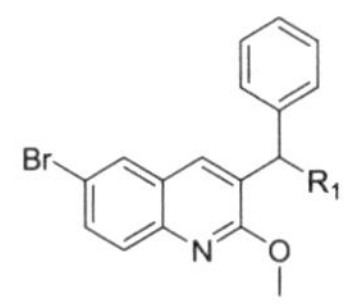

R_1 = Imidazolyl, Pyrazolyl, 1-(3-Trifluoromethyl-phenyl)-piperazinyl, 6-Amino-chromen-2-one

43

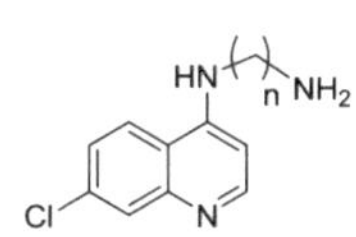

n = 8-10

44

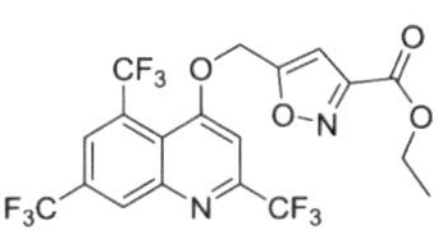

45

3.5 Antitumor

Some amido-anilinoquinolines, **46** were synthesized by Scott et al. that act as antitumor agents by inhibiting CSF-1R kinase [41]. Certain derivatives of 4-hydroxyquinolines, **47** were synthesized by Mai et al. that showed histone acetyltransferase (HAT) inhibitory activity [42]. A few 3-cyanoquinolines, **48** were developed by Miller et al. as inhibitors of growth factor receptors (IGF-1R) for treating cancer [43]. 4-Anilinoquinolines, **49** were synthesized by Assefa et al. which were found to contain tyrosine kinase inhibitors [44]. Quinoline carboxylic acids, **50** have been synthesized by Chen et al. that act as antitumor compounds by inhibiting insulin-like growth factors [45]. A few c-Met kinase inhibitory quinolones, **51** were developed by Wang et al. with IC_{50} < 1 nM. These derivatives were found to show the inhibition of c-Met phosphorylation in c-Met-dependent cell lines [46]. Marganakop et al. developed few 6,7,8-substituted thiosemicarbazones of 2-chloro-3-formyl-quinoline derivatives, **52** which exhibit excellent anticancer activities [47]. Recently, some quinoline derivatives, **53** were synthesized as novel Raf kinase inhibitors with potent and selective antitumor activities. These derivatives were synthesized by modifying the structure of sorafenib [48].

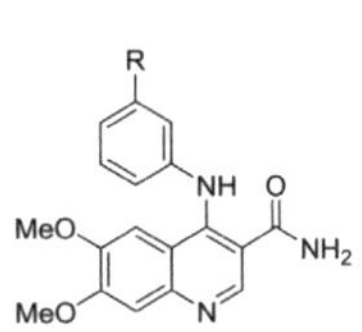

R = F, Cl, Br, CH_3

46

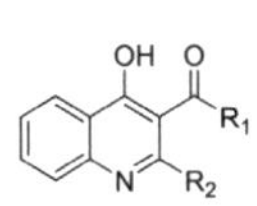

R_1 = OH, OEt
R_2 = CH_3, C_5H_{11}, $C_{10}H_{21}$, $C_{15}H_{31}$

47

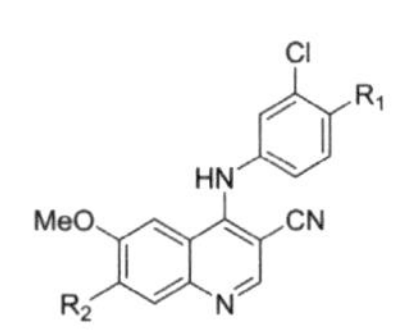

R_1 = Substituted-2-thioimidazole
R_2 = Substituted nitrogen heterocyclic

48

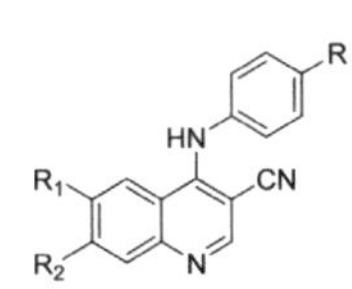

R = 3'-Br, 3'-Cl, 3'-CF_3, 3'-CN
R_1 = R_2 = OMe, OEt

49

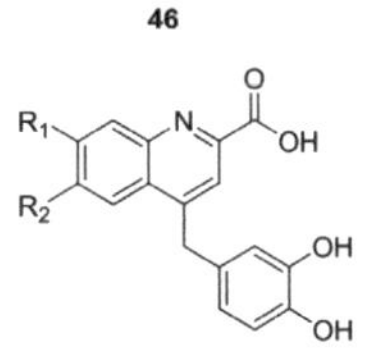

R_1 = OH, H, COOH, F, Cl, NH_2
R_2 = OH, OMe, COOH

50

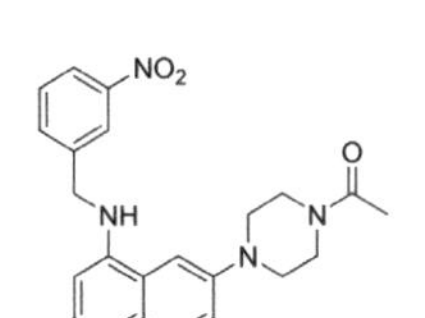

51

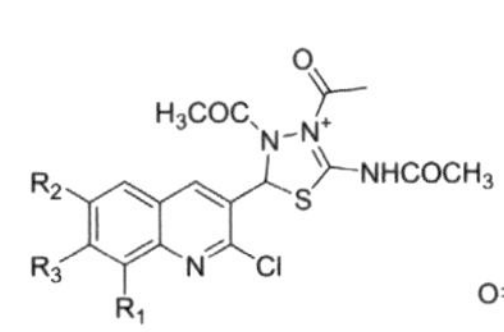

R_1 = R_2 = R_3 = H, CH_3, OCH_3

52

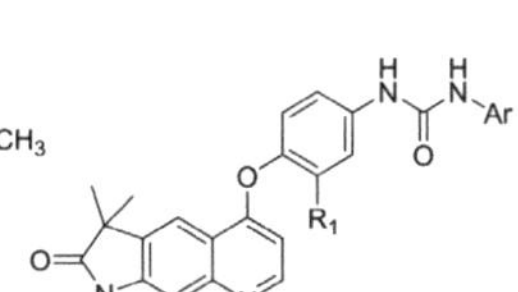

53

3.6 Antifungal

Certain tetrahydroquinolines, **54** were synthesized by Gholap et al. which were found to possess good antifungal activity against *Candida albicans*, *Fusarium oxysporum*, and *Mucor* fungi [49]. Kharkar et al. synthesized few quinoline derivatives, **55** that show good antifungal properties [50]. Kumar et al. developed few non-azole antimycotic agents having secondary amine attached 2-chloroquinolines, **56** and evaluated their antifungal activity against *Penicillium citrinum*, *Aspergillus niger*, *Monascus purpureus*, and *A. flavus* sp. [51].

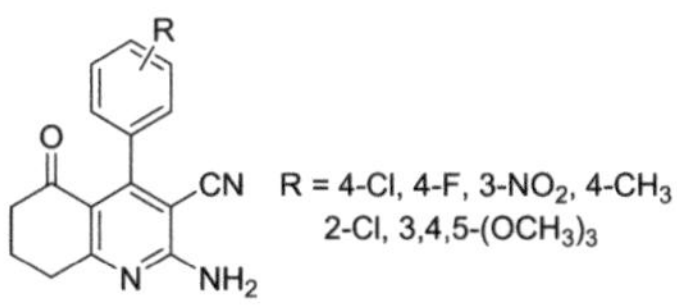

54

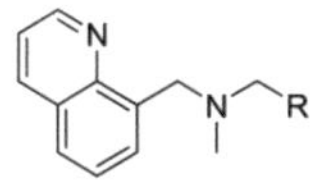

55

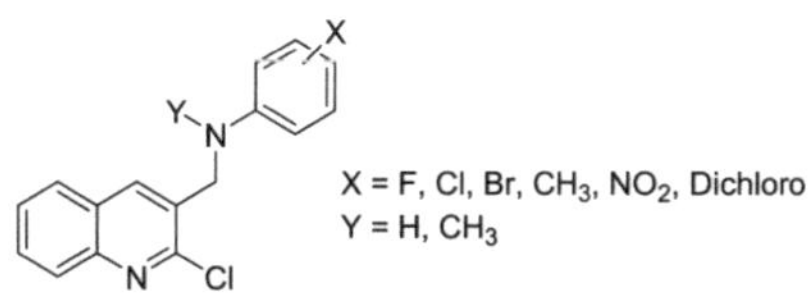

56

3.7 Antiviral

Several mono- and poly-substituted quinolones, **57–59** synthesized by Fakhfakh et al. were found to exhibit activity against HIV-1 [52]. Ghosh et al. synthesized anilidoquinoline derivatives, **60** which were found to possess an excellent antiviral activity against Japanese encephalitis virus [53]. A few quinoline derivatives, **61** possessing the behavior as HIV-1 Tat-TAR interaction inhibitors were synthesized by Chen et al. [45]. Massari et al. synthesized few desfluoroquinolones, **62** for treating HIV infection [54].

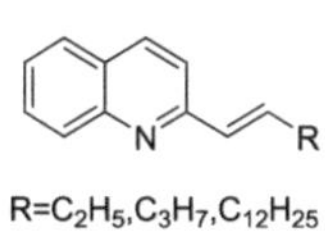

57

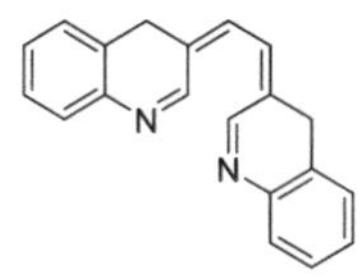

58

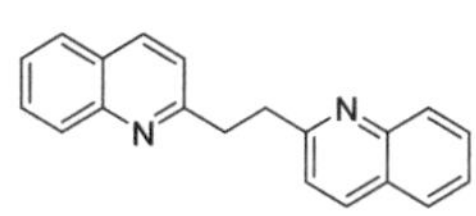

59

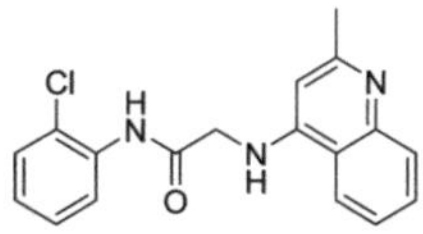

60

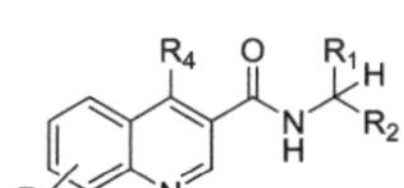

61

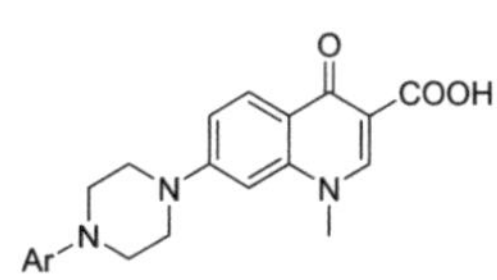

62

3.8 Anthelmintic

Substituted 2,4-arylquinolines, **63–66** have been synthesized by Rossiter et al. which possess good anthelmintic activity against levamisole-, ivermectin-, and thiabendazole-resistant strains of *H. contortus* [55].

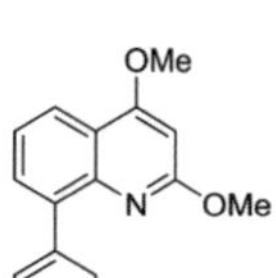

63

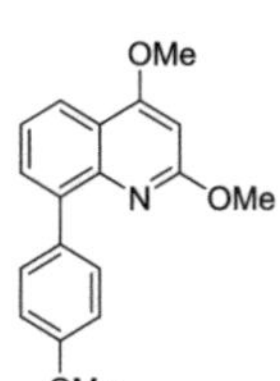

64

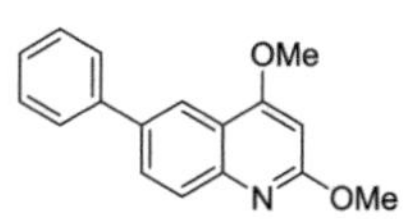

65

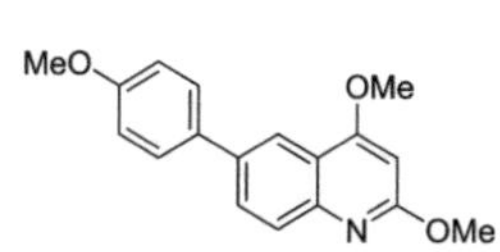

66

3.9 Antiprotozoal

2-Propyl quinoline and 2-(3-methyloxiran-2-yl)quinoline alkaloids **67, 68** isolated from *G. longiflora* plant were found to show antileishmanial activity against *Leishmania* spp. [56]. Alkenyl and alkynyl quinolones, **69, 70** reported by Fakhfakh et al. were found to have antiprotozoal activity against cutaneous leishmaniasis, African trypanosomiasis, Chagas disease, and visceral leishmaniasis [52]. Ma et al. developed a few quinolones, **71** that showed activity against *Trypanosoma cruzi* [35].

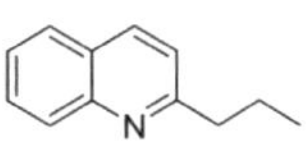

67

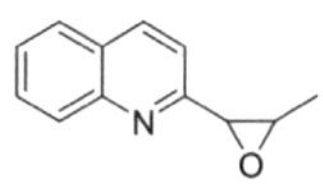

68

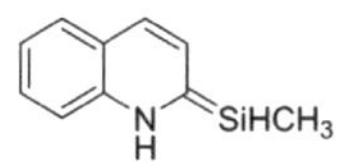

69

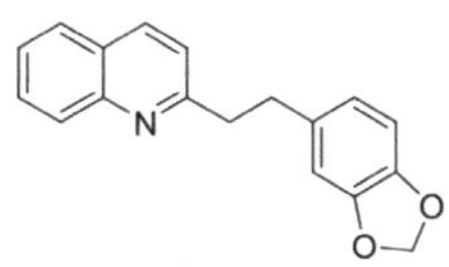

70

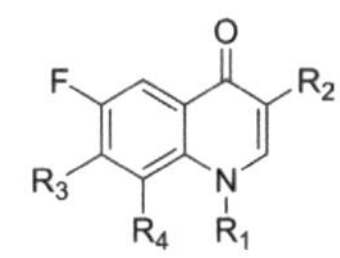

71

3.10 Cardiovascular activity

Srimal et al. demonstrated the hypotensive activity of centhaquin, **72**, and it was found to show the property of reducing the blood pressure in cat in a dose-dependent manner [57]. Quinoline-4-carboxylic acids, **73** have been synthesized by Lloyd et al. that are angiotensin II receptor antagonists and thereby act as hypotensive agents [58]. Certain biarylether amide quinolones, **74** have been developed by Bernotas et al. which act as liver X receptor agonists and are useful in the situation of dyslipidemia [59]. Phenyl acetic acid-based quinolones, **75** have been developed by Hu et al. which act as agonists at liver X receptors and found to have good binding affinity for LXRb and LXRa receptors [60]. A few 4-thiophenyl quinolones, **76** have been developed by Cai et al. that are HMG-CoA reductase inhibitors and useful as hypocholesterolemic agents [61]. Tetrahydroquinolines, **77** which inhibit the cholesteryl ester transfer protein have been synthesized by Rano et al. [62]. Certain tetrahydroquinolinamines, **78** have been developed by Ramos et al. which are found to inhibit platelet aggregation [63].

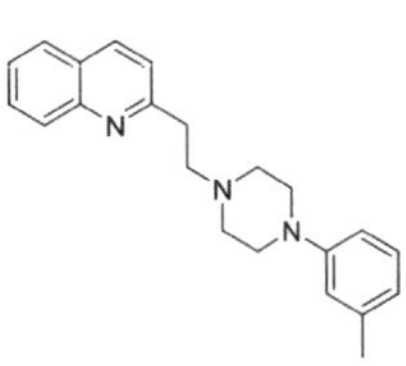

72

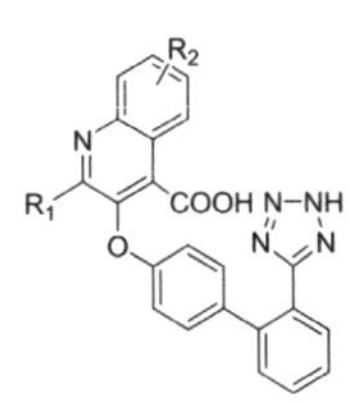

R_1 = Pr; R_2 = H

73

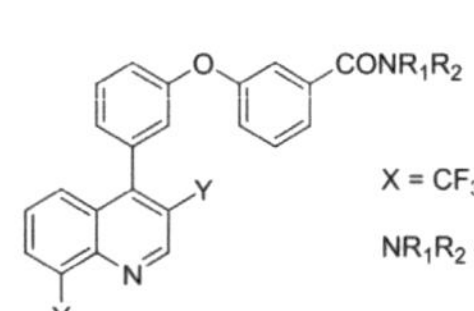

X = CF_3, Cl; Y = CH_2Ph

NR_1R_2 = Methyl ester, Pyrrolidine, Piperidine, Morpholine

74

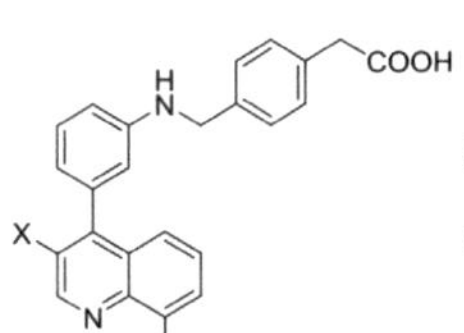

X = CH_2Ph, COPh, CN, $CONH_2$

Y = CF_3, CH_3, Cl

75

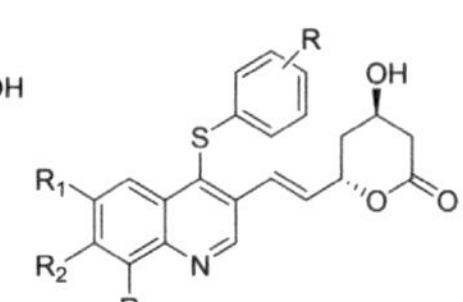

R = 4-$CH(CH_3)_2$, 4-F, 3-OCH_3; R_1 = H, F

R_2 = H, F, Cl or Substituted thiophenyl group

R_3 = H, F or Substituted thiophenyl group

76

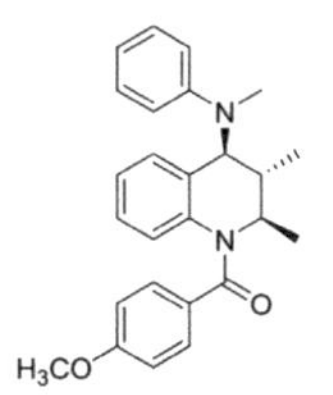

77

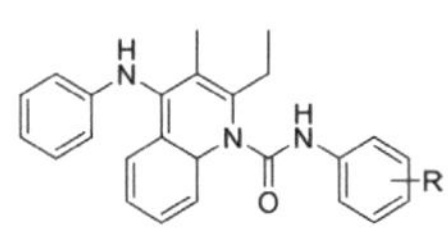

R = 3-Cl, 3-Br, 3-OCH_3

78

3.11 Reproductive system

Tetrahydroquinolines, **79** have been synthesized by Wallace et al. that are selective estrogen receptor modulators [64]. Bi et al. developed few quinolones, **80** which are potent PDE5 inhibitors thus are useful in the treating erectile dysfunction [65].

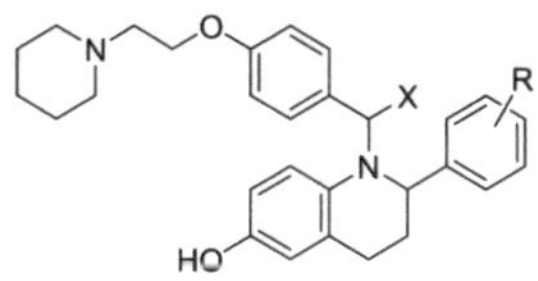

R = H, 3-OH, 4-OH; X = CH_2,O

79

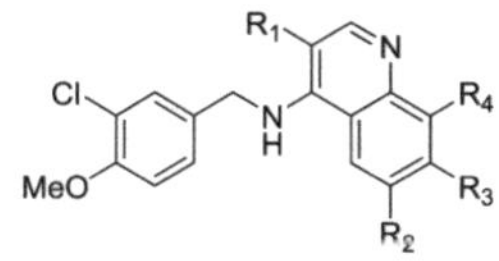

R_1 = COOEt; R_2 = H, CN; R_3 = H, CF_3; R_4 = H, Et

80

3.12 Miscellaneous

Quinolines and quinoline derivatives possess a number of miscellaneous biological activities also. Evans et al. synthesized few quinolones, **81** that are leukotriene synthesis inhibitors [66]. 1,2,3,4-Tetrahydroquinoline-2,2,4-trione oximes, **82** are developed by Cai et al. that act as antagonists of NDMA in glycine receptors and also found to be used as agents against neurodegenerative diseases (e.g., Alzheimer's disease) [61]. Lunniss et al. developed few selective PDE4 inhibitor quinolones **83, 84** which are useful in chronic obstructive pulmonary disorder [67]. Bachiller et al. have developed few tacrine–8-hydroxyquinoline hybrids, **85** that show activity against Alzheimer's [68]. Tetrahydroquinoline-6-yloxy propanes, **86** have been developed by Shakya et al. which show the β-3 agonists [69]. Few aminoalkoxy-quinolines, **87** which act as somatostatin receptor subtype-2 agonists have been developed by Wolkenberg et al. which are useful in proliferative diabetic retinopathy and also found utility in exudative age-related macular degeneration [70].

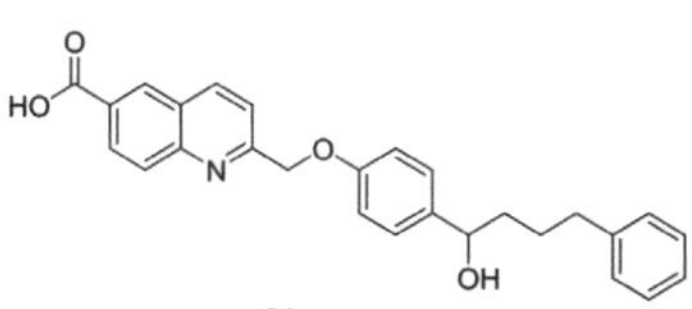

81

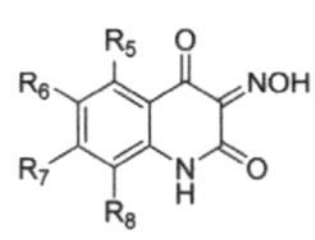

R = H, Cl, CH_3, F

82

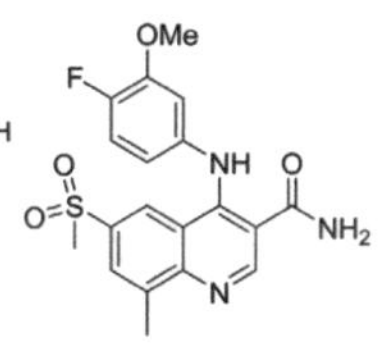

83

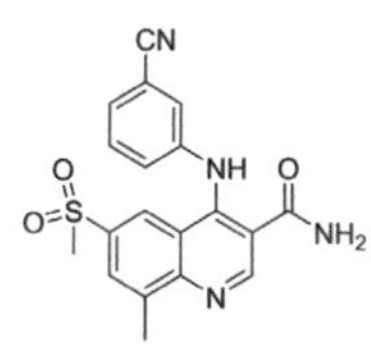

84

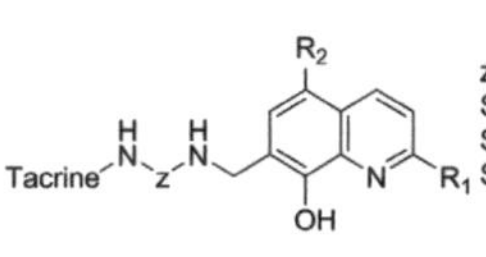

z = Alkyl chain
Str-I:R_1 = R_2 = H
Str-II:R_1 =CH_3; R_2 = H
Str-III:R_1 = H; R_2 = Cl

85

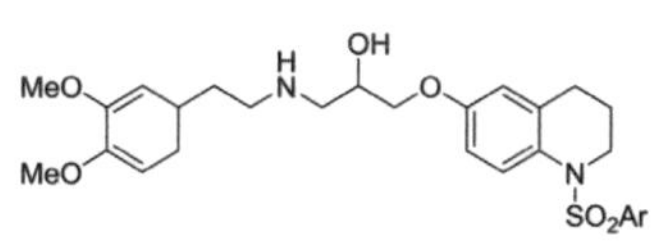

86

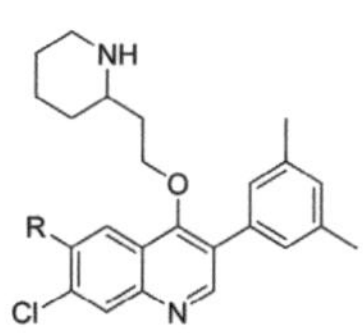

R = Aromatic ring

87

4. Conclusion

Since quinoline and its derivatives are known for their wide spectrum of pharmacological activities, a number of synthetic methods have been developed from time to time for their synthesis by conventional, homogeneous, and heterogeneous acid-catalyzed methods; rare-earth-catalyzed, transition metal-catalyzed, radical-catalyzed, microwave-assisted, ultrasound-promoted, or solvent-free conditions, and many more. This book chapter will be very useful to the researcher working in this field, and it would help them to develop new synthetic methods for the potent quinoline derivatives with good or enhanced biological activities for the future.

Acknowledgements

The author is grateful to DST-SERB for the Early Career Research Award grant (ECR/2016/001041). The author is thankful to SASTRA Deemed University, Thanjavur, for their encouragement and support.

Conflict of interest

The authors declare that there is no conflict of interest.

Author details

Yamajala B.R.D. Rajesh
School of Chemical and Biotechnology, SASTRA Deemed University, Thanjavur, India

*Address all correspondence to: ybrdrajesh@gmail.com

References

[1] Kouznetsov VV, Mendez LY, Gomez CM. Recent progress in the synthesis of quinolones. Current Organic Chemistry. 2005;**9**:141-161. DOI: 10.2174/1385272053369196

[2] Chen Y, Zhao Y, Lu C, Tzeng C, Wang JP. Synthesis, cytotoxicity, and antiinflammatory evaluation of 2-(furan-2-yl)-4-(phenoxy)quinoline derivatives. Bioorganic & Medicinal Chemistry. 2006;**14**:4373-4378. DOI: 10.1016/j.bmc.2006.02.039

[3] Sarma R, Prajapati D. Ionic liquid—An efficient recyclable system for the synthesis of 2,4-disubstituted quinolines via Meyer–Schuster rearrangement. Synlett. 2008;**19**:3001-3005. DOI: 10.1055/s-0028-1087340

[4] Lekhok KC, Prajapati D, Boruah RC. Indium(III) trifluoromethanesulfonate: An efficient reusable catalyst for the alkynylation–cyclization of 2-aminoaryl ketones and synthesis of 2,4-disubstituted quinolines. Synlett. 2008;**5**:655-658. DOI: 10.1055/s-2008-1042800

[5] Hegedus A, Hell Z, Vargadi T, Potor A, Gresits I. A new, simple synthesis of 1,2-dihydroquinolines via cyclocondensation using zeolite catalyst. Catalysis Letters. 2007;**117**:99-101. DOI: 10.1007/s10562-007-9127-4

[6] Zhou T, Lin J, Chen Z. A convenient synthesis of quinolones via ionic liquid-catalysed Friedlander annulation. Letters in Organic Chemistry. 2008;**5**:47-50. DOI: 10.2174/157017808783330261

[7] Ghassamipour S, Sardarian AR. Friedländer synthesis of poly-substituted quinolines in the presence of dodecylphosphonic acid (DPA) as a highly efficient, recyclable and novel catalyst in aqueous media and solvent-free conditions. Tetrahedron Letters. 2009;**50**:514-519. DOI: 10.1016/j.tetlet.2008.09.097

[8] Martinez R, Ramon DJ, Yus M. Transition metal free indirect Friedlander synthesis of quinolines from alcohols. The Journal of Organic Chemistry. 2008;**73**:9778-9780. DOI: 10.1021/jo801678n

[9] Horn J, Marsden SP, Nelson A, House D, Weingarten GG. Convergent, regiospecific synthesis of quinolines from o-aminophenylboronates. Organic Letters. 2008;**10**:4117-4120. DOI: 10.1021/ol8016726

[10] Zhou W, Zhang L, Jiao N. The tandem reaction combining radical and ionic processes: An efficient approach to substituted 3,4-dihydroquinolin-2-ones. Tetrahedron. 2009;**65**:1982-1987. DOI: 10.1016/j.tet.2009.01.027

[11] Wang LM, Hu L, Chen HJ, Sui YY, Shen W. One-pot synthesis of quinoline-4-carboxylic acid derivatives in water: Ytterbium perfluorooctanoate catalysed Doebner reaction. Journal of Fluorine Chemistry. 2009;**130**:406-409. DOI: 10.1016/j.jfluchem.2009.01.002

[12] Kouznetsov VV. Recent synthetic developments in a powerful imino Diels–Alder reaction (Povarov reaction): Application to the synthesis of N-polyheterocycles and related alkaloids. Tetrahedron. 2009;**65**:2721-2750. DOI: 10.1016/j.tet.2008.12.059

[13] Wang Y, Peng C, Liu L, Zhao J, Su L, Zhu Q. Sulfuric acid promoted condensation cyclization of 2-(2-(trimethylsilyl)ethynyl)anilines with arylaldehydes in alcoholic solvents: An efficient one-pot synthesis of 4-alkoxy-2-arylquinolines. Tetrahedron Letters. 2009;**50**:2261-2265. DOI: 10.1016/j.tetlet.2009.02.206

[14] Qi C, Zheng Q, Hua R. A domino three-component condensation of ortho-haloacetophenones with urea or amines: A novel one-pot synthesis of halogen-substituted quinolines. Tetrahedron. 2009;**65**:1316-1320. DOI: 10.1016/j.tet.2008.12.039

[15] Iraj MB, Shahram T, Majid M, Valiollah M, Salma A, Arsalan M. Microwave-promoted alkynylation-cyclization of 2-aminoaryl ketones: A green strategy for the synthesis of 2,4-disubstituted quinolines. Synlett. 2010;**20**:3104-3112. DOI: 10.1055/s-0030-1259065

[16] Kowsari E, Mallakmohammadi M. Ultrasound promoted synthesis of quinolines using basic ionic liquids in aqueous media as a green procedure. Ultrasonics Sonochemistry. 2011;**18**:447-454. DOI: 10.1016/j.ultsonch.2010.07.020

[17] Zhao J, Peng C, Liu L, Wang Y, Zhu Q. Synthesis of 2-alkoxy(aroxy)-3-substituted quinolines by DABCO promoted cyclization of o-alkynylaryl isocyanides. The Journal of Organic Chemistry. 2010;**75**:7502-7504. DOI: 10.1021/jo1017525

[18] Gao GL, Niu YN, Yan ZY, Wang HL, Wang GW, Shaukat A, et al. Unexpected domino reaction via Pd-catalysed sonogashira coupling of benzimidoyl chlorides with 1,6-enynes and cyclization to synthesize quinoline derivatives. The Journal of Organic Chemistry. 2010;**75**:1305-1308. DOI: 10.1021/jo9026116

[19] Huo Z, Gridnev ID, Yamamoto Y. A method for the synthesis of substituted quinolines via electrophilic cyclization of 1-azido-2-(2-propynyl)benzene. The Journal of Organic Chemistry. 2010;**75**:1266-1270. DOI: 10.1021/jo902603v

[20] Raynes K, Foley M, Tilley L, Deady LW. Novel bisquinoline antimalarials: Synthesis, antimalarial activity and inhibition of haem polymerisation. Biochemical Pharmacology. 1996;**52**:551-559. DOI: 10.1016/0006-2952(96)00306-1

[21] Chibale K, Moss JR, Blackie M, Schalkwyk D, Smith PJ. New amine and urea analogs of ferrochloroquine: Synthesis, antimalarial activity in vitro and electrochemical studies. Tetrahedron Letters. 2000;**41**:6231-6235. DOI: 10.1016/S0040-4039(00)01036-4

[22] Modapa S, Tusi Z, Sridhar D, Kumar A, Siddiqi MI, Srivastava K, et al. Search for new pharmacophores for antimalarial activity. Part I: Synthesis and antimalarial activity of new 2-methyl-6-ureido-4-quinolinamides. Bioorganic & Medicinal Chemistry. 2009;**17**:203-221. DOI: 10.1016/j.bmc.2008.11.021

[23] Mahajan A, Yeh S, Nell M, Rensburg CEJ, Chibale K. Synthesis of new 7-chloroquinolinyl thioureas and their biological investigation as potential anti-malarial and anticancer agents. Bioorganic & Medicinal Chemistry Letters. 2007;**17**:5683-5685. DOI: 10.1016/j.bmcl.2007.07.049

[24] Kovi KE, Yearick K, Iwaniuk DP, Natarajan JK, Alumasa J, de Dois AC, et al. Search for new pharmacophores for antimalarial activity. Part I: Synthesis and antimalarial activity of new 2-methyl-6-ureido-4-quinolinamides. Bioorganic & Medicinal Chemistry. 2009;**17**:270-283. DOI: 10.1016/j.bmc.2008.11.009

[25] Acharya BN, Thavaselvam D, Kaushik MB. Synthesis and antimalarial evaluation of novel pyridine quinoline hybrids. Medicinal Chemistry Research. 2008;**17**:487-494. DOI: 10.1007/s00044-008-9092-5

[26] Shiraki H, Kozar MP, Melendez V, Hudson TH, Ohrt C, Magill AJ, et al.

Antimalarial activity of novel 5-aryl-8-aminoquinoline derivatives. Journal of Medicinal Chemistry. 2011;**54**:131-142. DOI: 10.1021/jm100911f

[27] Singh B, Chetia D, Puri SK, Srivastava K, Prakash A. Synthesis and in vitro and in vivo antimalarial activity of novel 4-anilinoquinoline Mannich base derivatives. Medicinal Chemistry Research. 2011;**20**:1523-1529. DOI: 10.1007/s00044-010-9397-z

[28] Bhat HR, Singh UP, Gahtori P, Ghosh SK, Gogoi K, Prakash A, et al. Antimalarial activity and docking studies of novel bi-functional hybrids derived from 4-aminoquinolineand 1,3,5-triazine against wild and mutant malaria parasites as *pf*-DHFR inhibitor. RSC Advances. 2013;**3**:2942-2952. DOI: 10.1039/C2RA21915H

[29] McNulty J, Vemula R, Bordon C, Yolken R, Jones-Brando L. Synthesis and anti-toxoplasmosis activity of 4-arylquinoline-2-carboxylate derivatives. Organic & Biomolecular Chemistry. 2014;**12**:255-260. DOI: 10.1039/C3OB41539B

[30] Baba A, Kawamura N, Makino H, Ohta Y, Taketomi S, Sohda T. Studies on disease modifying antirheumatic drugs: Synthesis of novel quinoline and quinazoline derivatives and their anti-inflammatory effect. Journal of Medicinal Chemistry. 1996;**39**:5176-5182. DOI: 10.1021/jm9509408

[31] Gilbert AM, Bursavich MG, Lombardi S, Georgiadis KE, Reifenberg E, Flannery C, et al. N-((8-Hydroxy-5-substituted-quinolin-7-yl)(phenyl)methyl)-2-phenyloxy/amino-acetamide inhibitors of ADAMTS-5 (Aggrecanase-2). Bioorganic & Medicinal Chemistry Letters. 2008;**18**:6454-6457. DOI: 10.1016/j.bmcl.2008.10.065

[32] Abadi AH, Hegazy GH, Zaher AAE. Synthesis of novel 4-substituted-7-trifluoromethylquinoline derivatives with nitric oxide releasing properties and their evaluation as analgesic and anti-inflammatory agents. Bioorganic & Medicinal Chemistry. 2005;**13**:5759-5765. DOI: 10.1016/j.bmc.2005.05.053

[33] Gomtsyan A, Bayburt EK, Schmidt RG, Zheng GZ, Perner PJ, Didomenico S, et al. Novel transient receptor potential vanilloid 1 receptor antagonists for the treatment of pain: Structure–activity relationships for ureas with quinoline, isoquinoline, quinazoline, phthalazine, quinoxaline, and cinnoline moieties. Journal of Medicinal Chemistry. 2005;**48**:744-752. DOI: 10.1021/jm0492958

[34] Manera C, Cascio MG, Benetti V, Allara M, Tuccinardi T, Martinelli A, et al. New 1,8-naphthyridine and quinoline derivatives as CB2 selective agonists. Bioorganic & Medicinal Chemistry Letters. 2007;**17**:6505-6510. DOI: 10.1016/j.bmcl.2007.09.089

[35] Ma X, Zhou W, Brun R. Synthesis, in vitro antitrypanosomal and antibacterial activity of phenoxy, phenylthio or benzyloxy substituted quinolones. Bioorganic & Medicinal Chemistry Letters. 2009;**19**:986-989. DOI: 10.1016/j.bmcl.2008.11.078

[36] Sanchez JP, Domagala JM, Hagen SE, Heifetz CL, Hutt MP, Nichols JB, et al. Quinolone antibacterial agents. Synthesis and structure-activity relationships of 8-substituted quinoline-3-carboxylic acids and 1,8-naphthyridine-3-carboxylic acids. Journal of Medicinal Chemistry. 1988;**31**:983-991. DOI: 10.1021/jm00400a016

[37] Upadhayaya RS, Vandavasi JK, Vasireddy NR, Sharma V, Dixit SS, Chattopadhyaya J. Design, synthesis, biological evaluation and molecular modelling studies of novel quinoline derivatives against Mycobacterium

tuberculosis. Bioorganic & Medicinal Chemistry. 2009;**17**:2830-2841. DOI: 10.1016/j.bmc.2009.02.026

[38] de Souza MVN, Pais KC, Kaiser CR, Peralta MA, Ferreira MDL, Lourenco MCS. Synthesis and in vitro antitubercular activity of a series of quinoline derivatives. Bioorganic & Medicinal Chemistry. 2009;**17**:1474-1480. DOI: 10.1016/j.bmc.2009.01.013

[39] Lilienkampf A, Mao J, Wan B, Wang Y, Franzblau SG, Kozikowski AP. Structure activity relationships for a series of quinoline-based compounds active against replicating and nonreplicating mycobacterium tuberculosis. Journal of Medicinal Chemistry. 2009;**52**:2109-2118. DOI: 10.1021/jm900003c

[40] Bhat HR, Gupta SK, Singh UP. Discovery of potent, novel antibacterial hybrid conjugates from 4-aminoquinoline and 1,3,5-triazine: Design, synthesis and antibacterial evaluation. RSC Advances. 2012;**2**:12690-12695. DOI: 10.1039/C2RA22353H

[41] Scott DA, Balliet CL, Cook DJ, Davies AM, Gero TW, Omer CA, et al. Identification of 3-amido-4-anilinoquinolines as potent and selective inhibitors of CSF-1R kinase. Bioorganic & Medicinal Chemistry Letters. 2009;**19**:697-700. DOI: 10.1016/j.bmcl.2008.12.046

[42] Mai A, Rotili D, Tarantino D, Nebbioso A, Castellano S, Sbardella G, et al. Identification of 4-hydroxyquinolines inhibitors of p300/CBP histone acetyltransferases. Bioorganic & Medicinal Chemistry Letters. 2009;**19**:1132-1135. DOI: 10.1016/j.bmcl.2008.12.097

[43] Miller LM, Mayer SC, Berger DM, Boschelli DH, Boschelli F, Di L, et al. Lead identification to generate 3-cyanoquinoline inhibitors of insulin-like growth factor receptor (IGF-1R) for potential use in cancer treatment. Bioorganic & Medicinal Chemistry Letters. 2009;**19**:62-66. DOI: 10.1016/j.bmcl.2008.11.037

[44] Assefa H, Kamath S, Buolamwini JK. 3D-QSAR and docking studies on 4-anilinoquinazoline and 4-anilinoquinoline epidermal growth factor receptor (EGFR) tyrosine kinase inhibitors. Journal of Computer-Aided Molecular Design. 2003;**17**:475-493. DOI: 10.1023/B:JCAM.0000004622.13865.4f

[45] Chen S, Chen R, He M, Pang R, Tan Z, Yang M. Design, synthesis, and biological evaluation of novel quinoline derivatives as HIV-1 Tat–TAR interaction inhibitors. Bioorganic & Medicinal Chemistry. 2009;**17**:1948-1956. DOI: 10.1016/j.bmc.2009.01.038

[46] Wang Y, Ai J, Wang Y, Chen Y, Wang L, Liu G, et al. Synthesis and c-Met kinase inhibition of 3,5-disubstituted and 3,5,7-trisubstituted quinolines: Identification of 3-(4 acetylpiperazin-1-yl)-5-(3-nitrobenzylamino)-7 (trifluoromethyl) quinoline as a novel anticancer agent. Journal of Medicinal Chemistry. 2011;**54**:2127-2142. DOI: 10.1021/jm101340q

[47] Marganakop SB, Kamble RR, Taj T, Kariduraganvar MY. An efficient one-pot cyclization of quinoline thiosemicarbazones to quinolines derivatized with 1,3,4-thiadiazole as anticancer and anti-tubercular agents. Medicinal Chemistry Research. 2012;**21**:185-191. DOI: 10.1007/s00044-010-9522-z

[48] Li Y, Shi X, Xie N, Zhao Y, Li S. 3,3-Dimethyl-1H-pyrrolo[3,2-g]quinolin-2(3H)-one derivatives as novel Raf kinase inhibitors. Medicinal Chemistry Communications. 2013;**4**:367. DOI: 10.1039/C2MD20275A

[49] Gholap AR, Toti KS, Shirazi F, Kumari R, Bhat MK, Deshpande MV, et al. Synthesis and evaluation of antifungal properties of a series of the novel 2-amino-5-oxo-4-phenyl-5,6,7,8-tetrahydroquinoline-3-carbonitrile and its analogues. Bioorganic & Medicinal Chemistry. 2007;**15**:6705-6715. DOI: 10.1016/j.bmc.2007.08.009

[50] Kharkar PS, Deodhar MN, Kulkarni VM. Design, synthesis, antifungal activity, and ADME prediction of functional analogues of terbinafine. Medicinal Chemistry Research. 2009;**18**:421-432. DOI: 10.1007/s00044-008-9138-8

[51] Kumar S, Bawa S, Drabu S, Panda BP. Design and synthesis of 2-chloroquinoline derivatives as non-azoles antimycotic agents. Medicinal Chemistry Research. 2011;**20**:1340-1348. DOI: 10.1007/s00044-010-9463-6

[52] Fakhfakh MA, Fournet A, Prina E, Mouscadet JF, Franck X, Hocquemiller R, et al. Synthesis and biological evaluation of substituted quinolines: Potential treatment of protozoal and retroviral co-infections. Bioorganic & Medicinal Chemistry. 2003;**11**:5013-5023. DOI: 10.1016/j.bmc.2003.09.007

[53] Ghosh J, Swarup V, Saxena A, Das S, Hazra A, Paira P, et al. Therapeutic effect of a novel anilidoquinoline derivative, 2-(2-methyl-quinoline-4ylamino)-N-(2-chlorophenyl)-acetamide, in Japanese encephalitis: Correlation with in vitro neuroprotection. International Journal of Antimicrobial Agents. 2008;**32**:349-354. DOI: 10.1016/j.ijantimicag.2008.05.001

[54] Massari S, Daelemans D, Manfroni G, Sabatini S, Tabarrini O, Pannecouque C, et al. Studies on anti-HIV quinolones: New insights on the C-6 position. Bioorganic & Medicinal Chemistry. 2009;**17**:667-674. DOI: 10.1016/j.bmc.2008.11.056

[55] Rossiter S, Peron SJ, Whitfield PJ, Jones K. Synthesis and anthelmintic properties of arylquinolines with activity against drugresistant nematodes. Bioorganic & Medicinal Chemistry Letters. 2005;**15**:4806-4808. DOI: 10.1016/j.bmcl.2005.07.044

[56] Fournet A, Barrios AA, Munoz V, Hocquemiller R, Cave A, Bruneton J. 2-substituted quinolinc alkaloids as potential antileishmanial drugs. Antimicrobial Agents and Chemotherapy. 1993;**37**:859-863. DOI: 10.1128/AAC.37.4.859

[57] Srimal RC, Gulati K, Nityanand S, Dhawan BN. Pharmacological studies on 2-(2-(4-(3-methylphenyl)-1-piperazinyl) ethyl) quinoline (centhaquin). I. Hypotensive activity. Pharmacological Research. 1990;**22**:319-329. DOI: 10.1016/1043-6618(90)90729-W

[58] Lloyd J, Ryono DE, Bird JE, Buote J, Delaney CL, Dejneka T, et al. Quinoline-4 carboxylic acids as angiotensin II receptor antagonists. Bioorganic & Medicinal Chemistry Letters. 1994;**4**:195-200. DOI: 10.1016/S0960-894X(01)81146-X

[59] Bernotas RC, Singhaus RR, Kaufman DH, Ullrich J, Fletcher H, Quinet E, et al. Biarylether amide quinolines as liver X receptor agonists. Bioorganic & Medicinal Chemistry. 2009;**17**:1663-1670. DOI: 10.1016/j.bmc.2008.12.048

[60] Hu B, Jette J, Kaufman D, Singhaus R, Bernotas R, Unwalla R, et al. Further modification on phenyl acetic acid based quinolines as liver X receptor modulators. Bioorganic & Medicinal Chemistry. 2007;**15**:3321-3333. DOI: 10.1016/j.bmc.2007.03.013

[61] Cai Z, Zhou W, Sun L. Synthesis and HMG CoA reductase inhibition of 4-thiophenyl quinolines as potential hypocholesterolemic agents. Bioorganic & Medicinal Chemistry.

2007;**15**:7809-7829. DOI: 10.1016/j.bmc.2007.08.044

[62] Rano TA, McMaster ES, Pelton PD, Yang M, Demarest KT, Kuo GH. Design and synthesis of potent inhibitors of cholesteryl ester transfer protein (CETP) exploiting a 1,2,3,4-tetrahydroquinoline platform. Bioorganic & Medicinal Chemistry Letters. 2009;**19**:2456-2460. DOI: 10.1016/j.bmcl.2009.03.051

[63] Ramos AIM, Mecom JS, Kiesow TJ, Graybill TL, Brown GD, Aiyar NV, et al. Tetrahydro-4-quinolinamines identified as novel P2Y1 receptor antagonists. Bioorganic & Medicinal Chemistry Letters. 2008;**18**:6222-6226. DOI: 10.1016/j.bmcl.2008.09.102

[64] Wallace OB, Lauwers KS, Jones SA, Dodge JA. Tetrahydroquinoline based selective estrogen receptor modulators (SERMs). Bioorganic & Medicinal Chemistry Letters. 2003;**13**:1907-1910. DOI: 10.1016/S0960-894X(03)00306-8

[65] Bi Y, Stoy P, Adam L, He B, Krupinski J, Normandin D, et al. Quinolines as extremely potent and selective PDE5 inhibitors as potential agents for treatment of erectile dysfunction. Bioorganic & Medicinal Chemistry Letters. 2004;**14**:1577-1580. DOI: 10.1016/j.bmcl.2003.12.090

[66] Evans JF, Leveille C, Mancini JA, Prasit P, Therien M, Zamboni R, et al. 5-lipooxygenase-activating protein is the target of a quinoline class of leukotriene synthesis inhibitors. Molecular Pharmaceutics. 1991;**40**:22-27

[67] Lunniss CJ, Cooper AWJ, Eldred CD, Kranz M, Lindvall M, Lucas FS, et al. Quinolines as a novel structural class of potent and selective PDE4 inhibitors: Optimisation for oral administration. Bioorganic & Medicinal Chemistry Letters. 2009;**19**:1380-1385. DOI: 10.1016/j.bmcl.2009.01.045

[68] Bachiller MIF, Perez C, Munoz GCG, Conde S, Lopez MG, Villarroya M, et al. Novel tacrine–8-hydroxyquinoline hybrids as multifunctional agents for the treatment of Alzheimer's disease, with neuroprotective, cholinergic, antioxidant, and copper complexing properties. Journal of Medicinal Chemistry. 2010;**53**:4927-4937. DOI: 10.1021/jm100329q

[69] Shakya N, Roy KK, Saxena AK. Substituted 1,2,3,4-tetrahydroquinolin-6-yloxypropanes as b3-adrenergic receptor agonists: Design, synthesis, biological evaluation and pharmacophore modeling. Bioorganic & Medicinal Chemistry. 2009;**17**:830-847. DOI: 10.1016/j.bmc.2008.11.030

[70] Wolkenberg SE, Zhao Z, Thut C, Maxwell JW, McDonald TP, Kinose F, et al. Design, synthesis, and evaluation of novel 3,6-diaryl-4-aminoalkoxyquinolines as selective agonists of somatostatin receptor subtype 2. Journal of Medicinal Chemistry. 2011;**54**:2351-2358. DOI: 10.1021/jm101501b

Chapter 11

Potent Antibacterial Profile of 5-Oxo-Imidazolines in the New Millennium

Roshan D. Nasare, Mohammad Idrees, Satish S. Kola and Rajendra S. Dongre

Abstract

Pharmaceutics and therapeutics industries enforced chemists to seek/discover antibacterial novel heterocycles owing specific bioactivity and innate characteristics significance. This chapter summarized potent antibacterial profile of 5-oxo-imidazolines in the new millennium as an antibacterial against Gram-positive and Gram-negative bacteria viz. *B. thuringiensis*, *S. aureus*, *E. coli*, and *E. aerogenes* is presented in this chapter. 5-(H/Br benzofuran-2-yl)-1-phenyl 1H-pyrazole-3-carbohydrazides are condensed with 4-(arylidene)-2 phenyloxazol-5(4H)-one in acetic acid at elevated temperature to yield product 5-(H/Br benzofuran-2-yl)-N-(4-arylidene-5-oxo-2-phenyl-4,5-dihydroimidazol-1-yl)-1-phenyl-1H-pyrazole-3-carboxamides. Different substrates like 4-(arylidene)-2-phenyloxazol-5(4H)-one allowed to react with benzaldehyde hippuric acid to yield 5-oxo-imidazolines/5-oxo-4,5-dihydroimidazole. All synthesized 5-oxo-imidazolines were characterized via elemental analysis and FT-IR, ^{1}H-NMR and mass spectra techniques. All 5-oxo-imidazolines assayed in vitro for inherent antimicrobial activity at different concentration against stated bacterial strains and compared with standard chloramphenicol. 5-Oxo-imidazolines (**3a** and **3c**) with 125 μg/mL concentration showed excellent antibacterial profile against Gram-positive bacteria, *B. thuringiensis*, while other derivatives at different concentrations showed moderate antibacterial activity against Gram-positive bacteria, *S. aureus* and *B. thuringiensis*. Gram-negative bacteria like *E. coli* and *E. aerogenes* are tested at higher concentration (1000, 500, and 125 μg/mL) and found good-to-moderate antibacterial activity. Tested products found non-active against *E. aerogenes* for 125, 61, and 31 μg/mL concentration also inactive at conc. 31 μg/mL against *E. coli*.

Keywords: antibacterial, Gram positive/negative, *B. thuringiensis*, *S. aureus*, *E. coli*, *E. aerogenes*, 5-oxo-imidazoline, azlactones, medicinal

1. Introduction

Imidazole is a planer five-member ring with molecular formula $\mathbf{C_3N_2H_4}$, containing three carbon atoms and two nitrogen atoms in 1 and 3 skeletal positions as depicted in **Figure 1**. This is an aromatic heterocyclic ring that's classified as a diazole family owing non-adjacent nitrogens in its skeleton.

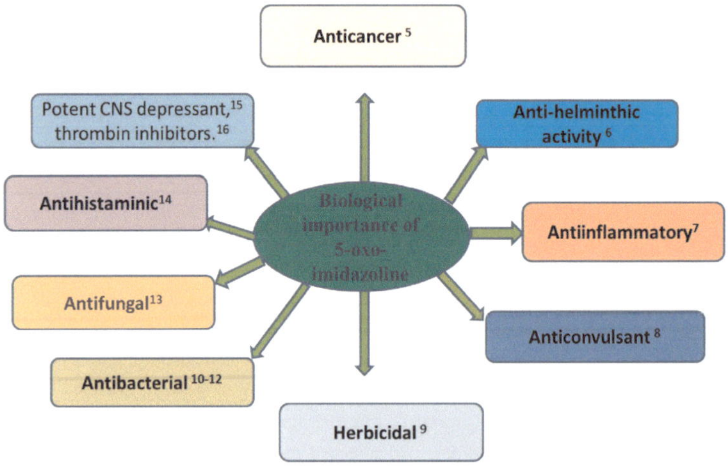

Figure 1.
Certain potent antibacterial profile of 5-oxo-imidazolines in the new millennium [5–16].

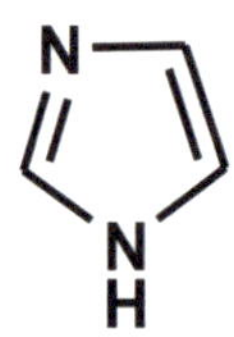

Imidazole
(1)

Assorted naturally occurring alkaloids own this imidazole moiety as vital biological building blocks viz; histidine and related hormone histamine. Various synthetic drugs are based on imidazole rings like antifungal, antibiotics: nitroimidazole and sedative: midazolam etc. Oxo-imidazoline derivatives are keto-dihydroimidazoles too, known as **imidazolinone** a five member ring system having 2-nitrogen situated at 1 and 3-positions and —C=O at various positions like 2, 4 and 5 of ring. Three possible isomers of imidazolinone observed based on position of C=O substituent at skeleton namely: 2-oxo-imidazoline **(2)**, 4-oxo-imidazoline **(3)** and 5-oxo-imidazoline **(4).**

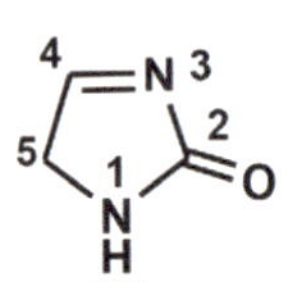
(2)

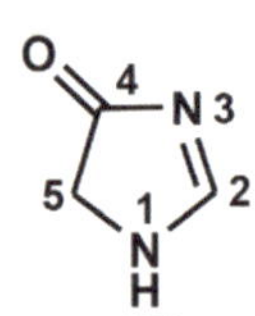
(3)

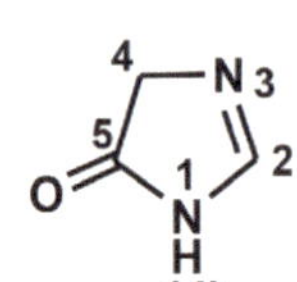
(4)

5-Oxo-4,5-dihydroimidazole derivative is called as 5-oxo-imidazoline, unsaturated system, in fact nitrogen analogues of azlactone/oxazolone can be converted into amino acids [1, 2] and also employed active pharmaceutical ingredient/API component in drugs [3]. 5-Oxo-imidazoline holds biological as well as chemical aspects for a long time; among the various heterocycles, it is preferred due to its wide antimicrobial profile. Certain imidazolines are useful intermediates in synthesis of many natural products as well as common building blocks in many biologically active moieties [4].

Biological importance of 5-oxo-imidazoline: Literature survey indicated that the synthetic drugs/molecules incorporated with 5-oxo-imidazoline found to owe assorted biological/clinical significance and wide range of pharmacological activities as mention below:

Solankee et al. [17] synthesized some 5-imidazolinones (**5**) and evaluated as anticancer agent.

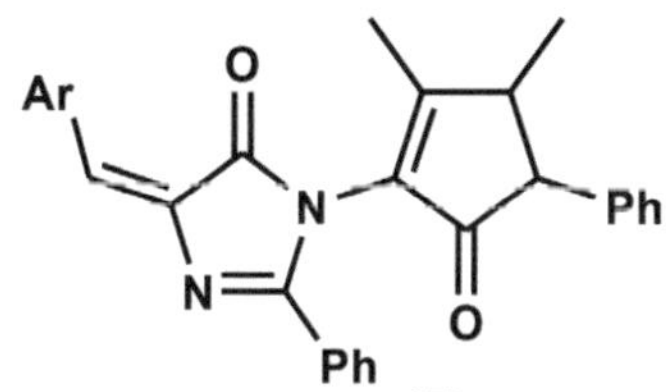

(5)

Mistry et al. [18] have synthesized imidazolinone (**6**) and studied antibacterial, antifungal activities.

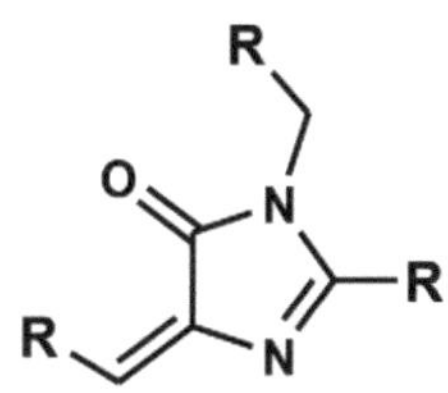

(6)

Kathrotiya et al. [19] and co-workers reported a series of some new quinoline based imidazole-5-one derivatives (**7**) and evaluated them as antibacterial and antifungal agent.

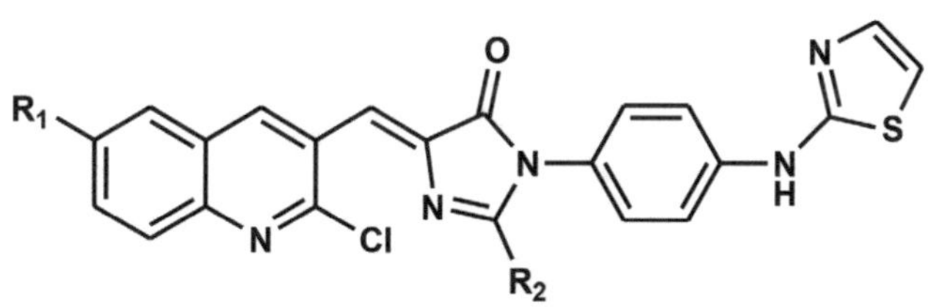

(7)

Desai et al. [20] also reported the synthesis of 5-oxo-imidazole amides derivatives including quinoline unit (**8**) and assessed their antibacterial and antifungal agent.

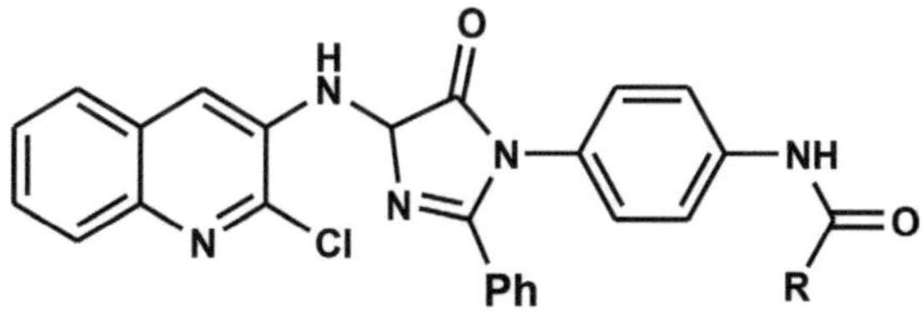

(8)

Mohammad and coworkers [21] have prepared some new imidazolinones and investigated their antimicrobial activities. Khan et al. [22] have also reported antibacterial and fungicidal activity of 5-oxo-imidazolines. Herbicidal activity of imidazolinone derivatives have been reported by Andreani et al. [23]. Moreover Zhou et al. [24] and Pai et al. [25] have reported anticancer active analogues of 5-oxo-imidazolines. Imidazolinone derivatives which possess antifungal activities have been reported by Shah et al. [26]. Some new 5-oxo-imidazolines as antimicrobial agents have been investigated by Patel et al. [27]. Rao [28] have prepared substituted imidazolone derivatives and reported their pharmaceutical use as inhibitors of p38 MAP Kinase and ERK-2 inhibitors. Xue et al. [29] have synthesized and evaluated imidazole-2-one derivatives as potential antitumor agents. Parekh and co-workers [30] have synthesized 5-oxo-imidazolines as novel bioactive compounds derived from benzimidazole. Kanjaria and co-workers [31] have described imidazolinones as potential antimicrobial agents. Joshi et al. [3] have synthesized imidazolinones as potent anticonvulsant agents. Acharya et al. [32] tested the imidazolinone (**9**) having quinolone nucleus for their antibacterial activity toward Gram-positive and Gram-negative bacteria and antifungal activity toward *Aspergillus niger* at a concentration of 40 μg, they found active against microorganism.

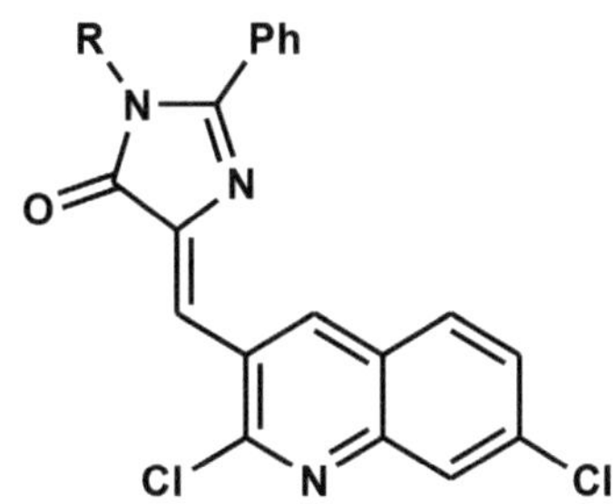

(9)

In view of potent antimicrobial and other pharmacological activities exhibited by 5-oxo-imidazolines, a variety of novel imidazolone analogs (**3a**-**g**) were synthesized by the condensation of different substituted oxazolines (**2a**-**g**) with hetero-aromatic amines (**1a**-**b**). All the synthesized compounds were screened for in vitro activities against a panel of Gram-positive and Gram-negative bacteria.

2. Materials and method

Melting points of all synthesized compounds were recorded in open capillary tube and are uncorrected. IR was recorded on a Shimadzu IR Spectrophotometer in KBr pellets. 1H-NMR recorded on a Bruker AM 400 model (400 MHz) using tetramethylsilane (TMS) as an internal reference and DMSO-d6 as solvent. Chemical shifts are given in parts per million (ppm). Positive-ion electrospray ionization (ESI) mass spectra were obtained with a Waters MicromassQ–TOF Micro, Mass Spectrophotometer. Elemental analysis was done on Vario EL III Elemental Analyzer, all compounds showed satisfactory elemental analysis. Reactions were monitored by E. Merck TLC aluminum sheet silica gel 60F254 and seen spot in UV light and iodine chamber.

3. Experimental

(I) **Synthesis of benzoyl glycine [33]:** A solution of glycine (0.33 mol) in 10% NaOH (250 mL) of was prepared and benzoyl chloride (45 mL, 0.385 mol) was added to the above solution in portions. The mixture was shaken vigorously after each addition until all the chlorides have been reacted. The mixture was cooled by adding few grams of crushed ice and was acidified by adding conc. HCl slowly with constant stirring. The resulting crystalline precipitate of benzoyl glycine was filtered and washed with cold water and dried. The solid was treated with hot CCl_4 in order to remove benzoic acid. The dried product was recrystallized with boiling water.

(II) **Synthesis of 4-(arylidene)-2-phenyloxazol-5(4*H*)-ones [33] (2a-g):** Benzoyl glycine (0.0476 mmol), aryl aldehydes (0.0476 mol), acetic anhydride (14 mL, 0.146 mmol) and anhydrous sodium acetate (0.0476 mmol) were placed in a 250 mL conical flask. It was heated on electric hot plate with constant shaking until the mixture liquefies completely. Then it was refluxed for 2 h on water bath. Then ethanol (10 mL) was added and mixture was allowed to stand overnight. The crystalline precipitate was filtered, washed with ice-cold alcohol and boiling water. The product was dried and recrystallized using benzene.

(III) **Preparation of 5-(5-H/Br benzofuran-2-yl)-1-phenyl-1*H*-pyrazole carbohydrazide (1a-b):** Synthesis of (5-(5-H/Br benzofuran-2-yl)-1-phenyl-1*H*-pyrazole-3-carbohydrazides (**1a-b**) were prepared in laboratory in quantitative yield according to reference method [34].

(IV) **General procedure for the synthesis of 5-(5-H/Br benzofuran-2-yl)-*N*-(4-arylidene-5-oxo-2-phenyl-4,5-dihydroimidazol-1-yl)-1-phenyl-1*H*-pyrazole-3 carboxamide (3a-g):** To a mixture of 4-benzylidene-2-phenyloxazol-5(4*H*)-one, **2a** (0.002 mol) and 5-(benzofuran-2-yl)-1-phenyl-1*H*-pyrazole-3-carbohydrazide **1a** (0.002 mol), acetic acid (20 mL) were added and the contents were refluxed for 9 h. Resulting mass was poured onto crushed ice, filtered and the product was recrystallized from ethanol to give **3a**.

Similarly, other 5-(bromobenzofuran-2-yl)-*N*-(4-arylidene-5-oxo-2-phenyl-4,5-dihydro imidazole-1-yl)-1-phenyl-1*H*-pyrazole-3-carboxamide **3b-g** were synthesized from **1b** and **2b-g** by extending the same procedure followed for **3a**.

- **Reaction scheme:**

See **Figures 2** and **3**.

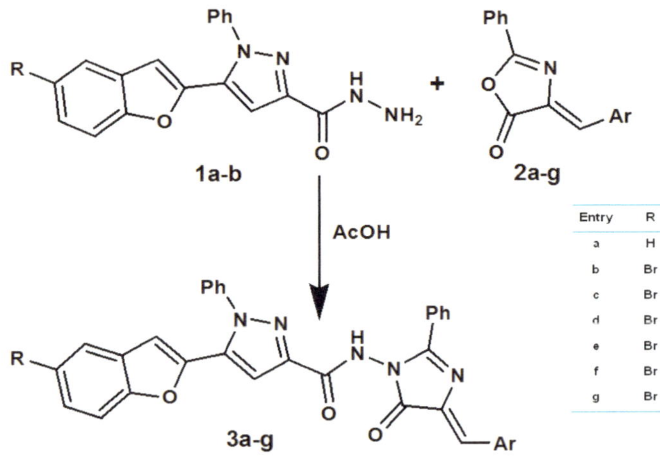

Entry	R	Ar
a	H	-C_6H_5
b	Br	4-OMe-C_6H_4
c	Br	2-Cl-C_6H_4
d	Br	1- $C_{10}H_7$
e	Br	4-$OCH_2C_6H_5$ -C_6H_4
f	Br	-CH=CH-C_6H_5
g	Br	2-Furfuryl

Figure 2.
Reaction scheme for 5-oxo-imidazoline derivatives.

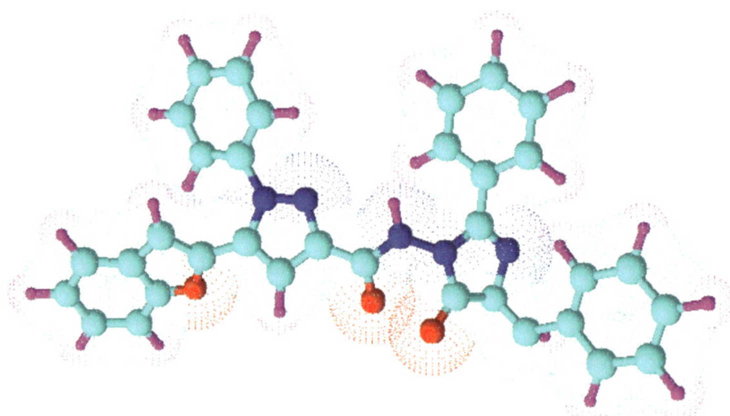

Figure 3.
*3D representation of 5-(benzofuran-2-yl)-*N*-(4-benzylidene-5-oxo-2-phenyl-4,5-dihydroimidazol-1-yl)-1-phenyl-1*H*-pyrazole-3-carboxamide (compound **3a**).*

4. Results and discussion

4.1 Spectral, elemental and physical data of synthesized compounds

5-(Benzofuran-2-yl)-*N*-(4-benzylidene-5-oxo-2-phenyl-4,5-dihydroimidazol-1-yl)-1-phenyl-1*H*-pyrazole-3-carboxamide (3a): Yellow crystalline solid; mp. 200–204°C; yield, 90%.

IR (KBr, ν max in cm^{-1}): 3197 (NH), 3062 (ArH), 1793, 1719 (C=O imidazole), 1597, 1525, 1496, 1448, (C=C), 1207, 1292, 1028 (C—O—C), 1164 (C—N—C stretch), 1640 (C=O in amide group), 1525 (C=N), 1110 (C—N).

^{1}H-NMR (DMSO-d_6): δ (ppm) 6.53 (s, 1H, C_4 of pyrazole ring), 11.65 (s, 1H, NH of amide group), 7.22–8.37 (m, 21H, ArH + benzofuran ring).

MS: *m/z*550 $[M+H]^+$, 551 $[M+2]^+$, 572 $[M+Na]^+$, 573 $[(M+H)+Na]^+$.

Elemental analysis: Calcd: for $C_{34}H_{23}N_5O_3$; calculated: C, 74.30; H, 4.22; N, 12.74; found: C, 74.16; H, 4.05; N, 12.37.

4-(4-Methoxybenzylidene)-5-oxo-2-phenyl-4,5-dihydroimidazol-1-yl)-5-(5-bromo benzofuran-2-yl)-1-phenyl-1*H*-pyrazole-3-carboxamide (3b): Yellow crystalline solid; recrystallization solvent, Ethanol; mp. 132–135°C; yield, 78%; IR (KBr, v max in cm^{-1}): 3315 (NH), 3063 (ArH), 1779, 1720 (C=O imidazole), 1502, 1438 (C=C), 1257, 998 (C—O—C), 1159 (C—N—C stretch), 1649 (C=O in amide group), 1595 (C=N), 1106 (C—N). Elemental anal. calcd: for $C_{35}H_{24}BrN_5O_4$; calculated: N, 10.64; found: N, 10.03.

4-(2-Chlorobenzylidene)-5-oxo-2-phenyl-4,5-dihydroimidazol-1-yl)-5-(5-bromobenzofuran-2-yl)-1-phenyl-1*H*-pyrazole-3-carboxamide (3c): Yellow crystalline solid; re-crystallization solvent, ethanol; mp. 155–158°C; yield, 82%; IR (KBr, v max in cm^{-1}): 3417 (NH), 1786, 1715 (C=O imidazole), 1501, 1433 (C—C), 1243, 1060 (C—O—C), 1155 (C—N—C stretch), 1643 (C=O in amide group), 1595 (C=N), 1106 (C—N). Elemental anal. calcd: for $C_{34}H_{21}BrClN_5O_3$; calculated: N, 10.56; found: N, 10.11.

5-(5-Bromobenzofuran-2-yl)-N-(4-(naphthalen-1-ylmethylene)-5-oxo-2-phenyl-4,5-dihydroimidazol-1-yl)-1-phenyl-1*H*-pyrazole-3-carboxamide (3d): Yellow crystalline solid; recrystallization solvent, ethanol; mp. 136–138°C; yield, 76%; IR (KBr, v max in cm^{-1}): 3378 (NH), 3005 (ArH), 1778, (C=O imidazole), 1489, 1431 (C=C), 1236, 1069 (C—O—C), 1151 (C—N—C stretch), 1689 (C=O in amide group), 1593 (C=N), 1151 (C—N). Elemental anal. calcd: for $C_{38}H_{24}BrN_5O_3$; calculated: N, 10.32; found: N, 10.40.

4-(4-(Benzyloxy)benzylidene)-5-oxo-2-phenyl-4,5-dihydroimidazol-1-yl)-5-(5 bromobenzofuran-2-yl)-1-phenyl-1*H*-pyrazole-3-carboxamide (3e): Yellow crystalline solid; recrystallization solvent, ethanol; mp. 155–157°C; yield, 80%; IR (KBr, v max in cm^{-1}): 3432 (NH), 3062, 2986 (ArH), 1786, 1716 (C=O imidazole), 1501, 1438 (C=C), 1249, 998 (C—O—C), 1160 (C—N—C stretch), 1642 (C=O in amide group), 1595 (C=N), 1110 (C—N). Elemental anal. calcd: for $C_{41}H_{28}BrN_5O_4$; calculated: N, 9.53; found: N, 9.07.

5-(5-Bromobenzofuran-2-yl)-N-(–5-oxo-2-phenyl-4-((E)-3-phenylallylidene)-4,5-dihydroimidazol-1-yl)-1-phenyl-1*H*-pyrazole-3-carboxamide (3f): Yellow crystalline solid; recrystallization solvent, ethanol; mp. 158–160°C; yield, 84%; IR (KBr, v max in cm^{-1}): 3342 (NH), 3034 (ArH), 1783 (C=O imidazole), 1493, 1439 (C=C), 1237, 1068 (C—O—C), 1158 (C—N—C stretch), 1627 (C=O in amide group), 1597 (C=N), 1105 (C—N). Elemental anal. calcd: for $C_{36}H_{24}BrN_5O_3$; calculated: N, 10.70; found: N, 10.25.

5-(5-Bromobenzofuran-2-yl)-N-(4-(furan-2-ylmethylene)-5-oxo-2-phenyl-4,5-dihydroimidazol-1-yl)-1-phenyl-1*H*-pyrazole-3-carboxamide(3g): Yellow crystalline solid; recrystallization solvent, ethanol; mp. 148–150°C; yield, 83%; IR (KBr, v max in cm^{-1}): 3431 (NH), 3062(ArH), 1783 (C=O imidazole), 1496, 1450 (C=C), 1231, 1008 (C—O—C), 1153 (C—N—C stretch), 1641 (C=O in amide group), 1525 (C=N in imidazole), 1079 (C—N). Elemental anal. calcd: for $C_{32}H_{20}BrN_5O_4$; calculated: N, 11.32; found: N, 10.96.

4.2 Common examination of the product

The newly synthesized compounds are soluble in following solvents which are listed in table also identification of newly synthesized compounds has been further confirmed by Lassaigne's test for nitrogen, all compound gives positive test. **Table 1** represents the structure of all derivatives along with solubility solvent and Lassaigne's test.

Sr. no.	Structure and compound code	Solubility	Lassaigne's test for nitrogen
1.	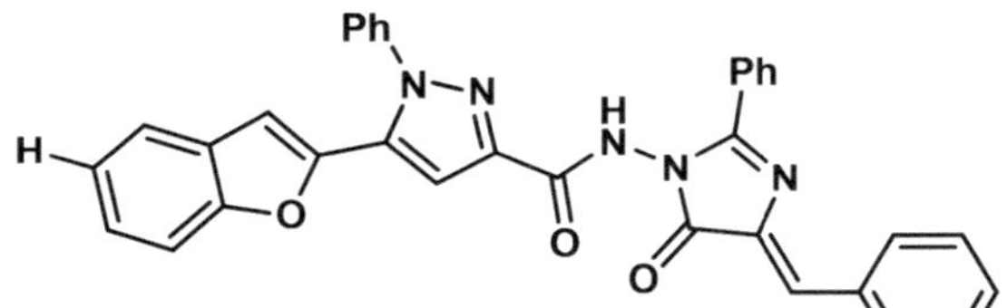 **3a**	1,4-Dioxane, DMSO, THF	Prussian blue coloration
2.	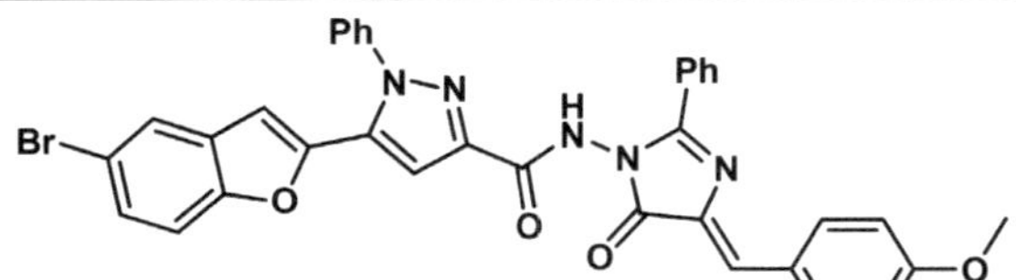 **3b**	1,4-Dioxane, DMSO, THF	Prussian blue coloration
3.	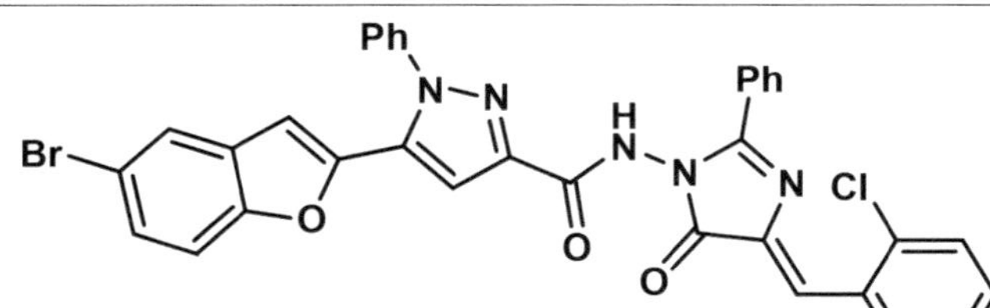 **3c**	1,4-Dioxane, DMSO, THF	Prussian blue coloration
4.	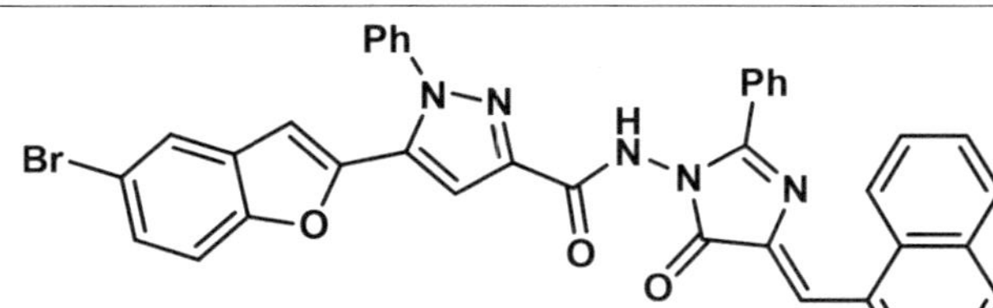 **3d**	1,4-Dioxane, DMSO, THF	Prussian blue coloration

DOI: http://dx.doi.org/10.5772/intechopen.81269

Sr. no.	Structure and compound code	Solubility	Lassaigne's test for nitrogen
5.	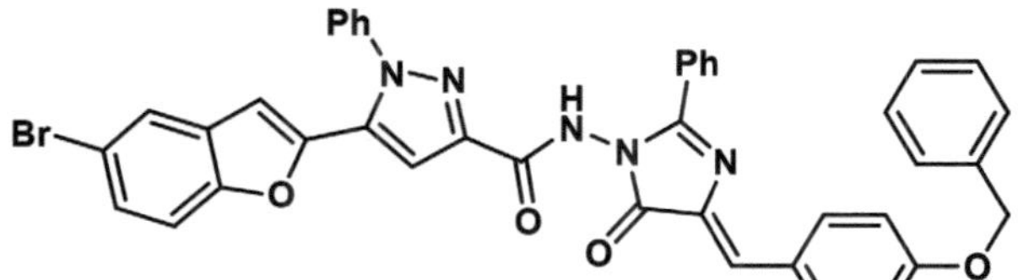 3e	1,4-Dioxane, DMSO, THF	Prussian blue coloration
6.	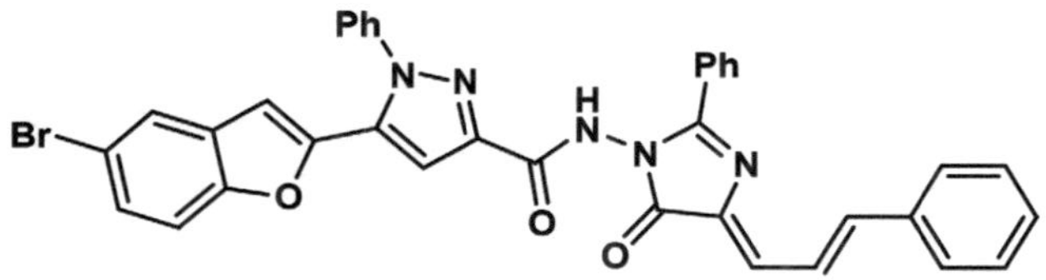 3f	1,4-Dioxane, DMSO, THF	Prussian blue coloration
7.	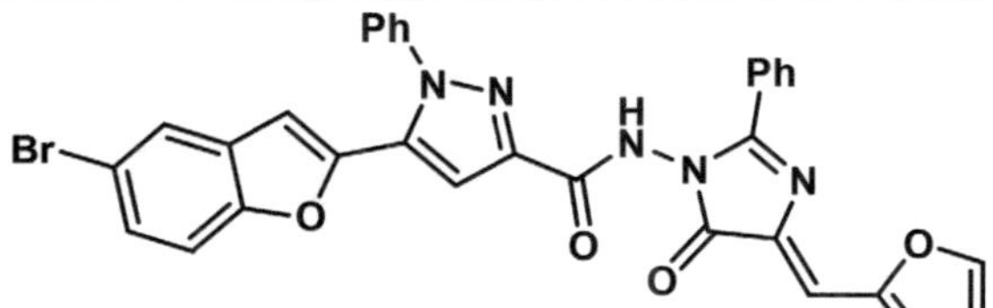 3g	1,4-Dioxane, DMSO, THF	Prussian blue coloration

Table 1.
Analysis characteristics.

4.3 Physico-chemical characterization

The synthesis of the novel compounds **3a-g** is described in the reaction schemes. Purity of the compounds was monitored by TLC technique. The structures of the newly synthesized compounds were confirmed using chemical transformation reaction, physical data, elemental analysis and different spectroscopic techniques such as IR, ^{1}H NMR and mass. The synthesis of the starting compound, 5-(5-H/Br benzofuran-2-yl)-1-phenyl-1*H*-pyrazole-3-carbohydrazides **(1a-b)** and 4-(arylidene)-2-phenyloxazol-5(4*H*)-ones **(2a-g)** achieved in quantitative yields according to the reference method. The reaction of **1a-b** with **2a-g** (4-(arylidene)-2-phenyloxazol-5(4*H*)-ones) in acetic acid solvent yields compounds **3a-g**.

IR spectrum of this **3a** showed absorption bands at 3197 cm^{-1} due to —NH stretching, disappearance of absorption band due to —NH_2 stretching and two absorption bands at 1719 and 1640 cm^{-1} for two carbonyl groups of imidazoline and aryl amide respectively indicated that 4-(arylidene)-2-phenyloxazol-5(4*H*)-ones has condensed with 5-(5-H/Br benzofuran-2-yl)-1-phenyl-1*H*-pyrazole-3-carbohydrazides to form **3a**. In addition, ^{1}H NMR spectrum of **3a** showed singlet at δ 10.65 ppm for —NH group and disappearance of signal due to —NH_2 group in the synthesized compound **3a** which is expected in carbohydrazide **1a** and also exhibited multiplet at δ 7.22–8.37 ppm due to 21 aromatic protons is in consistent with aromatic protons of **3a**. The % of elements in **3a** was C 74.16, H 4.05 and N 12.37, while its mass spectrum shows molecular ion peaks at *m/z* 550 $[M+H]^+$, 551 $[M+2]^+$, 572 $[M+Na]^+$, 573 $[(M+H)+Na].^+$ which is in good agreement with the proposed structure and molecular formula $C_{34}H_{23}N_5O_3$.

Similarly other imidazolinones (**3b-g**) were also identified on the basis of chemical transformation reaction, physical data, IR and elemental detection. IR spectra of each compound showed characteristics absorption bands for —NH stretching and disappearance of absorption band due to —NH_2 stretching, also showed corresponding band for carbonyl group. Elemental analysis was carried for nitrogen and sulfur of all compounds is found to be in good agreement with the calculated values.

4.4 Antimicrobial activity/profile

Antimicrobial activity means activity of any agent or drug against microbial organism. Microbial organism includes bacteria, viruses, fungi and protozoa. On the basis of their activity against specific microbial organism they termed as like antibacterial (against bacteria) that means they are capable to inhibit the growth of bacteria or to kill the bacteria. Other term is antifungal (against fungi), antiviral (against virus), antiprotozoal (against protozoa). Heterocyclic entities possess different antimicrobial activity. Activity changes by changing structural unit. It is very interesting thing to check out antimicrobial activity of newly synthesized compound. We carried out antibacterial activity of the novel compound (**Figure 4**).

4.5 Potent antibacterial/inhibition profile of 5-oxo-imidazolines (at different concentration) by agar disc-diffusion method

Test solutions were prepared with known weight of compound in DMSO and half diluted suitably to give the resultant concentration of 31–1000 μg/mL [35]. Whatman No. 1 sterile filter paper discs (6 mm) were impregnated with solution and allowed to dry at room temperature. *In-vitro* antibacterial activity was determined by using Mueller Hinton Agar obtained from Himedia Ltd., Mumbai. Petri plates were prepared by pouring 10 mL of Mueller Hinton Agar for bacteria

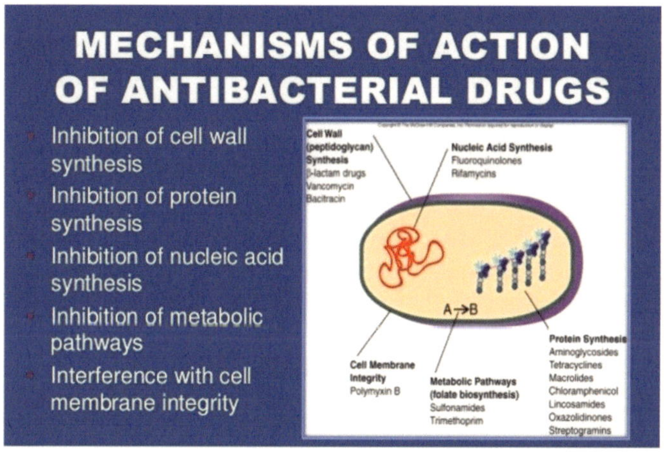

Figure 4.
Antibacterial drug mechanism in cell wall of microbe. Source: Google image.

containing microbial culture was allowed to solidify. The discs were then applied and the plates were incubated at 37°C for 24 h (bacteria) and the inhibition zone was measured as diameter in four directions and expressed as mean. The results were compared using chloramphenicol as a standard antibacterial agent. The results of antibacterial activity (i.e. zone of inhibition in mm) are given in the **Tables** 2 and 3.

4.6 Inhibition profile zone for Gram-positive bacterial strains of tested compound-3a-g

The synthesized compounds **3a-g** were screened for their *in vitro* antimicrobial activity using agar disc-diffusion method against two Gram-positive bacterial

	Zone of inhibition (mm)											
	Gram-positive bacteria											
	B. thuringiensis						*S. aureus*					
Compd. code	Conc. (μg/mL)						Conc. (μg/mL)					
	1000	500	250	125	63	31	1000	500	250	125	63	31
3a	15	14	12	**18**	12	10	14	12	16	12	**18**	15
3b	16	15	11	**16**	9	9	16	11	11	8	12	—
3c	14	15	12	**17**	11	8	16	14	16	11	14	13
3d	16	13	12	14	8	—	13	12	14	10	12	11
3e	13	11	13	12	11	—	14	13	13	10	11	10
3f	14	14	11	10	10	8	13	12	10	—	9	—
3 g	17	16	13	11	12	10	15	15	14	17	9	7
Std. drug	22	20	21	16	15	16	26	30	27	21	18	20

Standard drug: chloramphenicol.
Bold value indicates activity of tested compound is equal or high than standard drug.

Table 2.
Antibacterial activity of 3a-g.

strains such as *B. thuringiensis, S. aureus.* Chloramphenicol was used as standard drug for bacteria. According to antibacterial data obtained the test compounds **3a–c** at 125 µg/mL conc. showed excellent activity i.e. equal or higher than the standard drug and other derivatives viz. **3d–g** at 125 µg/mL conc. showed good inhibitory activity against *B. thuringiensis*. At conc. 1000, 500, and 250 µg/mL imidazolinone derivatives 3a-g showed good to moderate activity against *B. thuringiensis*, whereas **3d** and **3e** are found to be inactive at 31 µg/mL against Gram-positive bacteria, *B. thuringiensis*. In case of *S. aureus* **3a** exhibit with excellent activity at 63 µg/mL conc. While at 1000, 500, 250 µg/mL concentrations **3a-g** possesses good to moderate activity. Whereas **3b & 3f** are found to be inactive at 31 µg/mL also **3f** found to be inactive at 125 µg/mL against *S. aureus*. Obtained results of *in-vitro* antimicrobial activities of **3a-h** are summarized in **Table 2**.

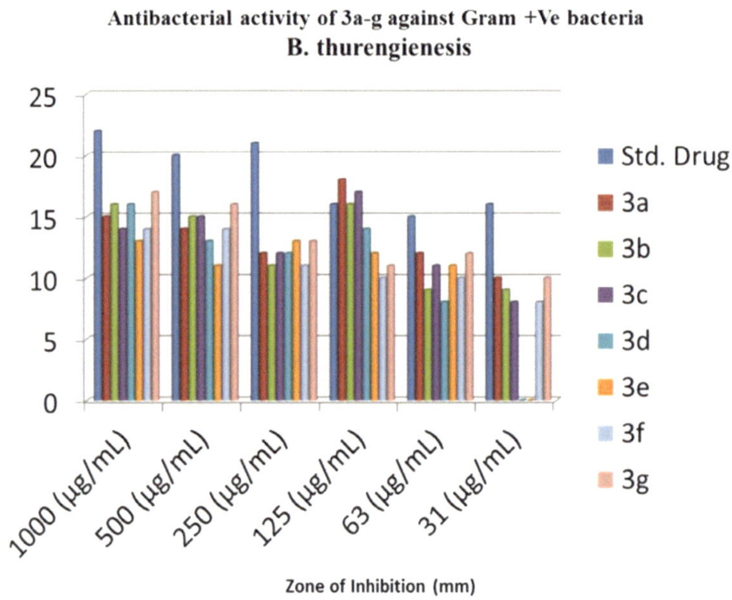

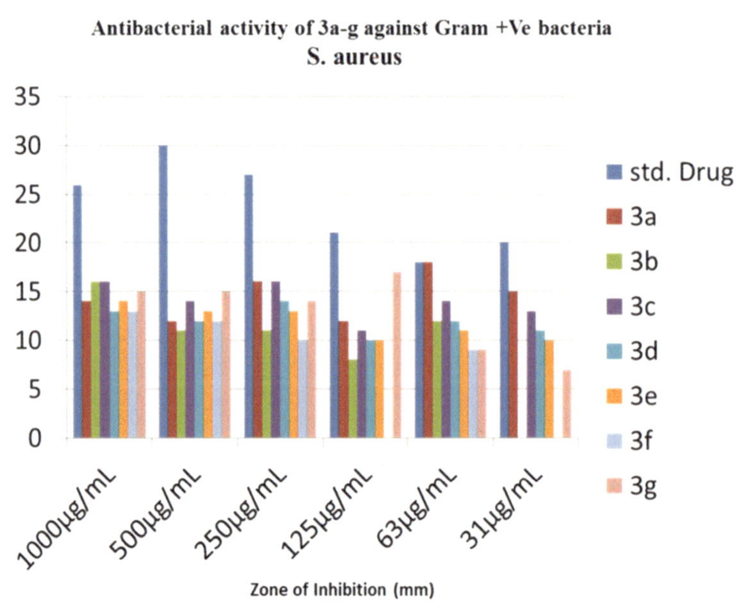

4.7 Inhibition profile zone for Gram-negative bacterial strains of tested compound-3a-g

The synthesized compounds **3a-g** were screened for their *in vitro* antimicrobial activity using agar disc-diffusion method against two Gram-negative bacterial strains such as *E. coli, E. aerogenes.* Chloramphenicol was used as standard drug for bacteria. According to antibacterial data obtained the test compounds **3a-g** possesses good to moderate activity at higher concentrations, i.e. 1000, 500, 250 and 125 μg/mL against Gram-negative bacteria *E. coli.* At conc. 63 μg/mL **3a-g** showed good activity while **3b** & **3e** found to be inactive against *E. coli*. At conc. 31 μg/mL **3d, 3f & 3g** showed moderate activity whereas **3a, 3b, 3c & 3e** found to be inactive against *E. coli*. In case of *E. aerogenes,* tested compounds showed moderate activity at higher concentrations while poor activity at lower concentrations. At conc. 125 μg/mL **3a, 3d, 3e** and at 63 μg/mL **3a, 3b, 3d, 3e, 3f** found to be inactive. All the compounds were inactive at a concentration of 31 μg/mL against *E. aerogenes.* Obtained results of *in-vitro* antimicrobial activities of synthesized 5-oxo-imidazolines (**3a-g**) are summarized in **Table 3**.

Compd. code	Zone of inhibition (mm)											
	Gram-negative bacteria											
	E. coli						*E. aerogenes*					
	Conc. (μg/mL)						Conc. (μg/mL)					
	1000	500	250	125	63	31	1000	500	250	125	63	31
3a	14	10	12	15	12	—	10	12	10	—	—	—
3b	11	11	9	12	—	—	12	11	13	8	—	—
3c	18	14	11	13	8	—	13	13	12	10	8	—
3d	13	12	10	12	10	8	10	9	10	—	—	—
3e	15	12	8	10	—	—	9	10	11	—	—	—
3f	14	10	12	13	9	9	11	12	8	9	—	—
3 g	15	13	12	14	10	8	12	10	9	7	8	—
Std. drug	**24**	**20**	**18**	**17**	**17**	**21**	**16**	**16**	**17**	**16**	**15**	**15**

Standard drug: chloramphenicol.
Bold value indicates activity of tested compound is equal or high than standard drug.

Table 3.
Antibacterial activity of 5-oxo-imidazoline compounds (3a-g).

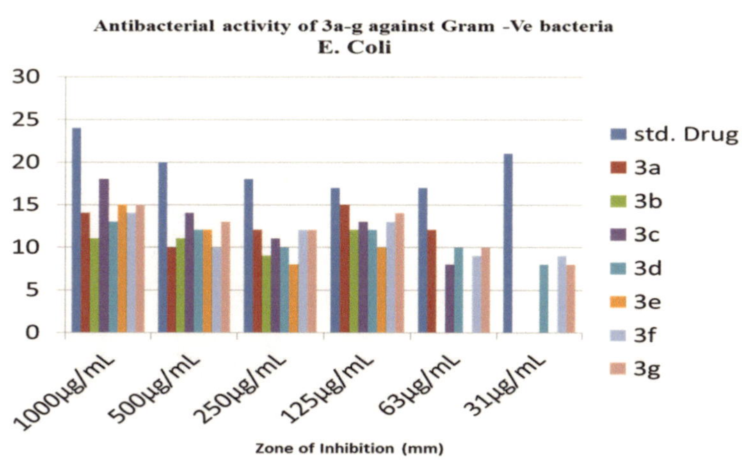

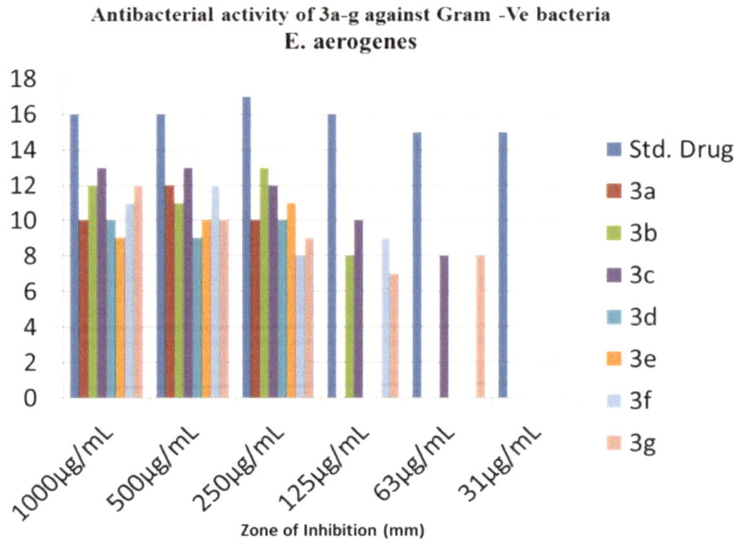

4.8 Mechanism of inhibition/prohibition

Gram-negative bacteria habitually owe low susceptibility as outer membrane of their cell wall not gets blocked/penetrated by drugs easily and factors like amount of peptidoglycan, receptors, and lipids availability, nature of cross-linking, autolytic enzymes activity greatly influence the bio-activity, permeation, and incorporation of the antibacterial drugs. 5-Oxo-imidazoles showed their specificity for polysaccharides, thats present in the outer membrane of many Gram-negative bacteria and so acted selectively toxic for series of *B. thuringiensis*, *S. aureus* bacteria. Mechanistically, once alliance with lipopolysaccharide substrate in outer membrane of *B. thuringiensis*, *S. aureus* bacteria, synthesized imidazolinone: potent antibacterial agent changes their membrane structure, thus enhances permeability and disruption of osmotic balance that ultimately results higher physiological effects.

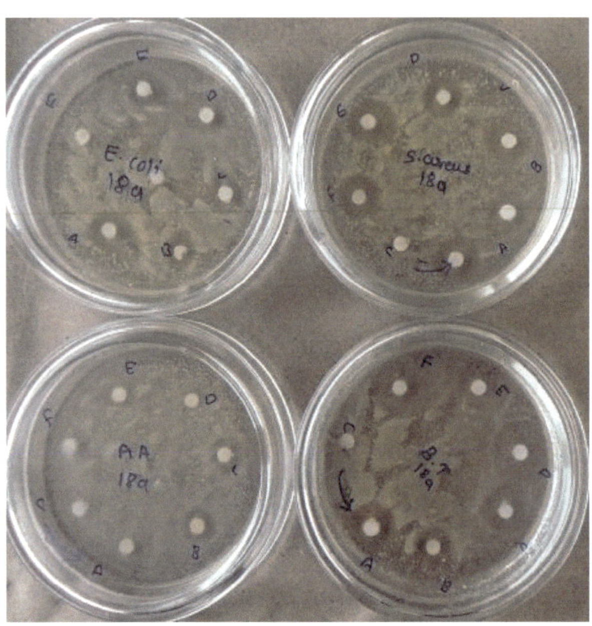

Figure 5.
Inhibition profile zone for four different bacterial strains of tested compound-3a (at different concentration).

Also, alteration like discharge of molecules from interior of *B. thuringiensis*, *S. aureus* bacterial cell inhibits respiration and increased water uptake may leads to cell death. Gram-positive bacteria own too thick cell wall and deny easy access of 5-oxo-imidazoles via their bacterial cell membrane, thus less effective on *E. coli* and *E. aerogenes* series of bacteria. The inhibition profile zone for four different bacterial strains of tested compound-**3a** (at different concentration) 5-oxo-imidazoline compounds are shown in **Figure 5**.

5. Conclusion

Assorted antibacterial agents own certain limitations viz.; resistance/potency, vast types and numbers own different structures besides slightly dissimilar pattern of activity which made it necessary to discover/explore the existing class and functions of almost all the antibacterial agents. Thus, futuristic pathogenic bacterial infections/diseases can be easily cured via promising antibacterial chemotherapeutic agents derive from 5-oxo-Imidazole skeleton. Pursued chapter described antibacterial resistance of 5-oxo-imidazoles mostly against Gram-positive series like *B. thuringiensis*, *S. aureus*. 5-Oxo-imidazoles can act onto simple one-celled bacterial organism that could kill, inhibit, or at least slower down their growth and ultimately can inhibit concern diseases/infections. This chapter focused on helping futuristic researchers, clinicians, and academicians involved in synthesizing and corresponding biological screening of innate activity of certain novel imidazolinone heterocycles. These synthesized 5-oxo-imidazoles restrain potent antibacterial activity may own prospective different therapeutic behavior if developed as advanced drug moiety. Therefore, chapter focus on the basis of chemical structure of 5-oxo-imidazoles. Gram-positive and Gram-negative bacteria showed varied response/susceptibility toward 5-oxo-imidazoles.

Targeted 5-oxo-imidazolines (**3a-g**) a class of imidazolinones are successfully synthesized in good yields and purity checked by physical, analytical and spectral data. Antibacterial screening of 5-oxo-imidazolines (**3a-g**) exhibited a potent bactericidal. Thus, 5-oxo-imidazolines could be powerfully stimulates major advances in remarkable significant chemotherapeutics in medicine, biology and pharmacy. Overall these imidazolinones disturb macromolecules like cytoplasmic membrane covering cytoplasm which acts selective barrier to control internal composition of cell. 5-Oxo-imidazoles in particular interrupted such functional roles of cytoplasmic membrane and ionic outflow that resulted cell destruction/death. Synthesized potent bioactive 5-oxo-imidazoles may open new possibilities in the successful treatment of several diseases due to promising antibacterial profile. So, ample scope exists in further research of imidazolinones especially innate selectivity of 5-oxo-imidazoles needs to carry out their chemotherapy as potent antibacterial aims to target cell membrane of range of Gram-negative bacteria as to derive novel drugs of new millennium.

Acknowledgements

The authors are thankful to The Principal, Government Science College, Gadchiroli and Dr. N. J. Siddiqui, for his support and cooperation. The authors are also thankful to Dr. S. D. Narkhede, Head, Department of Botany, GSC, Gadchiroli for permitting to carry out the antimicrobial activity, similarly the authors are also thankful to the Director, SAIF, Punjab University, Chandigarh for providing CHN analysis, IR, ^{1}HNMR and mass spectra.

Author details

Roshan D. Nasare[1]*, Mohammad Idrees[2], Satish S. Kola[3] and Rajendra S. Dongre[4]

1 Department of Chemistry, G. H. Raisoni University, Chhindwara, M.P., India

2 Department of Chemistry, Government Institute of Science, Nagpur, M.H., India

3 Department of Chemistry, Government Science College, Gadchiroli, M.H., India

4 Department of Chemistry, R.T.M., Nagpur University, Nagpur, M.H., India

*Address all correspondence to: nasare.roshan17@gmail.com

IntechOpen

References

[1] Kidwai AR, Dewasia GMV. A New Method for the Synthesis of Amino Acids. Synthesis of Amino Acids and Their Derivatives through 2, 4-Disubstituted 2-Imidazolin-5-ones. Journal of the American Chemical Society. 1962;**27**:4527

[2] Devasia M, Pillai C. An improved method for the synthesis of cylamino acids through 2,4- disubstituted 2-imidazolin-5-ones. Tetrahedron Letters. 1975;**6**:4051

[3] Joshi H, Upadhyay P, Karia D, Baxi AG. Synthesis of some novel imidazolinones as potent anticonvulsant agents. European Journal of Medicinal Chemistry. 2003;**38**:837

[4] Fox SW. Chemistry of the biologically important Imidazoles. Chemical Reviews. 1943;**32**:479

[5] El-Hady HA, Abubshait SA. Synthesis and anticancer evaluation of imidazolinone and benzoxazole derivatives. Research on Chemical Intermediates. 2015;**41**:1833

[6] Maneshwar T, Vijethal N, Balakrishna V, Kumar CV, Suresh M. Synthesis, characterization and anthelmintic activity of novel isatin substituted imidazoline derivatives. International Journal of Pharmaceutics. 2014;**4**:437

[7] Kalluraya B, Gunaga P, Banji D, Isloor AM. Synthesis and biological studies of some imidazolinone derivatives. Bollettino Chimico Farmaceutico. 2001;**140**:428

[8] Moorthy NS, Saxena V, Karthikeyan C, Trivedi P. Synthesis, in silico metabolic and toxicity prediction of some novel imidazolinones derivatives as potent anticonvulsant agents. Journal of Enzyme Inhibition and Medicinal Chemistry. 2012;**27**:201

[9] Mehta P, Davadra P, Shah N, Joshi H. Synthesis and antimicrobial activity of some new imidazolinone derivatives containing benzimidazole, International Letters of Chemistry, Physics and Astronomy. 2014;**29**:74

[10] Biplab D, Jayanta KG, Venkatapuram SS. Synthesis of some oxazolinones and imidazolinones and their antimicrobial screening. Acta Pharmaceutica. 2005;**55**:287

[11] Desai NC, Bhavsar AM, Baldaniya BB. Synthesis and antimicrobial activity of 5-imidazolinone derivatives. Indian Journal of Pharmaceutical Sciences. 2009;**71**:90

[12] Omar AM, Nada MA, Hamdi MH, Ahmad S, Abu SM. Synthesis and Biological Activity Evaluation of Some New Heterocyclic Spirocompounds with Imidazolinone and Pyrazoline Moieties. International Journal of Chemistry. 2011;**3**:20

[13] Mohd A, Arun K, Israr A, Khan SA. Synthesis of pharmaceutically important 1,3,4-Thiadiazole and imidazolinone derivatives as antimicrobial. Indian Journal of Chemistry. 2009;**48**:1288

[14] Sidney WF. Chemistry of the biologically important imidazoles. Chemical Reviews, 2002;**32**(1)

[15] Wadekar MP, Raut AR, Murhekar GH. Synthesis of some novel substituted 5-oxo imidazolines containing azo linkages and their biological screening. Der Pharma Chemica. 2010;**2**:76

[16] He L-W, Dai W-C, Li N-g. Development of orally active thrombin inhibitors for the treatment of thrombotic disorder diseases. Molecules. 2015;**20**:11046

[17] Solankee A, Kapadiya K, Upadhyay K, Patel J. Synthesis and Anticancer

Activity of Some 1-(2′,3′-Dimethyl-1′-Phenyl-3′-Pyrazoline-5′-One-4′-Yl)-2-Phenyl-4-(Substituted Benzylidine)-5-Imidazolinones. Oriental Journal of Chemistry. 2001;**17**:315

[18] Mistry RN, Desa KR. Studies on synthesis of some novel heterocyclic azlactone derivatives and imidazolinone derivatives and their antimicrobial activity. E-Journal of Chemistry. 2005;**2**:42

[19] Kathrotiya HG, Patel NA, Patel RG, Patel MP. An efficient synthesis of 3′-quinolinyl substituted imidazole-5-one derivatives catalyzed by zeolite and their antimicrobial activity. Chinese Chemical Letters. 2012;**23**:273

[20] Desai NC, Maheta AS, Rajpara KM, Joshi VV, Vaghani HV. Green synthesis of novel quinoline based imidazole derivatives and evaluation of their antimicrobi alactivity. Journal of Saudi Chemical Society. 2018;**14**:963

[21] Mohammad I, Bodkhe YG, Siddiqui NJ. Synthesis of some novel 5-imidazolones and its antimicrobial activity. International Research Journal of Pharmacy. 2018;**9**:85

[22] Khan KM, Mughal UR, Khan S, Khan S, Perveen S, Choudhary MI. Synthesis and antibacterial and antifungal activity of 5-substituted imidazolones. Letters in Drug Design and Discovery. 2009;**6**:69

[23] Andreani A, Locatelli AY, Rambaldi M. Synthesis and herbicidal properties of 3-(2-oxo-1-imidazolidinyliminomethyl) indoles; Journal of Heterocyclic Chemistry. 1995;**32**:49-51

[24] Zhou Y, Li Y, Zhu J, Lv J, Li Y, Zheng C, et al. Faming Zhuanli Shenqing Gongkai Shuoming shu. Patent CN 101691357 A 20100407; 2010

[25] Pai A, Singla RK, Joseph A, Kedar T, Thomas AT. Synthesis, in vitro and in vivo anticancer activity of substituted imidazolones. Pharmacology. 2009;**2**:933

[26] Shah MD, Desai NC, Awasthi KK, Saxena AK. Synthesis and QSAR studies of 5-imidazolinone derivatives as potential antibacterial agent. Indian Journal of Chemistry. 2001;**40**:201

[27] Patel PS, Shah RA, Trivedi DK, Vyas P. Synthesis and characterization of 4-{4-(2-methyl-4-benzylidene-5-oxo-imidazol-1-yl)phenyl}-6-substitutedphenyl)-5,6 dihydropyrimi din -2-one and study of their antimicrobial actvities. Journal of Organic Chemistry: An Indian Journal. 2010;**6**:66

[28] Rao SN. Tracking binding modes of 1,2,4-trisubstituted imidazolinone P38 MAP kinase and ERK-2 inhibitors, Journal of Molecular Graphics & Modeling. 2017;**76**:161

[29] Xue N, Yang X, Wu R, Chen J, He Q, Yang B, Lu X, Hu Y. Synthesis and biological evaluation of imidazol-2-one derivatives as potential antitumor agents. Bioorganic & Medicinal Chemistry. 2008;**16**:2550

[30] Kagthara P, Shah N, Doshi R, Parekh H. Synthesis of some arylamides, sulphonamides and 5-oxo-imidazolines as novel bioactive compounds derived from benzimidazole. Heterocyclic Communications. 1998;**4**:6

[31] Knjariya H, Verma D, Parekh H. Oriental Journal of Chemistry. 2002;**18**: 583

[32] Acharya HH. Discovering the new heterocycles of therapeutic interest [PhD thesis]. Saurashtra University; 2004

[33] Furniss BS, Hanna Ford AJ, Smith PWG, Tatchell AR. Vogel's Text Book of Practical Organic Chemistry. 5th ed. ELBS; 1989

[34] Siddiqui NJ, Idrees M, Khati N, Dhonde M. Use of transesterified 1,3-diketoesters in the synthesis of trisubstituted pyrazoles and their biological screening. Bulletin of the Chemical Society of Ethiopia. 2013;**27**:85

[35] Bauer AW, Kirby WM, Sherris JC, Turck M. Antibiotic susceptibility testing by a standardized single disk method. American Journal of Clinical Pathology. 1966;**36**:493

Chapter 12

Azoles as Potent Antimicrobial Agents

Rohit Singh and Swastika Ganguly

Abstract

Imidazole analogs have proved to be a very good source of medicinal agents. The various activities associated with these moieties include antibacterial, antifungal, anthelmintic, Anti HIV activity, anticancer, antihypertensive, analgesic, anti-inflammatory, anticonvulsant, sedative and other pharmacological activities.

Keywords: imidazole, antibacterial, antifungal and antiviral

1. Introduction

Azoles are basically five member heterocyclic compounds containing one or more different hetero atom out of which at least one must be nitrogen and other heterocyclic may be nitrogen or other than nitrogen like sulfur or oxygen.

Some of the common five member azoles which consist the nitrogen hetero atom only are as follows as imidazole (1), pyrazole (2), 1,2,3-triazole (3), 1,2,4-triazole (4), tetrazole (5) and pentazole (6).

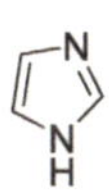
(1)
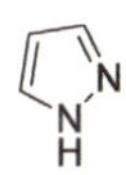
(2)
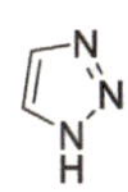
(3)
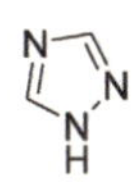
(4)
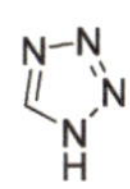
(5)
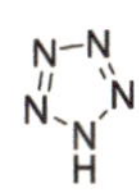
(6)

Some of the five member azoles which consist sulfur and oxygen as hetero atom along with nitrogen atom such as thiazole (7), isothiazole (8), 1,2,3-thiadiazole (9), 1,2,4-thiadiazole (10), 1,2,5-thiadiazole (11), 1,3,4-thiadiazole (12), oxazole (13), isoxazole (14), 1,2,3-oxadiazole (15), 1,2,4-oxadiazole (16), 1,2,5-oxadiazole (17) and 1,3,4-oxadiazole (18).

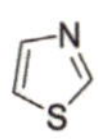
(7)

(8)
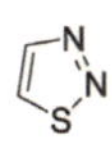
(9)
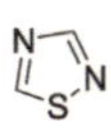
(10)
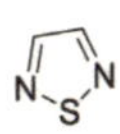
(11)
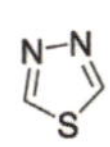
(12)

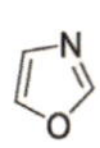
(13)
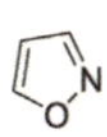
(14)
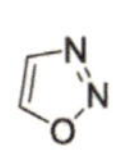
(15)
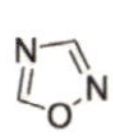
(16)
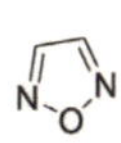
(17)
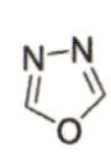
(18)

Some of the heterocyclic azoles are fused with benzene ring to form bicyclic azole derivatives such as benzimidazole (19), benzotriazole (20), benzothiazole (21) and benzoxazole (22).

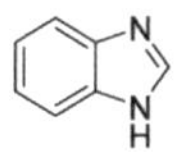
(19)
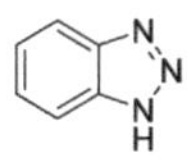
(20)
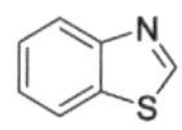
(21)
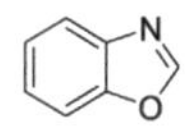
(22)

Among above mentioned class of azoles a brief review is presented focused on imidazoles and benzotriazoles as given below.

2. Imidazole analogs and their significance

In various oxidation states (23 and 24) imidazole has shown a number of interesting biological activities, like antiviral [1, 2], antibacterial [3] antifungal [4, 5], antiprotozoal [6, 7], antihypertensive [8, 9], antihistaminic [10], alpha-adrenergic agonist [11], alpha adrenergic blocking [12] and other activities [13, 14].

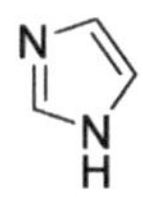
(23)
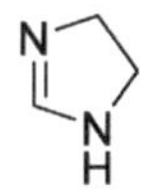
(24)

2.1 Antiprotozoals

Nitroimidazoles (25) with antitrichomonas activity were reported in year 1961 and then in 1966. Metronidazole (26) was among these compounds, it exhibited broad antiprotozoal activity and has found wide use in treating trichomoniasis orally.

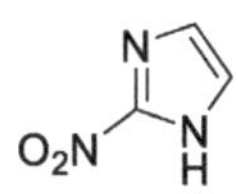
(25)
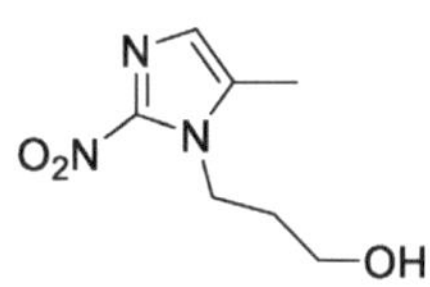
(26)

Structural variation of Metronidazole (26), mainly to improve trichomonacidal activity led to the discovery of Tinidazole (27), Nimorazole (28), and Panidazole (29). Tinadazole (27) is most potent towards *Entamoeba histolytica*, *in vitro*, cecal amoebiasis and hepatic amoebiasis in experimental animals [6–8].

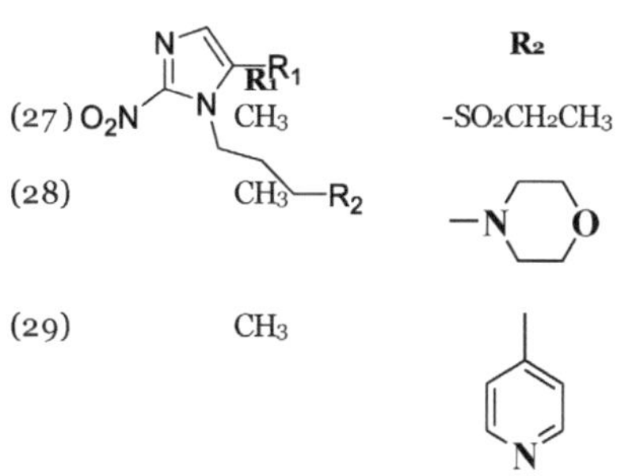

Giraldi et al. [15] in year 1967, synthesized a series of 1-aminoalkyl and 1-aminoalkyl-2-methyl-5(4)-nitro-4(5)-styrylimidazoles (30–37) and these compounds were tested to check the potency of synthesized compounds against various non-pathogenic bacteria and fungi.

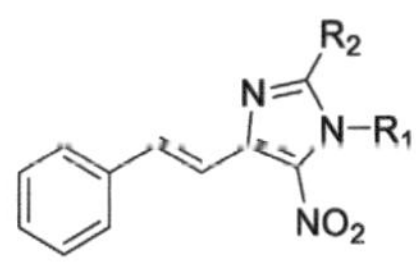

	R1	R2
(30)	$CH_2CH_2N(C_2H_5)_2$	H
(31)	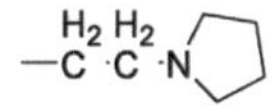	H
(32)	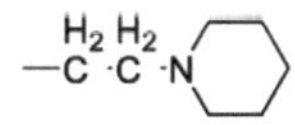	H
(33)	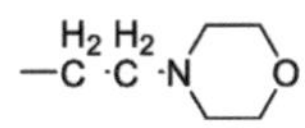	H
(34)	$CH_2CH_2N(C_2H_5)_2$	CH_3
(35)	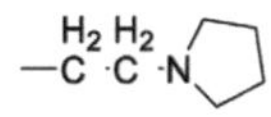	CH_3
(36)	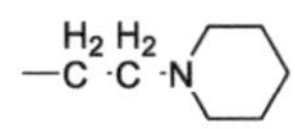	CH_3
(37)	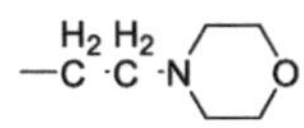	CH_3

In vitro activity against *Trichomonas vaginalis* of the compounds was found to be very potential, moderate against *Entamoeba histolytica* and least active against *Candida albicans.* It was found that those 5-nitroimidazoles in which the fourth position is free showed higher activity against *Trichomonas vaginalis*, whereas substituents at imidazole follow the order pyrrolidine > piperidine > diethylamine > morpholine.

In the year 1969, Lancini et al. [16] synthesized a various number of 1,5-disubstituted 2-nitro imidazoles (38–44) through diazotization reaction or Gattermann reaction.

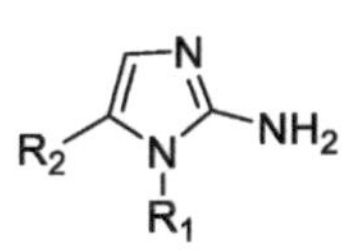

	R1	R2
(38)	CH_3	CH_3
(39)	CH_3	C_2H_5
(40)	CH_3	n-C_3H_7
(41)	CH_3	t-C_3H_7
(42)	CH_3	n-C_4H_9
(43)	C_2H_5	C_2H_5
(44)	C_2H_5	CH_3

All the synthesized compounds showed moderate *In vitro* activity against *Trichomonas vaginalis.*

Miller et al. [17] in the year 1970, synthesized a novel series of 2-methyl-5-nitroimidazoles (45–49) and evaluated their antiprotozoal activity. This series bore an aliphatic side chain incorporated with electronegative group. *In vitro* and *In vivo* evaluations were carried out against *Trichomonas foetus* and *Trichomonas vaginalis* as well as against *Entamoeba histolytica.*

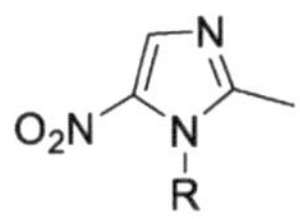

	R
(45)	$(CH_2)_2SO_2CH_2CH_3$
(46)	$(CH_2)_2SO_2CH_2(CH_3)_2$
(47)	$(CH_2)SO_2(CH_2)_2CH_3$
(48)	$(CH_2)_2SO_2CH_3$
(49)	$(CH_2)SO_2(CH_2)_3CH_3$

All the synthesized compounds showed mild activity against *Entamoeba histolytica* when compared to standard Tinidazole (27).

Nair et al. [18] in the year 1982, performed an acrylation of 1-methyl-5-nitroimidazole (50–58) with aroyl chlorides to form 2-aroyl-1-methyl-5-nitroimidazole through Beckmann rearrangement and yielded corresponding oximes and anilides as Beckmann product.

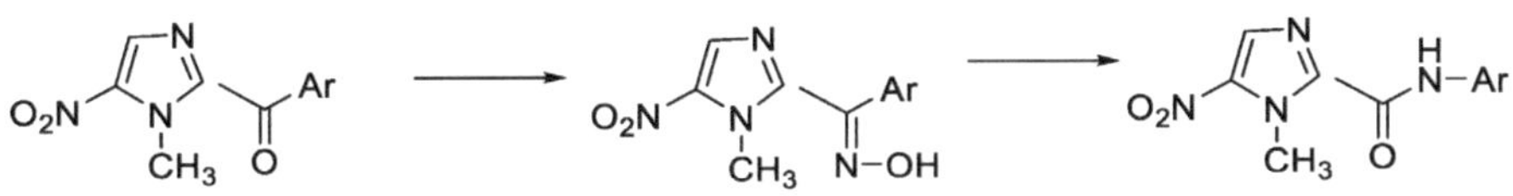

(50) Ar = C_6H_5,
(51) Ar = $C_6H_4Cl(p)$
(52) Ar = 2-theonyl

(53) Ar = C_6H_5,
(54) Ar = $C_6H_4Cl(p)$
(55) Ar = 2-theonyl

(56) Ar = C_6H_5,
(57) Ar = $C_6H_4Cl(p)$
(58) Ar = 2-theonyl

Walsh et al. [19] in the year 1986, synthesized a library of 5-nitroimidazole derivatives (59–65) which had ability to cause mutagenicity and these were also evaluated for their antitrichomonal activity.

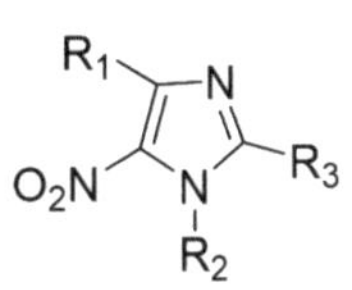

	R_1	R_2	R_3
(59)	CH_3	CH_3	CH_2OCONH_2
(60)	CH_3	CH_3	4-F-Ph
(61)	CH_3	CH_3CH_2OH	CH_3
(62)	CH_3	CH_3	CH_3
(63)	CH_3CH_2OH	CH_3	CH_3
(64)	CH_3CH_2OH	CH_3	CH_3
(65)	CH_3CH_2OH	CH_3	CH_3CH_2OH

Compounds 64 and 65 showed very less mutagenicity in comparison to standard drug ronidazole.

Foroumadi et al. [20] in the year 2005, synthesized a series of 2-(1-methyl-5-nitro imidazol-2-yl)-5-(1-piperazinyl, 1-piperidinyl and 1-morpholinyl)-1,3,4-thiadiazoles and estimated the antileishmanicidal activity for the synthesized compounds (77–83).

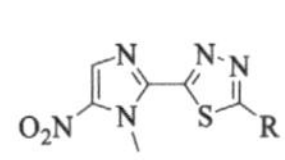

(66) 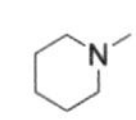

(70) 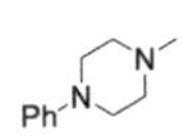

(67) 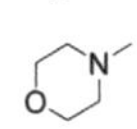

(71) 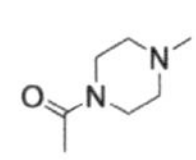

(68) 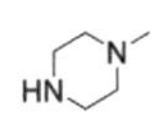

(72) 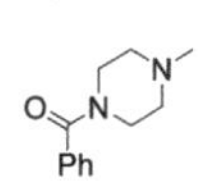

(69) 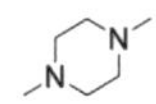

All compounds exhibited better activity against *Leishmania major* (IC_{50} < 1.744 μM).

Bhandari et al. [21] in the year 2010, synthesized a series of various substituted alkyl/aryl imidazoles (73–91) and estimated their activity against *Leishmania donovani* as antileishmanial agents.

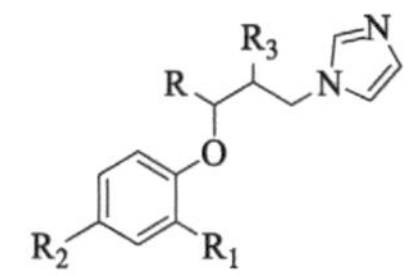

	R	R₁	R₂	R₃
(73)	Ph	H	CF_3	H
(74)	Ph	H	NO_2	H
(75)	Ph	NO_2	H	H
(76)	Ph	F	NO_2	H
(77)	Ph	H	CF_3	CH_3
(78)	Ph	NO_2	H	CH_3
(79)	CH_3	H	CF_3	H
(80)	CH_3	H	NO_2	H
(81)	CH_3	F	NO_2	H
(82)	CH_3	NO_2	H	H
(83)	CH_3	CF_3	NO_2	H

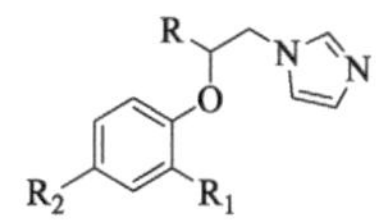

	R	R_1	R_2
(84)	Ph	H	CF_3
(85)	Ph	H	NO_2
(86)	Ph	F	NO_2
(87)	Ph	NO_2	H
(88)	$PhOCH_2$	H	NO_2
(89)	$PhOCH_2$	H	CF_3
(90)	$PhOCH_2$	F	NO_2
(91)	$PhOCH_2$	NO_2	H

Most of the synthesized compounds exhibited very significant activity up to 84–91% inhibition at the concentration of 10 μg/ml while some compounds showed high IC_{50} values ranging from 0.47–4.85 μg/ml against amastigotes.

Hernandez-Nunez et al. [22] in the year 2009, reported synthesis of novel imidazole derivatives (103–117). These compounds were biologically examined against various parasites namely *Giardia intestinalis*, *Trichomonas vaginalis* and *Entamoeba histolytica*.

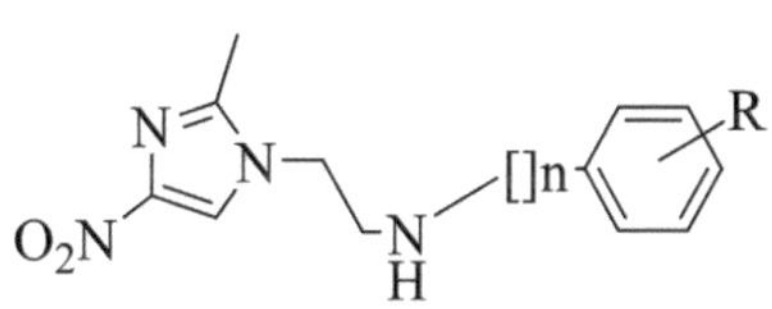

	n	R
(92)	1	H
(93)	0	H
(94)	0	4-F
(95)	0	4-Cl
(96)	0	4-CN
(97)	0	4-NO_2
(98)	0	2,4-di-Cl
(99)	0	3-CF_3

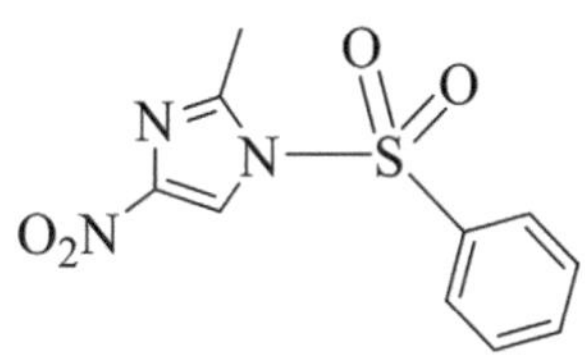

	R
(100)	$NHCOCH_3$
(101)	H
(102)	CH_3
(103)	Cl
(104)	F
(105)	NO_2
(106)	OCH_3

Compounds (100–106) exhibited two fold better activity in comparison to benzimidazole analogs against *Trichomonas* vaginalis and *Giardia intestinalis* and found to be more active analogs against *Entamoeba histolytica*.

2.2 Antibacterial and antifungal agents

During the last 30 years, antifungal azoles [23] such as clotrimazole (107), miconazole (108), ketoconazole (109), butoconazole (110), econazole (111), cloconazole (112), fenticonazole (113), oxiconazole (114) and sulcoconazole (115) have been introduced. In all these compounds N-1 atom of imidazole is linked to other aromatic rings. The other antimycotic azoles have a five membered ring with three nitrogen atoms.

The antifungal azoles inhibit the cytochrome P-450 which catalyzes the 14-α-demethylation of lanosterol to ergosterol [15]. The azole drugs are relatively of broad spectrum antifungal activity but there may be differences among the individual compounds.

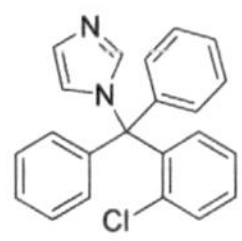
(107)

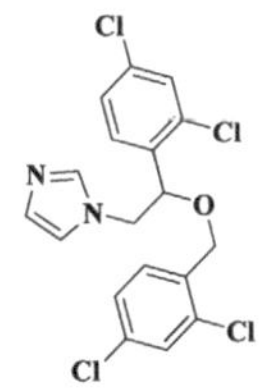
(108)

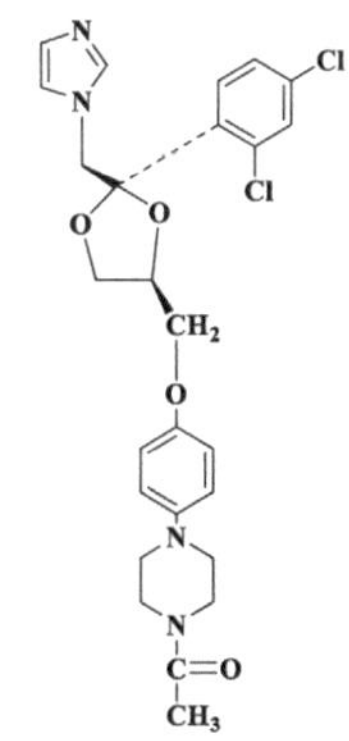
(109)

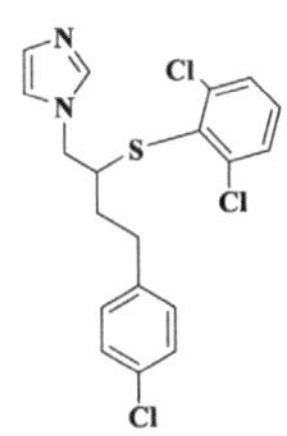
(110)

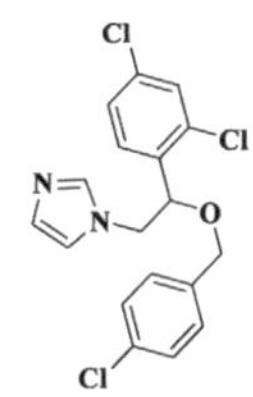
(111)

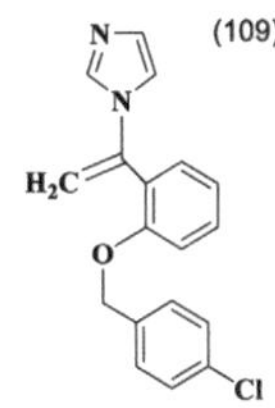
(112)

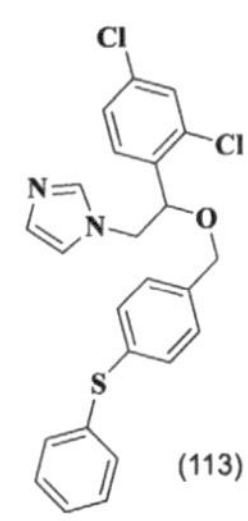
(113)

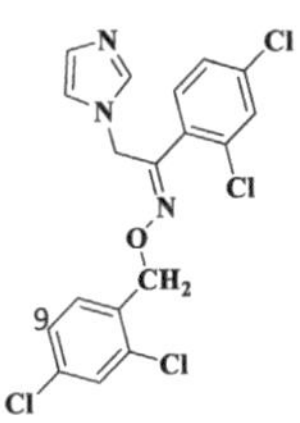
(114)

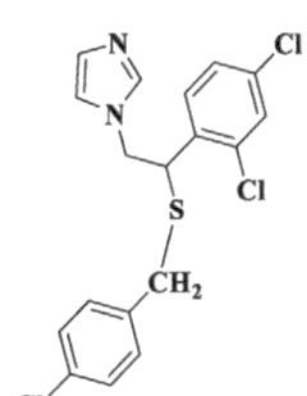
(115)

Castelli et al. [24] in the year 1997, synthesized two new compounds belonging to 5-nitroimidazole family: sulphuridazole and sulphonidazole derivatives (116–119) and compared their minimum inhibitory concentrations with metronidazole (26).

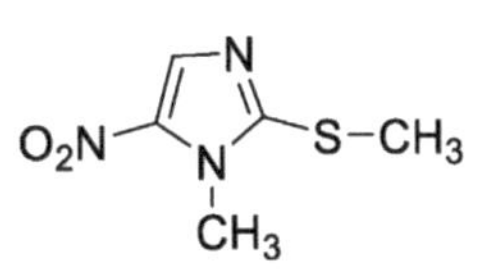
(116)

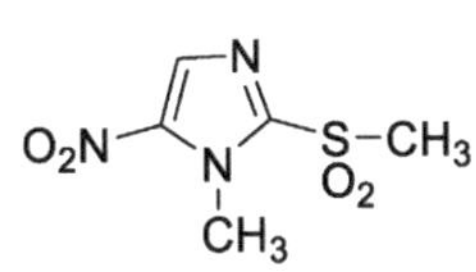
(117)

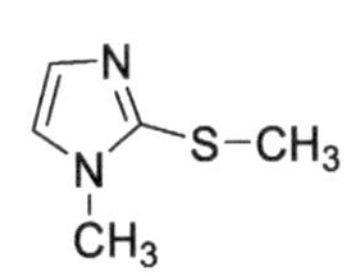
(118)

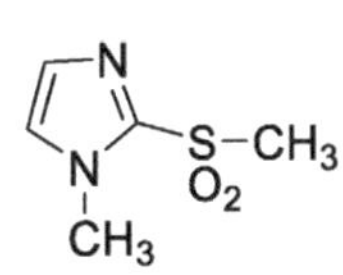
(119)

DOI: http://dx.doi.org/10.5772/intechopen.88547

Sulphonidazole (116–117) showed better activity than sulphuridazole (118–119) against all the bacterial and fungal strains.

Demirayak et al. [25] in the year 1999, synthesized a novel series of some pyrrole-nitroimidazole clubbed hybrid derivatives (120–125) and evaluated their antifungal activity.

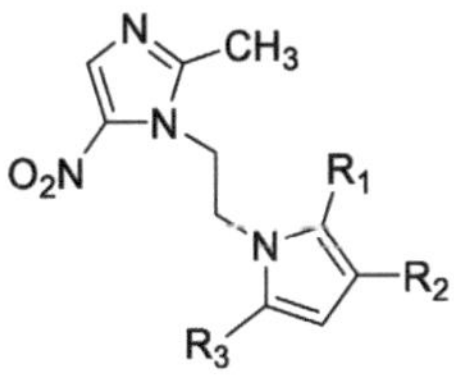

	R_1	R_2	R_3
(120)	H	H	H
(121)	CH_3	H	CH_3
(122)	CH_3	$COOC_2H_5$	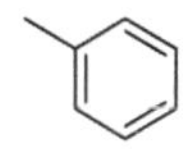
(123)	CH_3	$COOC_2H_5$	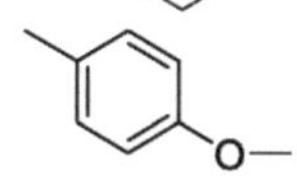
(124)	CH_3	$COOC_2H_5$	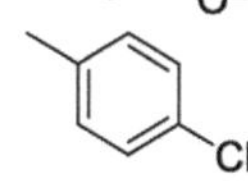
(125)	CH_3	$COOC_2H_5$	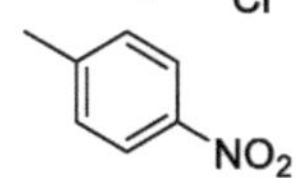

Compounds 120–124 showed excellent activity against *Staphylococcus aureus* at the dose of 8 mg/ml while all the synthesized compounds exhibited excellent activity against fungal strain *Candida albicans*.

Kolavi et al. [26] in the year 2006, synthesized a library of some imidazo thiadiazole derivatives (126–161) and evaluated their antibacterial activity.

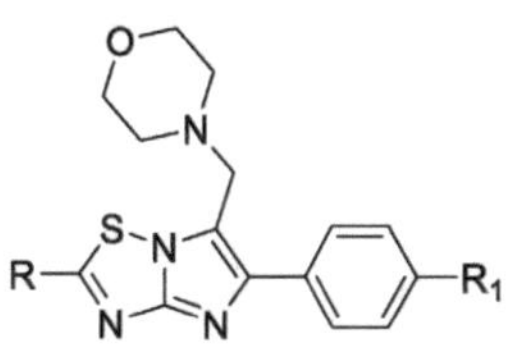

	R	R_1
(126)	Cyclohexyl	H
(127)	Cyclohexyl	Br
(128)	2-furfuryl	H
(129)	2-furfuryl	Br
(130)	2-thienyl	H
(131)	2-thienyl	Br

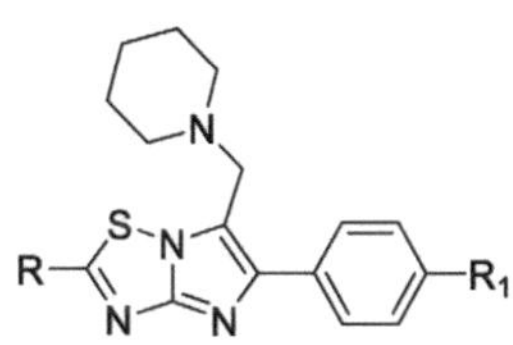

	R	R_1
(132)	Cyclohexyl	H
(133)	Cyclohexyl	Br
(134)	2-furfuryl	H
(135)	2-furfuryl	Br
(136)	2-thienyl	H
(137)	2-thienyl	Br

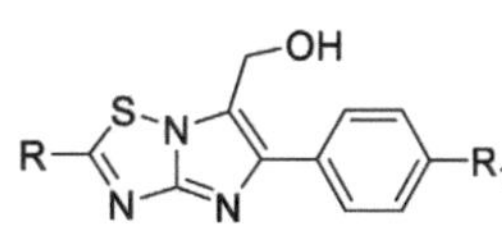

	R	R_1
(138)	Cyclohexyl	H
(139)	Cyclohexyl	Br
(140)	2-furfuryl	H
(141)	2-furfuryl	Br
(142)	2-thienyl	H
(143)	2-thienyl	Br

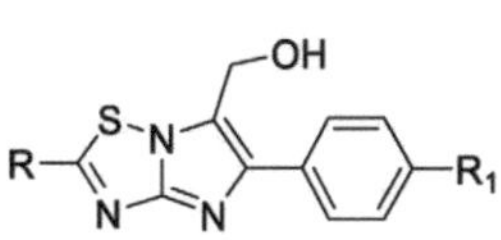

	R	R_1
(144)	Cyclohexyl	H
(145)	Cyclohexyl	Br
(146)	2-furfuryl	H
(147)	2-furfuryl	Br
(148)	2-thienyl	H
(149)	2-thienyl	Br

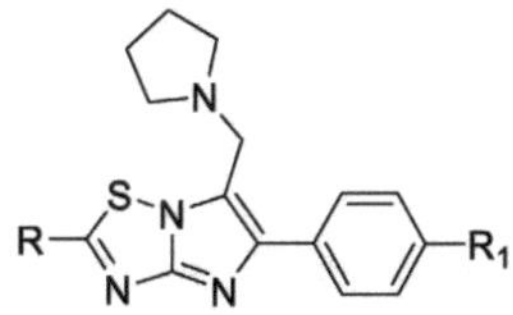

	R	R_1
(150)	Cyclohexyl	H
(151)	Cyclohexyl	Br
(152)	2-furfuryl	H
(153)	2-furfuryl	Br
(154)	2-thienyl	H
(155)	2-thienyl	Br

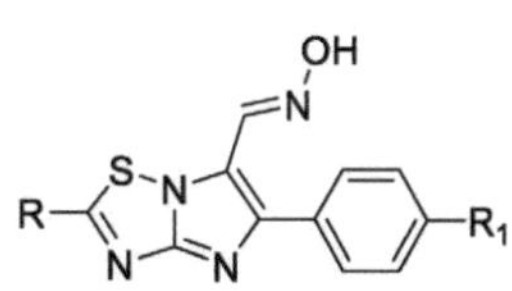

	R	R_1
(156)	Cyclohexyl	H
(157)	Cyclohexyl	Br
(158)	2-furfuryl	H
(159)	2-furfuryl	Br
(160)	2-thienyl	H
(161)	2-thienyl	Br

The antibacterial screening revealed that compounds 134 and 140 showed significant activity against *Escherichia coli*. Compounds 126 and 127 showed good inhibition of *Escherichia coli* at a concentration of 100 µg/ml.

The antifungal screening revealed that the compounds 132, 134, 142, 159 and 161, displayed good antifungal activity against *Penicillium wortmannii* and *Aspergillus niger*.

Banfi et al. [27] in the year 2006 synthesized and evaluated new imidazoles (162–166) and (167–171) for antifungal and antimycobacterial activity.

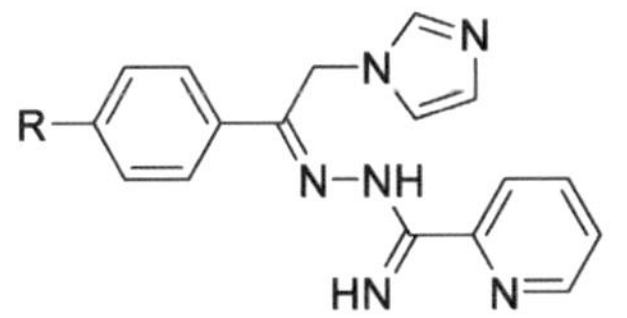

	R
(162)	H
(163)	Br
(164)	Cl
(165)	CH_3
(166)	C_6H_5

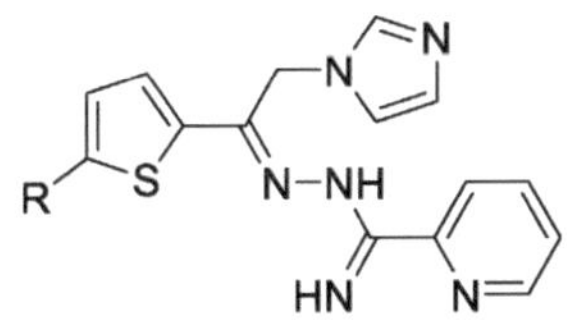

	R
(167)	H
(168)	Br
(169)	Cl
(170)	CH_3
(171)	C_6H_5

The results showed that the compounds 166 and 171 showed very good activity, while rest of the derivatives exhibited weak antifungal activity against *Candida* species.

Mamolo et al. [28] in the year 2007, reported the synthesis of novel bis-imidazole derivatives (172–175) and screened their antimycobacterial and antifungal activity.

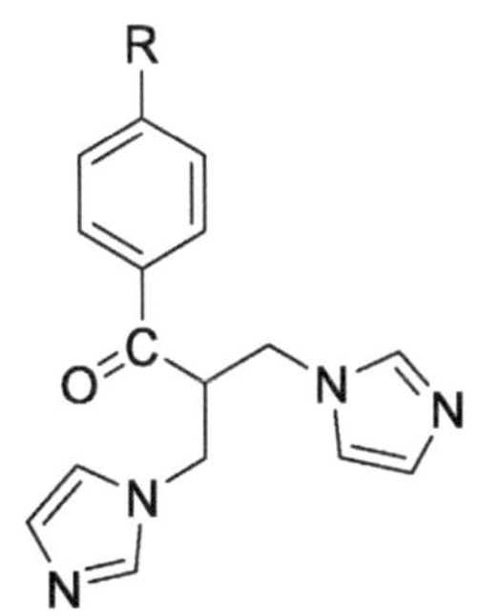

	R
(172)	H
(173)	Cl
(174)	Br
(175)	F

Maximum compounds exhibited weak activity towards *Mycobacterium tuberculosis* and *Candida* species. Compound 175 was considered to be a significant antifungal agent against *Candida* species.

Ganguly et al. [29] in the year 2009, synthesized a few compounds of the type (176–183) and these were evaluated for antibacterial, antifungal and anti-HIV activity.

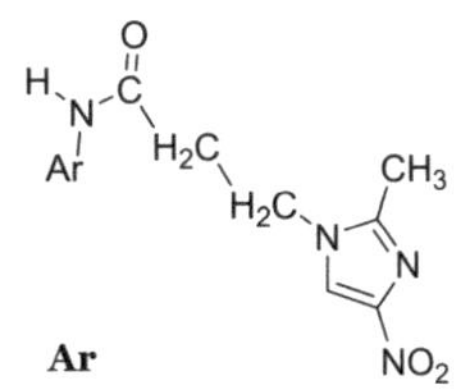

Ar **Ar**

Ar Ar

(176) 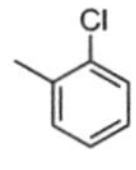 (180) 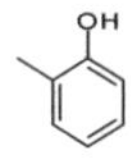

(177) 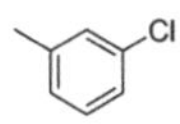 (181) 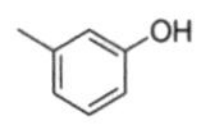

(178) 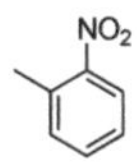 (182) 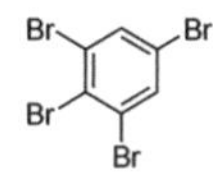

(179) 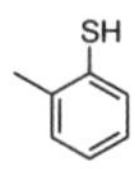 (183) 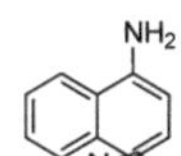

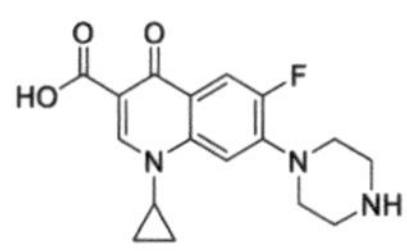

(184)

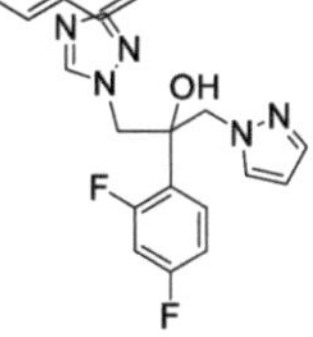

(185)

Compound (176) exhibited good activity against *Staphylococcus aureus* whereas compound (178) showed moderate activity towards *Escherichia coli*. However, all the compounds were less active than standard drug ciprofloxacin (184). Compounds (174) and (179) exhibited significant antifungal activities against *Candida albicans* comparable to the standard drug fluconazole (185). None of the compounds had appreciable anti-HIV activity.

Sharma et al. [30] in the year 2009, synthesized a series of novel 2-substituted benzimidazoles (186–189) and imidazoles (190–193) from long chain alkenoic acids and these were evaluated as antibacterial agents.

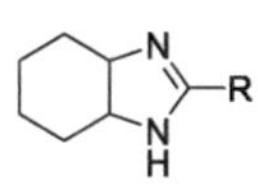

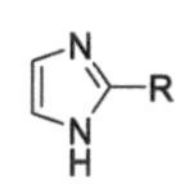

R

(186) 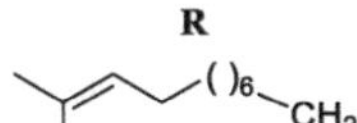

(187) 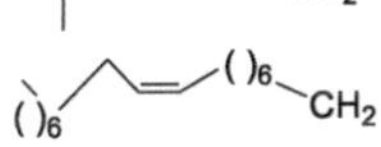

(188) 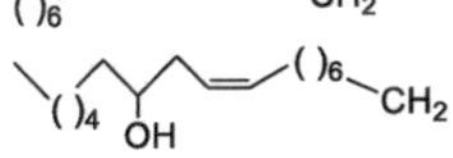

(189) 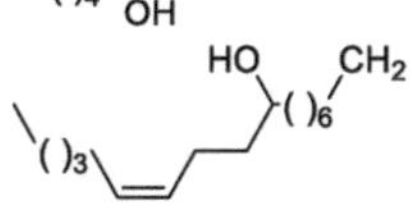

R

(190) 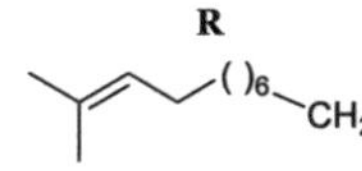

(191) 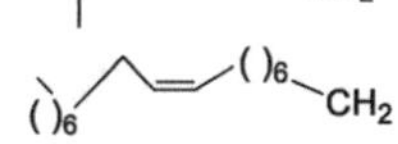

(192) 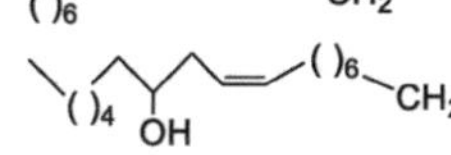

(193) 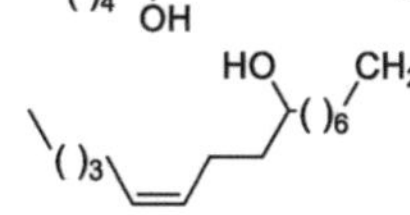

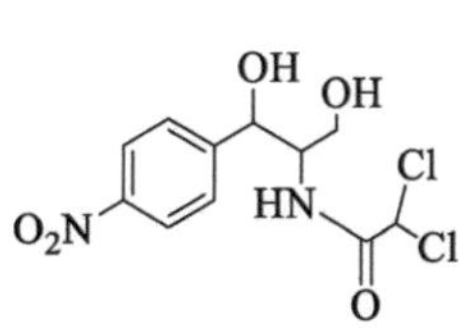

(194)

Compounds (190) and (193) were found to be most active against *Escherichia coli* and *Bacillus subtilis*. Whereas the imidazoles (186–189) exhibited moderate activity against the tested bacterial strains when compared to chloramphenicol (194) as standard.

Wu-Li-Ji et al. [31] in the year 2010, reported the synthesis of 2-benzyl thioimidazoles and 2-benzylthio sulfonyl-1H-imidazoles (195–208) and were evaluated for antibacterial, antifungal and antioxidant activity.

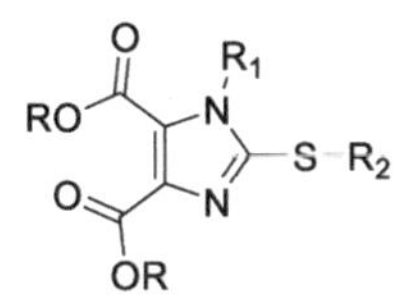

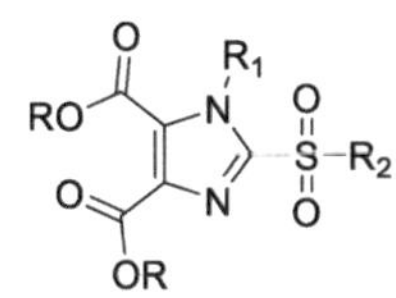

	R	R1	R2
(195)	H	$CH_3CH(CH_3)_2$	o-Me-$CH_3C_3H_4$
(196)	C_2H_5	H	o-Me-$CH_2C_6H_4$
(197)	H	H	$C_{10}H_{11}CH_2$
(198)	H	$COOC_2H_5$	o-Me-$CH_2C_6H_4$
(199)	H	o-Me-$CH_2C_6H_4$	o-Me-$CH_2C_6H_4$
(200)	H	$CH_2C_6H_5$	$CH_2C_6H_5$
(201)	H	$CH_2CH(CH_3)_2$	$CH_2CH(CH_3)_2$

	R	R1	R2
(202)	H	$CH_3CH(CH3)_2$	o-Me-$CH_3C_3H_4$
(203)	C_2H_5	H	o-Me-$CH_2C_6H_4$
(204)	H	H	$C_{10}H_{11}CH_2$
(205)	H	$COOC_2H_5$	o-Me-$CH_2C_6H_4$
(206)	H	o-Me-$CH_2C_6H_4$	o-Me-$CH_2C_6H_4$
(207)	H	$CH_2C_6H_5$	$CH_2C_6H_5$
(208)	H	$CH_2CH(CH_3)_2$	$CH_2CH(CH_3)_2$

All newly reported derivatives exhibited excellent antibacterial activity towards *Proteus vulgaris* and *Klebsiella pneumonia* while exhibiting excellent antifungal activity towards *Penicillium chrysogenum.*

Ganguly et al. [32] in the year 2010, synthesized some novel 2-methyl-5-nitroimidazole analogs (209–211). These were evaluated for antibacterial, antifungal and antidiarrheal activities.

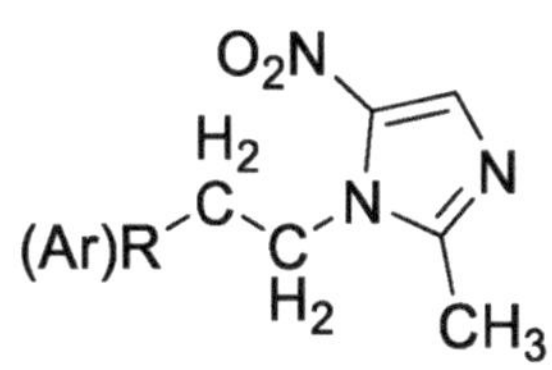

	R	Ar
(209)	$-NH-CH_2-CH_2-CH_3$	$-C_6H_5NH$
(210)	$-NH-CH(CH_3)_2$	o-$C_6H_5NHNO_2$
(211)	$-N-[CH(CH_3)_2]_2$	p-$C_6H_5NHNO_2$

Compounds 209–211 showed significant anti-diarrheal activity at a dose of 60 mg/kg body wt. while compound 209 exhibited significant antibacterial activity against *Staphylococcus aureus.*

Ganguly et al. [33] in the year 2011, reported some novel imidazole analogs of the type (261–274) and evaluated their antibacterial and anti-HIV activity.

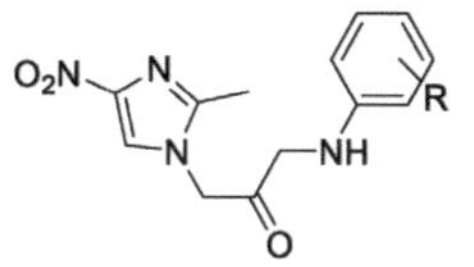

	R
(212)	H
(213)	p-CH3
(214)	p-CH3
(215)	m-Cl
(216)	p-Cl

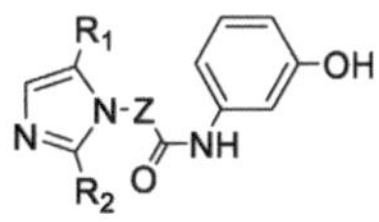

	R	R1	Z
(217)	CH_3	H	CH_2
(218)	CH_3	H	$CH(CH_3)$
(219)	CH_3	H	CH_2CH_2
(220)	CH_3	NO_2	CH_2
(221)	CH_3	NO_2	$CH(CH_3)$

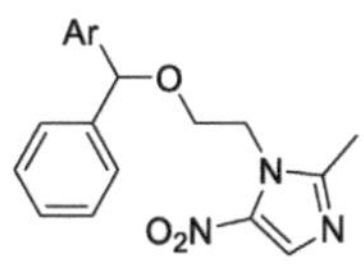

Ar

(222)
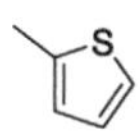

(223)
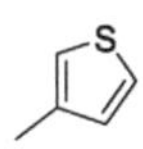

(224)
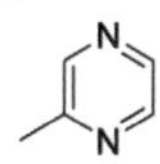

(225)
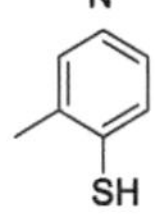

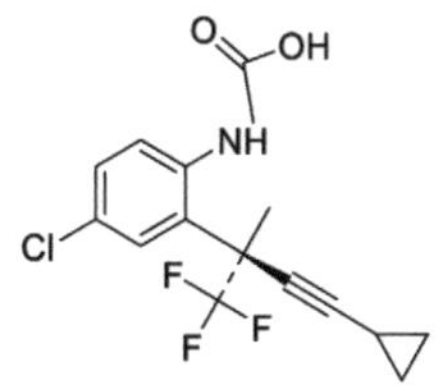

(226)

Compounds 212, 215 and 220 exhibited moderate activity as antibacterials, however compounds 212, 213 and 215 showed weak anti-HIV activity as compared to the standard efavirenz (226).

Ganguly et al. [34] in the year 2011 synthesized a few compounds of type (226–232) these were evaluated for antibacterial, antifungal and anti-HIV activity.

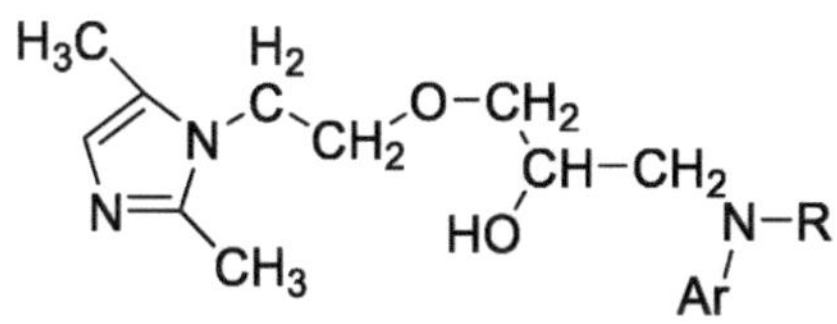

	R	Ar
(227)	H	aniline
(228)	H	m-Cl aniline
(229)	H	2,4-di-methyl aniline
(230)	H	1-naphylamine
(231)	H	p-toludine
(232)	H	m-toludine

Compounds 228 exhibited 44% inhibitory activity against HIV-1 RT.

Pathan and Rahatgoankar [35] in the year 2011, synthesized a series of substituted 4,5-diphenyl imidazolyl-pyrimidine hybrids (233–239).

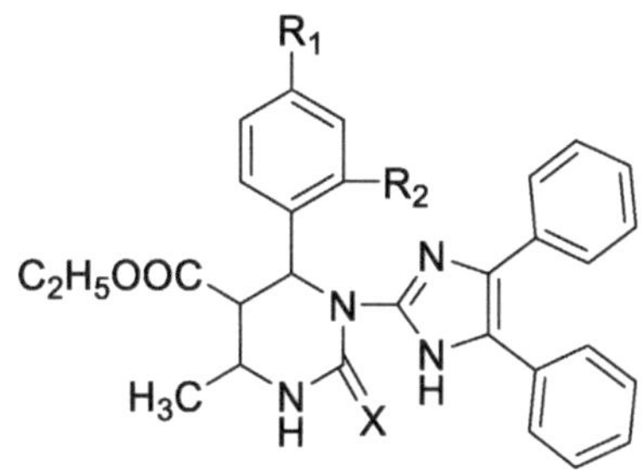

	X	R_1	R_2
(233)	O	H	N
(234)	O	H	NO_2
(235)	O	Cl	H
(236)	O	O	H
(237)	S	H	H
(238)	S	Cl	H
(239)	S	H	NO_2

Compound 236 was found to be most active against *Staphylococcus aureus* among all tested compounds.

Zhang and Zhou [36] in the year 2011 reported the synthesis of naphthalimide derived azoles (240–248) as novel anti-microbial agents and evaluated their efficiency *in vitro* against bacteria and fungi.

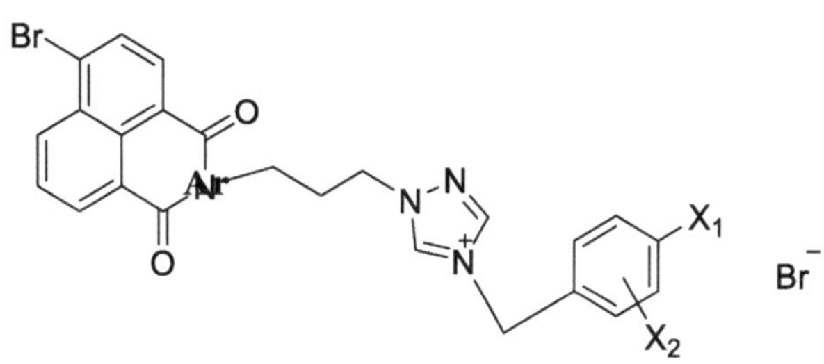

	n	X1	X2
(240)	1	F	2-F
(241)	1	Cl	2-Cl
(242)	1	Cl	3-Cl
(243)	1	H	2-Cl
(244)	1	Cl	H
(245)	1	NO_2	H
(246)	2	F	2-F
(247)	3	F	2-F
(248)	4	F	2-F

It was found that compounds 246–248 with different alkyl linkers were synthesized selectively and gave antibacterial profiles, especially compounds 246 and 247 showed prominent activity against *Pseudomonas aeruginosa* being eight fold more efficient than chloromycin.

Vijesh et al. [37] in the year 2011, synthesized a dual series containing imidazole-pyrazole combined derivatives (249–262).

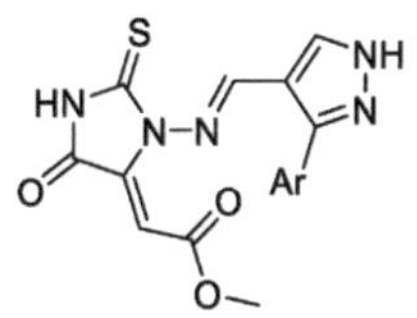

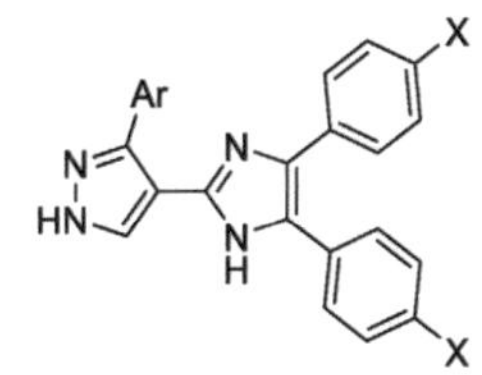

(249)	2,4-dichloro phenyl
(250)	2,5-dichloro thiophene
(251)	4-SCH_3-C_6H_4
(252)	4-SCH_3-C_6H_4

	Ar	X
(253)	4-SCH_3-C_6H_4	H
(254)	2,4-dichlorophenyl	H
(255)	Biphenyl	H
(256)	4-CH_3-C_6H_4	H
(257)	2,5-dichlorothiophene	H
(258)	4-SCH_3-C_6H_4	Br
(259)	2,4-dichlorophenyl	Br
(260)	Biphenyl	Br
(261)	4-CH_3-C_6H_4	Br
(262)	2,5-dichlorothiophene	Br

Among the tested compounds, compound 253 emerged as highly active against *Trychophyton rubrum* compared to standard fluconazole.

Zhu et al. [38] in year 2012 reported the design and synthesized oxadiazole derivatives (263–264) and evaluated their antibacterial activity.

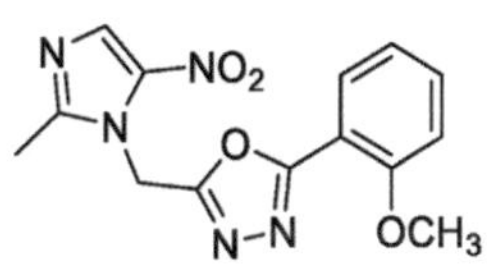

(263)

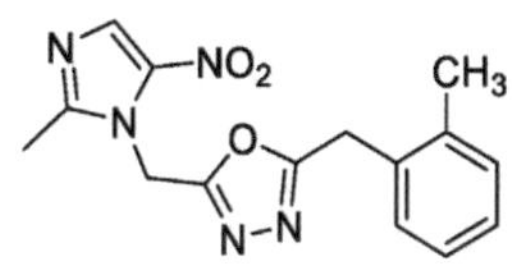

(264)

Compound 312 with MIC of 1.56–3.13 μg/ml and compound 313 with MIC of 1.56–6.25 μg/ml were the most potent inhibitors of FabH against *Escherichia coli.*

Desai et al. [39] in year 2012 synthesized a series of imidazole analogs (265–269) and reported their activity towards bacterial and fungal species.

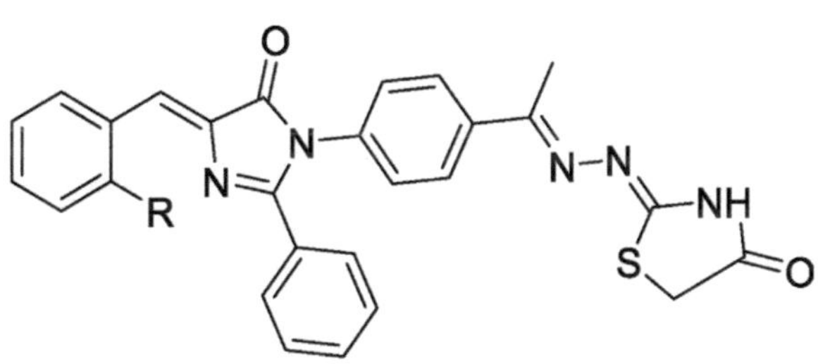

	R		R		R
(265)	H	(319)	2-NO_2	(324)	2-CH_3
(266)	2-Cl	(320)	4-NO_2	(325)	3-CH_3
(267)	4-Cl	(321)	2-OH	(326)	4-CH_3
(268)	2-F	(322)	3-OH	(327)	2-CH_3
(269)	2-F	(323)	4-OH	(328)	2-OCH_3

Compounds (265–269) were evaluated against *Gram-positive bacteria* mainly *Staphylococcus aureus*, *Staphylococcus pyogenes* and Gram-negative *bacteria* mainly *Escherichia coli*, *Pseudomonas aeruginosa* and fungi.

2.3. Antiviral agents

The first report on N-Amino imidazoles as antiHIV agents and particularly as NNRTIs came up with the discovery of Capravirine (S-1153) (270) in the year 2000 [1]. This also retained activity against HIV-1 strains carrying K103N mutation in RT structure.

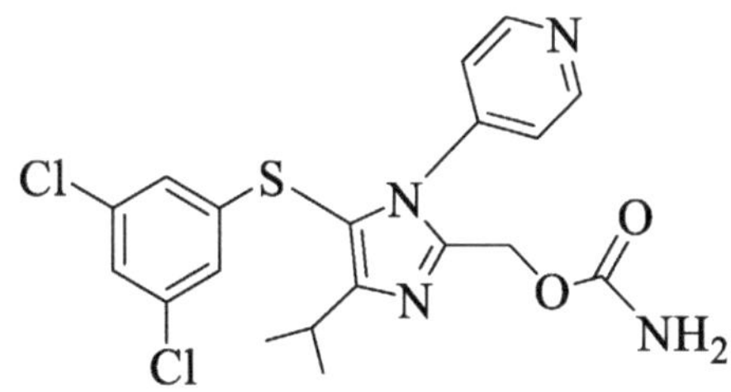

(270)

After this, anti-viral active N-amino imidazole (271) derivatives were reported by Lagoja et al. [2] in the year 2003, which exhibited considerable antiviral activity.

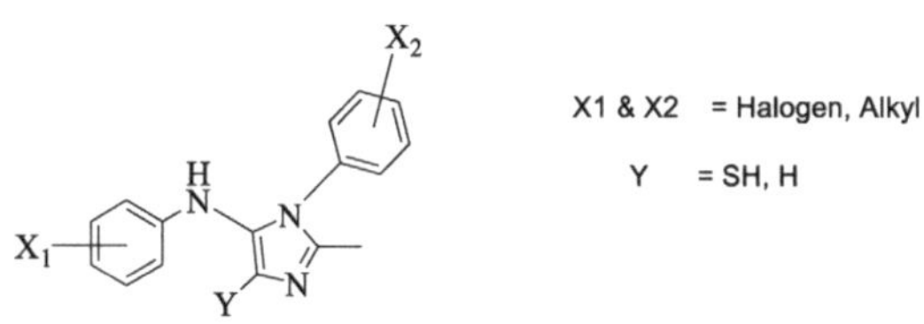

(271)

Methylation or benzoylation on sulfur group may demolish the anti HIV activity of compound, whereas compounds bearing alkyl/aryl substituents at para position to imidazole ring affected the antiHIV activity. Smaller the substituent higher the activity.

Silvestri et al. [40] in the year 2002 synthesized a novel series of 1-{2-(diarylmethoxy)ethyl]-2-methyl-5-nitroimidazoles (272) and evaluated their antiHIV activity.

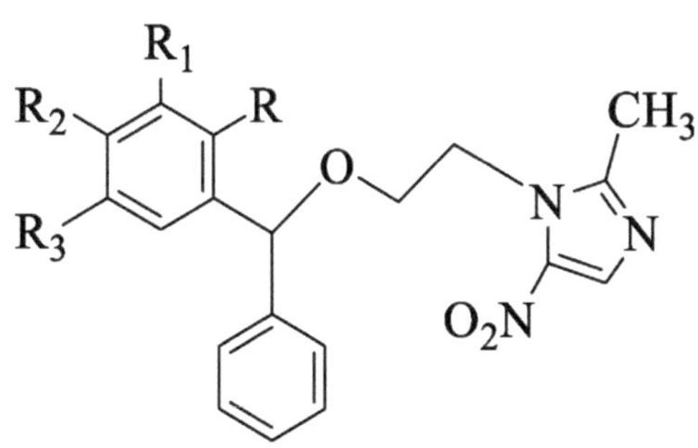

(272)

Substitution at meta position to the phenyl ring exhibited better anti-HIV activity while substituents like fluoro, chloro or methyl substituent enhances the activity than its prototype.

It was observed that substitution at meta position facilitates better activity rather than substitution at *ortho* and *para* position. Fluorine at meta position exhibited maximum potency among all derivatives.

De Martino et al. [40] replaced one phenyl ring of 1-[2-diarylmethoxy] ethyl) 2-methyl-5-nitroimidazoles (DAMNIs) with heterocyclic rings, such as 2-thienyl (273) or 3-pyridinyl ring (274), leading to novel DAMNIs with increased activity.

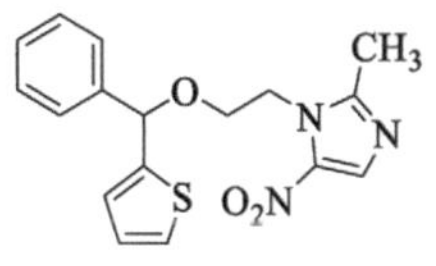

(273)

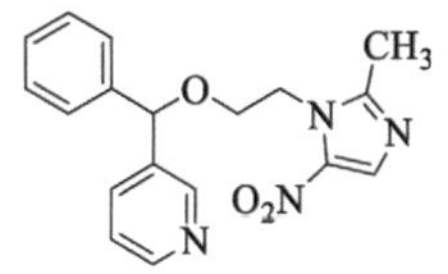

(274)

N-Alkylation of imidazole, 2-methyl imidazole and 2-methyl-4-nitroimidazole has been carried to achieve effective antiHIV agents.

Xu et al. [41] in the year 2008 synthesized some novel derivatives of N-arylindoles (275–282) and evaluated as HIV integrase inhibitors for first time.

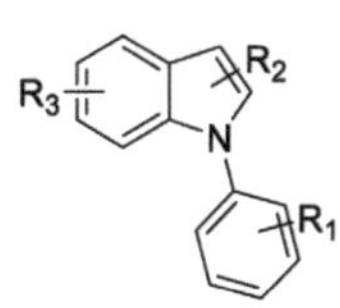

	R1	R2	R3
(275)	p-NO_2	H	H
(276)	o-NO_2	H	H
(277)	o-CN	H	H
(278)	p-NO_2	H	5-NO_2
(279)	o-NO_2	H	5-NO_2
(280)	o-NO_2	H	7-CH_3
(281)	p-NO_2	CH_3	H
(282)	o-NO_2	CH_3	H

Among all synthesized compounds, 276, 279 and 282 exhibited very significant anti HIV-1 integrase inhibitory activity. Especially, compound 277 showed highest activity with EC_{50} value 7.88 μg/ml and therapeutic index 24.61.

Serrao et al. [42] in the year 2013, reported a novel series of 5-carbonyl-1H-imidazole-4-carboxamides (283–299) capable of inhibiting HIV-1 integrase–LEDGF/p75 interaction.

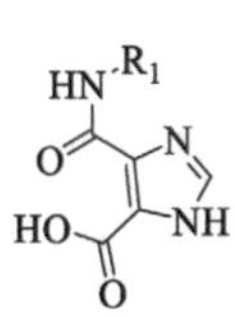

	R1
(283)	p-F-ph
(284)	p-F-Benzyl
(285)	Pyridine
(286)	p-NO_2-Ph
(287)	p-NH_2-Ph

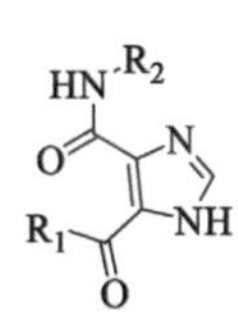

	R1	R2
(288)	Morpholine	Py
(289)	4-OH-pipreidine	Py
(290)	Piperazine	Py
(291)	$NHNH_2$	p-Me-ph
(292)	$NHNH_2$	Py
(293)	MeNH	Py

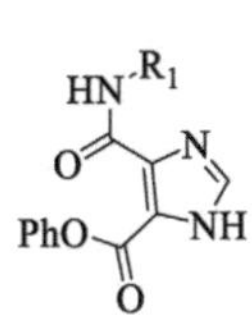

	R1
(294)	Py
(295)	p-Me-ph
(296)	2(Benzyloxy)phenyl

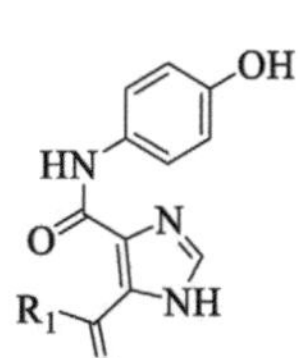

	R1
(297)	Morpholine
(298)	4-OH-pipreidine
(299)	Piperazine

All the synthesized compounds showed almost equivalent activity as their MTT/ MT-4 (CC_{50} and EC_{50}) values were same.

Huang et al. [43] in the year 2017, synthesized a series of diarylpyrimidines (300–310) and indolylarylsulphones (311–321) hybrids and showed their activity against HIV1-IIIB strain.

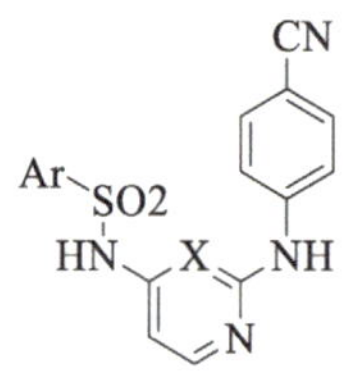

X=CH (300-310)
X=N (311-321)

	Ar		Ar
(300)	4-CH_3-Ph	(311)	4-CH_3-Ph
(301)	4-CN-Ph	(312)	4-CN-Ph
(302)	4-(tert-butyl)-Ph	(313)	4-(tert-butyl)-Ph
(303)	4-F-Ph	(314)	4-F-Ph
(304)	4-CF_3-Ph	(315)	4-CF_3-Ph
(305)	3,5-di-CH_3-Ph	(316)	3,5-di-CH_3-Ph
(306)	3,5-di-F-Ph	(317)	3,5-di-F-Ph
(307)	3,5-di-CF_3-Ph	(318)	3,5-di-CF_3-Ph
(308)	2,4,6-tri-CH_3-Ph	(319)	2,4,6-tri-CH_3-Ph
(309)	2,4,6-tri-isopropyl-Ph	(320)	2,4,6-tri-isopropyl-Ph
(310)	Napth-2-yl	(321)	Napth-2-yl

Compound 311 exhibited favorable selectivity index (SI = 80) which was the maximum above all synthesized compounds and determined by MTT method.

3. Conclusion

Compounds containing azole derivatives, exhibit a wide variety of activities such as antibacterial, antifungal, anthelmintic, antiprotozoal, antiviral, anticancer, antihistaminic, antiulcer, antipsychotic and various other biological activities.

Author details

Rohit Singh* and Swastika Ganguly
Department of Pharmaceutical Sciences, Birla Institute of Technology, Mesra, Ranchi, Jharkhand, India

*Address all correspondence to: rohitsingh20485bitmesra@gmail.com

References

[1] Ren J, Nichols C, Bird LE, Fujiwara T, Sugimoto H, Stuart DI, et al. Binding of the second generation non-nucleoside inhibitor S-1153 to HIV-1 reverse transcriptase involves extensive main chain hydrogen bonding. Journal of Biological Chemistry. 2000;**275**(19): 14316-14320

[2] Lagoja IM, Pannecouque C, Van Aerschot A, Witvrouw M, Debyser Z, Balzarini J, et al. N-aminoimidazole derivatives inhibiting retroviral replication via a yet unidentified mode of action. Journal of Medicinal Chemistry. 2003;**46**(8):1546-1553

[3] Maeda K, Osato T, Umezawa H. A new antibiotic, azomycin. The Journal of Antibiotics. 1953;**6**(4):182

[4] Cooper JS, Pajak TF, Forastiere AA, Jacobs J, Campbell BH, Saxman SB, et al. Postoperative concurrent radiotherapy and chemotherapy for high-risk squamous-cell carcinoma of the head and neck. New England Journal of Medicine. 2004;**350**(19): 1937-1944

[5] Wolff ME et al. Journal of Medicinal Chemistry. 1996;**39**(7):1567-1568

[6] Elslager EF, Thomson PE. Parasite chemotherapy. Annual Review of Pharmacology. 1962;**2**:193-214

[7] Elslager EF. Human antiparasitic agents. Annual Reports in Medicinal Chemistry. 1966;**1**:136-149

[8] Elslager EF. Human antiparasitic agents. Annual Reports in Medicinal Chemistry. 1967;**2**:131-146

[9] Bousquet P, Feldman J, Schwartz J. Central cardiovascular effects of alpha adrenergic drugs: Differences between catecholamines and imidazolines. Journal of Pharmacology and Experimental Therapeutics. 1984; **230**(1):232-236

[10] Urech E, Marxer A, Miescher K. 2-Aminoalkyl-imidazoline. Helvetica Chimica Acta. 1950;**33**(5):1386-1407

[11] Roberts LR, Bryans J, Conlon K, McMurray G, Stobie A, Whitlock GA. Novel 2-imidazoles as potent, selective and CNS penetrant α 1A adrenoceptor partial agonists. Bioorganic & Medicinal Chemistry Letters. 2008;**18**(24): 6437-6440

[12] Hong S-Y, Kwak K-W, Ryu C-K, Kang S-J, Chung K-H. Antiproliferative effects of 6-anilino-5-chloro-1H-benzo [d] imidazole-4, 7-dione in vascular smooth muscle cells. Bioorganic & Medicinal Chemistry. 2008;**16**(2): 644-649

[13] Lin N-H, Wang L, Wang X, Wang GT, Cohen J, Gu W-Z, et al. Synthesis and biological evaluation of 1-benzyl-5-(3-biphenyl-2-yl-propyl)-1H-imidazole as novel farnesyltransferase inhibitor. Bioorganic & Medicinal Chemistry Letters. 2004;**14**(20): 5057-5062

[14] Boström J, Berggren K, Elebring T, Greasley PJ, Wilstermann M. Scaffold hopping, synthesis and structure–activity relationships of 5, 6-diaryl-pyrazine-2-amide derivatives: A novel series of CB1 receptor antagonists. Bioorganic & Medicinal Chemistry. 2007;**15**(12):4077-4084

[15] Giraldi PN, Mariotti V, De Carneri I. Antiprotozoans. Synthesis and biological activity of some styrylimidazole derivatives. Journal of Medicinal Chemistry. 1968;**11**(1):66-70

[16] Lancini GC, Lazzari E, Arioli V, Bellani P. Synthesis and relationship between structure and activity of

2-nitroimidazole derivatives. Journal of Medicinal Chemistry. 1969;**12**(5): 775-780

[17] Miller MW, Howes HL Jr, Kasubick RV, English AR. Alkylation of 2-methyl-5-nitroimidazole. Some potent antiprotozoal agents. Journal of Medicinal Chemistry. 1970;**13**(5): 849-852

[18] Nair M, Sudarsanam V, Desai J. Nitroimidazoles. 12. Reaction of 1-methyl-5-nitroimidazoles with acid-chlorides. Indian Journal of Chemistry Section. 1982;**21**(11): 1027-1029

[19] Walsh JS, Wang R, Bagan E, Wang C, Wislocki P, Miwa GT. Structural alterations that differentially affect the mutagenic and antitrichomonal activities of 5-nitroimidazoles. Journal of Medicinal Chemistry. 1987;**30**(1):150-156

[20] Foroumadi A, Emami S, Hassanzadeh A, Rajaee M, Sokhanvar K, Moshafi MH, et al. Synthesis and antibacterial activity of N-(5-benzylthio-1, 3, 4-thiadiazol-2-yl) and N-(5-benzylsulfonyl-1, 3, 4-thiadiazol-2-yl) piperazinyl quinolone derivatives. Bioorganic & Medicinal Chemistry Letters. 2005;**15**(20): 4488-4492

[21] Bhandari K, Srinivas N, Marrapu VK, Verma A, Srivastava S, Gupta S. Synthesis of substituted aryloxy alkyl and aryloxy aryl alkyl imidazoles as antileishmanial agents. Bioorganic & Medicinal Chemistry Letters. 2010; **20**(1):291-293

[22] Hernández-Núñez E, Tlahuext H, Moo-Puc R, Torres-Gómez H, Reyes-Martínez R, Cedillo-Rivera R, et al. Synthesis and in vitro trichomonicidal, giardicidal and amebicidal activity of N-acetamide (sulfonamide)-2-methyl-4-nitro-1H-imidazoles. European Journal of Medicinal Chemistry. 2009;**44**(7): 2975-2984

[23] Abraham D. Burger's Medicinal Chemistry & Drug Discovery. 6th ed. Vol. 5

[24] Castelli M, Malagoli M, Ruberto A, Baggio A, Casolari C, Cermelli C, et al. In-vitro studies of two 5-nitroimidazole derivatives. The Journal of Antimicrobial Chemotherapy. 1997; **40**(1):19-25

[25] Demirayak S, Karaburun AÇ, Kiraz N. Synthesis and antibacterial activities of some 1-[2-(substituted pyrrol-1-yl) ethyl]-2-methyl-5-nitroimidazole derivatives. European Journal of Medicinal Chemistry. 1999; **34**(3):275-278

[26] Kolavi G, Hegde V, ahmed Khazi I, Gadad P. Synthesis and evaluation of antitubercular activity of imidazo [2, 1-b][1, 3, 4] thiadiazole derivatives. Bioorganic & Medicinal Chemistry. 2006;**14**(9):3069-3080

[27] Banfi E, Scialino G, Zampieri D, Mamolo MG, Vio L, Ferrone M, et al. Antifungal and antimycobacterial activity of new imidazole and triazole derivatives. A combined experimental and computational approach. Journal of Antimicrobial Chemotherapy. 2006; **58**(1):76-84

[28] Zampieri D, Mamolo MG, Vio L, Banfi E, Scialino G, Fermeglia M, et al. Synthesis, antifungal and antimycobacterial activities of new bis-imidazole derivatives, and prediction of their binding to P45014DM by molecular docking and MM/PBSA method. Bioorganic & Medicinal Chemistry. 2007;**15**(23): 7444-7458

[29] Ganguly S, Gopalakrishnan V, Chandra R. Synthesis and antibacterial

activity of some new 1-substituted 2-methyl-4-nitroimidazoles. International Journal of Chemical Sciences. 2009;7: 204-210

[30] Sharma S, Gangal S, Rauf A. Convenient one-pot synthesis of novel 2-substituted benzimidazoles, tetrahydrobenzimidazoles and imidazoles and evaluation of their in vitro antibacterial and antifungal activities. European Journal of Medicinal Chemistry. 2009;**44**(4): 1751-1757

[31] Wu-Li-Ji A, Qing-Hu W, Na-Yin-Tai D. The structural elucidation and antimicrobial activities of two new sesquiterpenes from Syringa pinnatifolia Hemsl. Chinese Journal of Natural Medicines. 2012;**10**(6): 477-480

[32] Ganguly S, Gupta PK, Dev A. Synthesis and antimicrobial evaluation of some 1-substituted 2-methyl-5-nitroimidazoles. Pharmbit. 2010;**21**(1): 29-32

[33] Ganguly S, Vithlani V, Kesharwani A, Kuhu R, Baskar L, Mitramazumder P, et al. Synthesis, antibacterial and potential anti-HIV activity of some novel imidazole analogs. Acta Pharmaceutica. 2011;**61**(2):187-201

[34] Ganguly S, Panigrahi N, Cremer SN, Dev A. Microwave assisted synthesis and evaluation of some 1-{2-(Hydroxyethyl)}-2-methyl-5-nitroimidazole analogs. Institution of Chemist India. 2011;**83**:65-71

[35] Pathan BN, Rahatgaonkar AM. Arabian Journal of Chemistry. 2011

[36] Zhang Y-Y, Zhou C-H. Synthesis and activities of naphthalimide azoles as a new type of antibacterial and antifungal agents. Bioorganic & Medicinal Chemistry Letters. 2011; **21**(14):4349-4352

[37] Vijesh A, Isloor AM, Telkar S, Peethambar S, Rai S, Isloor N. Synthesis, characterization and antimicrobial studies of some new pyrazole incorporated imidazole derivatives. European Journal of Medicinal Chemistry. 2011;**46**(8):3531-3536

[38] Li Y, Luo Y, Hu Y, Zhu D-D, Zhang S, Liu Z-J, et al. Design, synthesis and antimicrobial activities of nitroimidazole derivatives containing 1, 3, 4-oxadiazole scaffold as FabH inhibitors. Bioorganic & Medicinal Chemistry. 2012;**20**(14):4316-4322

[39] Desai N, Joshi V, Rajpara K, Makwana AH. A new synthetic approach and in vitro antimicrobial evaluation of novel imidazole incorporated 4-thiazolidinone motifs. Arabian Journal of Chemistry. 2017;**10**: S589-S599

[40] Silvestri R, Artico M, De Martino G, Ragno R, Massa S, Loddo R, et al. Synthesis, biological evaluation, and binding mode of novel 1-[2-(diarylmethoxy) ethyl]-2-methyl-5-nitroimidazoles targeted at the HIV-1 reverse transcriptase. Journal of Medicinal Chemistry. 2002;**45**(8): 1567-1576

[41] Xu H, Liu W-Q, Fan L-L, Chen Y, Yang L-M, Lv L, et al. Synthesis and HIV-1 integrase inhibition activity of some N-arylindoles. Chemical and Pharmaceutical Bulletin. 2008;**56**(5): 720-722

[42] Serrao E, Xu Z-L, Debnath B, Christ F, Debyser Z, Long Y-Q, et al. Discovery of a novel 5-carbonyl-1H-imidazole-4-carboxamide class of inhibitors of the HIV-1 integrase–LEDGF/p75 interaction. Bioorganic & Medicinal Chemistry. 2013;**21**(19): 5963-5972

[43] Huang B, Wang X, Liu X, Chen Z, Li W, Sun S, et al. Discovery of novel DAPY-IAS hybrid derivatives as potential HIV-1 inhibitors using molecular hybridization based on crystallographic overlays. Bioorganic & Medicinal Chemistry. 2017;**25**(16): 4397-4406